Advection and Diffusion in Random Media

Advection and Diffusion in Random Media

Implications for Sea Surface Temperature Anomalies

by

LEONID I. PITERBARG
Center for Applied Mathematical Sciences,
University of Southern California,
Los Angeles, California, U.S.A.

and

ALEXANDER G. OSTROVSKII
Research Institute for Applied Mechanics,
Kyushu University,
Kasuga, Japan

KLUWER ACADEMIC PUBLISHERS
DORDRECHT / BOSTON / LONDON

A C.I.P. Catalogue record for this book is available from the Library of Congress.

ISBN 978-1-4419-4773-4

Published by Kluwer Academic Publishers,
P.O. Box 17, 3300 AA Dordrecht, The Netherlands.

Kluwer Academic Publishers incorporates
the publishing programmes of
D. Reidel, Martinus Nijhoff, Dr W. Junk and MTP Press.

Sold and distributed in the U.S.A. and Canada
by Kluwer Academic Publishers,
101 Philip Drive, Norwell, MA 02061, U.S.A.

In all other countries, sold and distributed
by Kluwer Academic Publishers,
P.O. Box 322, 3300 AH Dordrecht, The Netherlands.

Printed on acid-free paper

CONTENTS

PREFACE

This book originated from our interest in sea surface temperature variability. Our initial, though entirely pragmatic, goal was to derive adequate mathematical tools for handling certain oceanographic problems. Eventually, however, these considerations went far beyond oceanographic applications partly because one of the authors is a mathematician. We found that many theoretical issues of turbulent transport problems had been repeatedly discussed in fields of hydrodynamics, plasma and solid matter physics, and mathematics itself. There are few monographs concerned with turbulent diffusion in the ocean (Csanady 1973, Okubo 1980, Monin and Ozmidov 1988).

While selecting material for this book we focused, first, on theoretical issues that could be helpful for understanding mixture processes in the ocean, and, second, on our own contribution to the problem. Mathematically all of the issues addressed in this book are concentrated around a single linear equation: the stochastic advection-diffusion equation. There is no attempt to derive universal statistics for turbulent flow. Instead, the focus is on a statistical description of a passive scalar (tracer) under given velocity statistics. As for applications, this book addresses only one phenomenon: transport of sea surface temperature anomalies.

Hopefully, however, our two main approaches are applicable to other subjects. The first idea is to establish a correspondence between different rescalings of the stochastic velocity field and different scale separations. For this purpose we use time scales such as the observation time, velocity correlation time, turnover time, and molecular diffusion time. This approach allows us to classify all limit cases and derive connections between the rigorous and non-rigorous results concerning turbulent diffusion. It should be noted that the language of scales used by physicists is not popular among mathematicians so far. Hopefully, mathematicians will not be confused by our scale language since all scales are rigorously defined.

The other idea addresses the problem of deriving velocity and diffusivity estimates from the tracer observations. Emphasis is placed on estimating the heat anomaly transport in the upper ocean. The estimation is treated as a statistical problem in stochastic partial differential equations. This statistical estimation takes advantage of powerful maximum likelihood and autoregressive techniques. It should be stressed that in the latter case we propose a rigorous way for deriving an autoregressive model from the stochastic partial differential equation instead of substituting finite differences for the partial derivatives, a procedure that is not reliable when dealing with stochastic forcing.

Admittedly, only a modest portion of the present results is proved with full mathematical rigor. Proved statements are referred to as 'lemmas' and 'theorems' while the term 'proposition' is reserved for the statements whose proofs contain additional plausible assumptions, which themselves need justification.

This book is organized as follows. Chapters 1-8 address general results of the advection and diffusion of a passive scalar (tracer) in random media, some important physical effects which follow from the obtained equations for the statistical characteristics of the tracer, and the inverse problem of extracting the velocity and diffusivity from passive scalar observations. In chapters 9-10 we apply the obtained results and the developed statistical techniques to analyze anomal heat processes in the upper ocean.

In chapter 1, we introduce the underlying equations and set up the forward and inverse problems. Consideration of the forward problem starts with some simple exactly solvable models and a description of general approaches. In particular, the case of constant flow and statistically homogeneous and stationary forcing is considered. Despite its simplicity, this case is of considerable practical importance concerning the properties of high level excursions of tracer fields, which are not well understood. Another simple case of random Gaussian velocities independent of the spatial variables offers significant guidance for understanding more complex situations.

In chapter 2, the main scales are introduced for the simplest model including only random advection and non-random constant diffusivity. Classification of limiting cases is proposed and the short-correlation approximation is specified.

Chapter 3 is devoted to the well-known approximation of δ-correlated velocity fields leading to the Fokker-Planck (FP) equation for the mean tracer field. In this chapter we concentrate on the less known equations for the second moment of the tracer and its gradient. Some important physical effects following from these equations are discussed. We also derive an equation for the mean tracer in the presence of a random source and a random potential which may be correlated with the velocity field. This procedure slightly differs from derivation of the FP equation, but we present it in detail because of its importance for the oceanographic applications.

The δ-correlated approximation is based on the assumption that the correlation time of velocity fluctuations is much less than the turnover time of a Lagrangian particle trapped by an eddy. This assumption can hold only for the top layer of the ocean. A more typical situation we face is either an approximate equality of these scales or a dominance of the velocity correlation time. These cases are considered in chapter 4.

The asymptotics in chapters 3 and 4 address the case of infinitesimally small velocity correlation time. In chapter 5, we briefly discuss consequences of removing this restriction. Substantial rigorous results in this field are only obtained in simple mathematical models. We also discuss some of the phenomenological equations elaborated by other authors.

As for the inverse problem (i.e., estimating velocity and dissipation parameters) we consider two methods, completely developed by the authors. In chapter 6, we present the maximum likelihood procedure for estimating unknown parameters in stochastic partial differential equations with a focus on the advection-diffusion equation. Necessary and sufficient conditions for the consistency, asymptotical normality and efficiency are given with full mathematical rigor. Properties of the estimates for small and moderate samples are studied by Monte-Carlo methods in chapter 7.

Chapter 8 is devoted to the autoregressive estimator for velocity and diffusivity. The advection-diffusion equation is discretized in time and space in order to derive the autoregressive model, which relates time series of the tracer fields. The formulas connecting the physical parameters of the advection-diffusion equation and the autoregressive model are presented. An estimate for error in the space approximation is given.

Chapter 9 serves to provide a physical rationale for application of the methods developed in chapters 6-8 to the heat anomaly balance equation for the upper ocean mixed layer. An extensive review of the oceanographic literature concerning this problem is given. Hasselmann's (1976) concept of stochastic climate models is introduced and extended to accommodate effects of the heat anomaly transport in the upper ocean mixed layer. The averaging technique developed in chapter 3 is applied in the derivation of a heat anomaly balance equation and an equation for the mean heat balance. The obtained model contains thorough definitions for advection and diffusion processes. In particular, it is shown that net advection of heat is not merely a result of mean currents but in fact consists of the two main parts: all motions with scales exceeding the given averaging scales and motion due to the joint effect of fluctuations in the feedback factor and the current velocity. Likewise, we described the diffusion, feedback, and source terms. In a wider sense, our inversion approach for advection and diffusion can be regarded as a diagnostic model for the tracer balance governed by a stochastic partial differential equation.

Chapter 10 is devoted to the practical application of the autoregressive and maximum likelihood inversion techniques to the problem of the heat anomaly transport in the midlatitude region of the North Pacific. The experimental time series are deduced from the Comprehensive Ocean Atmosphere Data Set. Major atmospheric forcing fields such as the net heat flux and the wind-driven ocean currents are computed. Some of their second order statistics, which are necessary for the inversions, are estimated and discussed. Analysis of the results obtained from the inversions reveal that the statistical models contain the essential physics of the heat anomaly balance. The autoregressive inversion provides us with reasonable estimates of heat anomaly advection in the North Pacific. The advection field exhibits certain features of ocean general circulation. The maximum likelihood inversion results in feasible estimates of the anisotropy of the heat diffusion in the upper ocean. By using the heat diffusivity estimates we come up with an evaluation of the size and lifetime of large-scale SST anomalies in the North Pacific.

Finally, we express our special thanks to colleagues who made this book possible. We are grateful to Boris Rozovskii for many thoughtful comments. Ken Owens carefully read the manuscript and suggested many improvements. Advice given by Toshio Yamagata was extremely important. Masaki Takematsu and Shiro Imawaki encouraged our research over a few years. Tsuyoshi Kondo helped to handle the tremendous amount of raw data. Victor Piterbarg helped us with the word processing. We appreciate the support of both the Office of Naval Research (USA) and the Ministry of Education, Science and Culture (Japan) which funded much of this research.

1

PROBLEM STATEMENT

We will use the following stochastic advection-diffusion equation to describe the transport and dissipation of a passive scalar (tracer) in the presence of a distributed source

$$\frac{\partial c}{\partial t} + \boldsymbol{u} \cdot \nabla c + \lambda c = \kappa \nabla^2 c + S. \tag{1.1}$$

where $c = c(t, \boldsymbol{r})$ is the tracer concentration at point $\boldsymbol{r}$ in d-dimensional Euclidean space $E = E^d$ at time t, $\boldsymbol{u} = \boldsymbol{u}(t, \boldsymbol{r})$ is a random velocity field, ∇ and ∇^2 stand for the gradient and Laplacean respectively, $\lambda = \lambda(t, \boldsymbol{r})$, and $S = S(t, \boldsymbol{r})$ are random fields, which represent the tracer dissipation and source, respectively, and, finally, κ is the constant nonrandom coefficient of molecular or, sometimes, small-scale diffusion. By passive scalar or tracer we mean an admixture which does not affect the fluid motion. In certain cases we will concentrate on 2- and 3-dimensional flows ($d = 2, 3$) because of their importance for applications. It is supposed that the field $\boldsymbol{u}$ is nondivergent, $\nabla \cdot \boldsymbol{u} = 0$, which means that the medium under consideration is incompressible.

Mostly, we will be interested in applications of eq.(1.1) for analysis of the sea surface temperature (SST) variability. In this case S will be the normalized total heat flux across the air-sea boundary and from the interior below, λ the atmosphere-ocean feedback factor (Newton's cooling coefficient) and $\boldsymbol{u}$ the near-surface ocean current vector. The meaning of these parameters will be discussed in detail in chapter 9.

Eq.(1.1) could be also used when describing phytoplankton variability in the upper ocean. In this case S is the phytoplankton production and λ is the depletion rate.

The forward problem for eq.(1.1) concerns modeling the statistics of c under given statistics of the parameters $\boldsymbol{u}, \lambda, S$ and the value of κ, while in the inverse problem $\boldsymbol{u}, \lambda, S$, and κ are to be estimated from given c. The inverse problem we will address has constant $\boldsymbol{u}, \kappa$, and λ, and in the forward problem we will concentrate on the effects of random velocities.

1.1 CONSTANT FLOW

To introduce the forward problem, we will start with the simplest case which nevertheless is of a great importance for practice. Namely, suppose that $\boldsymbol{u}, \lambda$, and κ are constants and $S(t, \boldsymbol{r})$ is a homogeneous in $\boldsymbol{r}$ and stationary in t Gaussian random field with zero mean. In this case the statistics of S are completely determined by its correlation function

$$R_S(s, \boldsymbol{r}) = \langle S(t, \boldsymbol{r}_0) S(t+s, \boldsymbol{r}_0 + \boldsymbol{r}) \rangle, \tag{1.2}$$

or by its spectral density $E_S(\omega, \boldsymbol{k})$ given by

$$R_S(s, \boldsymbol{r}) = \int_E \int_{-\infty}^{\infty} e^{i(\boldsymbol{k} \cdot \boldsymbol{r} - \omega s)} E_S(\omega, \boldsymbol{k}) d\omega d\boldsymbol{k}, \tag{1.3}$$

where ω is the angular frequency, $\boldsymbol{k}$ the wavevector, s the time lag, and the angle brackets denote averaging over the ensemble.

From the above assumptions it follows that the statistically stationary solution $c(t, \boldsymbol{r})$ of (1.1) is a Gaussian homogeneous random field with the spectral density

$$E_c(\omega, \boldsymbol{k}) = \frac{E_S(\omega, \boldsymbol{k})}{(\omega - \boldsymbol{k} \cdot \boldsymbol{u})^2 + (\lambda + \kappa k^2)^2}, \tag{1.4}$$

where $k = |\boldsymbol{k}|$ is the magnitude of the wavevector.

Thus, in this case, (1.4) gives a complete solution of the forward problem since the statistics of a Gaussian field are completely determined by its first and second moments.

One of the most important issues in the analysis of a coupled ocean and atmosphere system is the relationship between the space and time scales. In the simplest case, this problem can be posed as finding the connection between space and time correlation scales of the fields $S(t, \boldsymbol{r})$ and $c(t, \boldsymbol{r})$ whose spectra are related by (1.4). There are different ways to define these scales. First, let us consider an isotropic random field $\xi(t, \boldsymbol{r})$ with zero mean that is differentiable in all variables. The isotropy of ξ implies that its spectrum $E_\xi(\omega, \boldsymbol{k}) = E_\xi(\omega, k)$ is a function of the wavenumber magnitude k only. For such fields the appropriate definitions of the correlation radius and the correlation time are as follows

$$l_\xi = \frac{\sigma_\xi}{\langle |\nabla \xi|^2 \rangle^{1/2}} = \left(\frac{\int\int E_\xi(\omega, k) k^{d-1} d\omega dk}{\int\int E_\xi(\omega, k) k^{d+1} d\omega dk} \right)^{1/2} \tag{1.5}$$

and

$$\tau_\xi = \frac{\sigma_\xi}{\langle \xi_t^2 \rangle^{1/2}} = \left(\frac{\int E_\xi(\omega, k) k^{d-1} dk}{\int\int E_\xi(\omega, k) k^{d-1} \omega^2 d\omega dk} \right)^{1/2}, \tag{1.6}$$

where σ_ξ^2 is the variance, $\sigma_\xi^2 = \langle \xi^2 \rangle$, and the integration with respect to the frequency, ω, is done over the real line and with respect to the wavenumber magnitude, k, from zero to infinity unless otherwise specified. Let us note that the variance is given by

$$\sigma_\xi^2 = \vartheta_d \int\int E_\xi(\omega, k) k^{d-1} d\omega dk, \tag{1.7}$$

where ϑ_d is the volume of a unit ball, e.g. $\vartheta_2 = 2\pi$, $\vartheta_3 = 4\pi$.

We also will consider random fields which are not differentiable but instead quickly decorrelate in time. In this case a convenient definition of the correlation time is as follows

$$\tilde{\tau}_\xi = \frac{\int E_\xi(0, k) k^{d-1} dk}{\int\int E_\xi(\omega, k) k^{d-1} d\omega dk}. \tag{1.8}$$

The introduced scales can be defined for both the input and the output in the stochastic system described by (1.1).

Let us check the simplest two-scale spectrum of the input which is of interest for applications

$$E_S(\omega, k) = \frac{\sigma_S^2 \tau_S l_S^2}{\pi^2 (1 + \tau_S^2 \omega^2)(1 + l_S^2 k^2)^2}, \tag{1.9}$$

where the source is assumed to be Markovian in time with the correlation time τ_S and homogeneous in space with the correlation radius l_S, σ_S^2 is the variance of S which does not influence the scale relation. With future applications in mind we restrict ourselves to the $2D$ case. Additionally, assume that $\boldsymbol{u} = 0$ for the sake of simplicity. After some computations we obtain from (1.5)-(1.9) the following expressions for the space and time correlation scales of the tracer field

$$l_c = l_S \varphi_1(\alpha, \beta), \quad \tau_c = \lambda^{-1} \varphi_2(\alpha, \beta), \quad \tilde{\tau}_c = \lambda^{-1} \varphi_3(\alpha, \beta), \tag{1.10}$$

where

$$\varphi_1(\alpha, \beta) = \sqrt{\frac{p(\alpha, \beta)}{(\alpha - 1)(1 - \beta)(1 - \alpha\beta) - (1 - \beta)(1 - \alpha^2\beta) \log \alpha - \beta(1 - \alpha)^2 \log \beta}},$$

$$\varphi_2(\alpha, \beta) = \frac{\sqrt{\beta p(\alpha, \beta)}}{|(\alpha - 1)(\beta - 1)| \sqrt{1 - \alpha\beta + \alpha\beta \log(\alpha\beta)}},$$

$$\varphi_3(\alpha, \beta) = \frac{(\alpha^2 - 2\alpha \log \alpha - 1)(1 - \alpha\beta)^2}{2(\alpha - 1) p(\alpha, \beta)},$$

$$p(\alpha, \beta) = (1 - \alpha)(1 - \beta)(1 - \alpha\beta) + \alpha(1 - \beta)$$

$$\times (1 + \beta - 2\alpha\beta) \log \alpha - \alpha\beta^2 (1 - \alpha)^2 \log \beta, \tag{1.11}$$

the dimensionless parameters α and β are given by

$$\alpha = \frac{\kappa}{\lambda l_s^2}, \quad \beta = \frac{\lambda \tau_S}{1 + \lambda \tau_S}. \tag{1.12}$$

The corresponding variances are related by

$$\sigma_c^2 = \frac{\sigma_S^2 \beta p(\alpha, \beta)}{2\pi \lambda^2 (1 - \alpha)^2 (1 - \beta)(1 - \alpha\beta)^2}. \tag{1.13}$$

The plots of functions $\varphi_1(\alpha, \beta)$, $\varphi_2(\alpha, \beta)$, $\varphi_3(\alpha, \beta)$ (Figs. 1.1-1.3) show that for a reasonable range of parameters the space scale of c is determined mostly by the space scale of the source, while the time scale of the tracer depends mainly on the feedback parameter. Note that in the case of white noise forcing

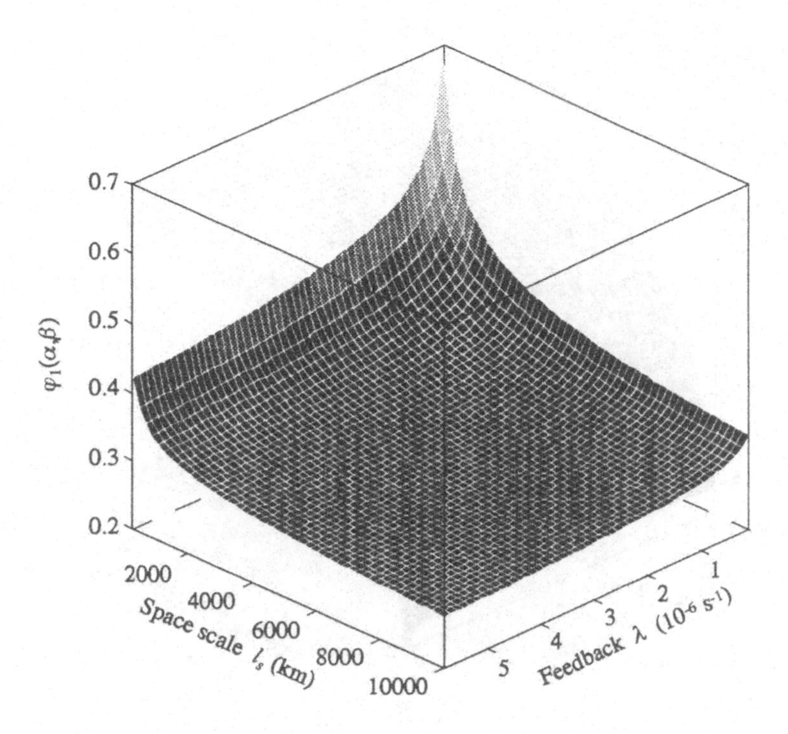

Figure 1.1 The space scale transfer function for the stochastic system with constant diffusivity and feedback and no advection.

($\tau_S = 0$) we have

$$\varphi_1(\alpha, 0) = \sqrt{\frac{\alpha \log \alpha - \alpha + 1}{\alpha - 1 - \log \alpha}},$$

$$\varphi_2(\alpha, 0) = 0, \tag{1.14}$$

$$\varphi_3(\alpha, 0) = \frac{\alpha^2 - 2\alpha \log \alpha - 1}{(\alpha - 1)(1 - \alpha + \alpha \log \alpha)},$$

The plots of $\varphi_1(\alpha, 0)$, $\varphi_3(\alpha, 0)$ are shown in Figs. 1.4 and 1.5.

In this case we can progress towards a solution of the forward problem by describing excursion of the output field c above a high level in terms of the input scales and the parameters of the underlying equation. Studying high level excursions is important for such problems as large scale SST anomalies.

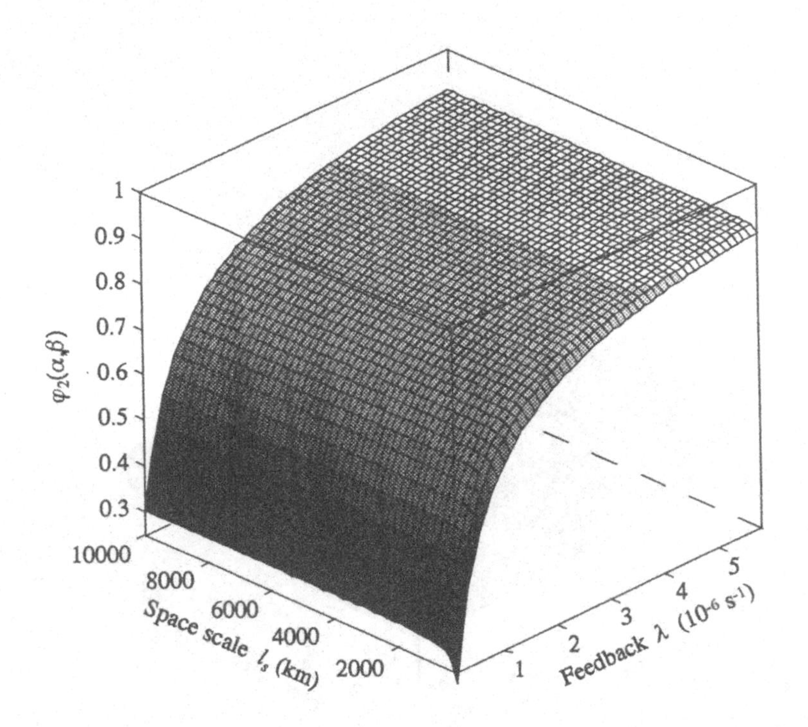

Figure 1.2 The time scale transfer function for the stochastic system with constant diffusivity and feedback and no advection when the input is a function differentiable in time.

First, recall the main issues of the theory of high excursions for Gaussian homogeneous fields (Nosko 1969, 1985). Let $\xi(x, y)$ be a random field with zero mean and u a fixed level. Consider the random surface $z = \xi(x, y)$ in the (x, y, z)-space. The connected components of the set

$$\{(x, y, z) : \xi(x, y) \geq u\} \tag{1.15}$$

are called the excursions above level u. Assume that $\xi(x, y)$ is Gaussian, homogeneous and differentiable in both variables and define the horizontal scales as follows

$$l_1^2 = \sigma_\xi^2 / \langle \xi_x^2 \rangle, \quad l_2^2 = \sigma_\xi^2 / \langle \xi_y^2 \rangle, \tag{1.16}$$

where the subscripts indicate the corresponding derivatives. Note that in the isotropic case $l_1 = l_2 = l_\xi$, where the latter is defined by (1.5). The key point of the mentioned theory is that an individual high excursion of ξ for a big u reproduces the shape of its correlation function. More exactly, if (x_0, y_0) is a

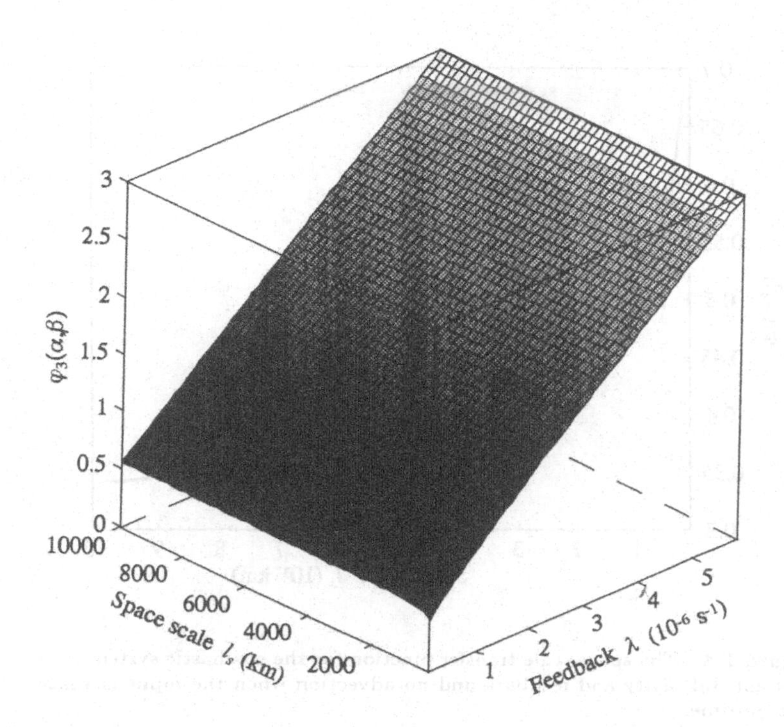

Figure 1.3 The same as Fig. 1.2 but for the case when the input is not differentiable.

point of a local maximum of ξ and this maximum is greater than u, then

$$\frac{\xi(x_0, y_0) - \xi(x, y)}{h_u} \sim \frac{(x - x_0)^2}{l_1^2} + \frac{(y - y_0)^2}{l_2^2}, \tag{1.17}$$

where h_u is the random value determining the excursion height. Here the sign '$\sim$' means that the quotient of both sides of the equality is bounded when $u \to \infty$. In other words, a high excursion looks like an elliptical paraboloid. Its bottom (intersection with the plane $z = u$) is an ellipse similar to the correlation ellipse of ξ (Fig. 1.6). Certainly, this ellipse is random. Its area S_u is described by the exponential distribution with the mean

$$\langle S_u \rangle = 2\pi l_1 l_2 (\sigma/u)^2, \tag{1.18}$$

where $\sigma = \sigma_\xi$. The centers of excursions form a Poisson process in the plane. Its density can also be expressed in terms of σ, l_1, and l_2 only.

Moreover, the discussed theory is able to describe excursions of time dependent random fields. Suppose now that $\xi(t, x, y)$ is a smooth Gaussian random field

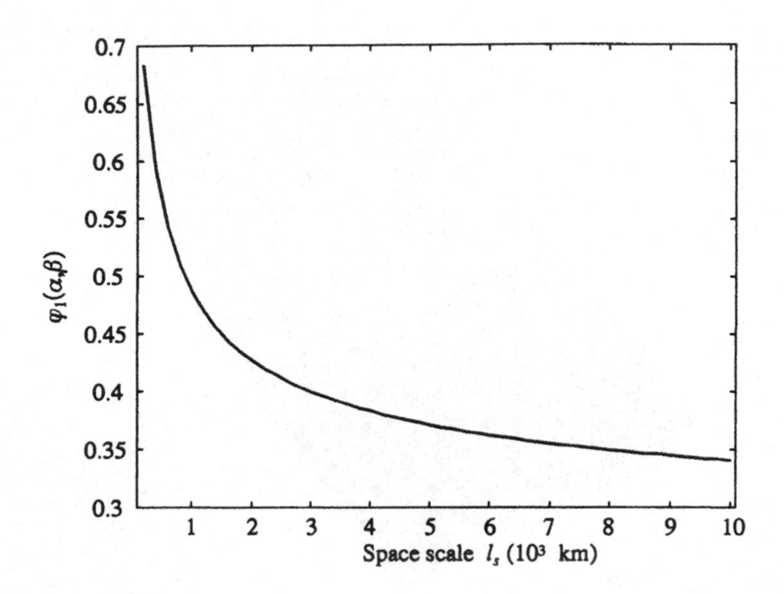

Figure 1.4 The space scale transfer function for the stochastic system with constant diffusivity and feedback and no advection when the input is white noise in time.

with zero mean, homogeneous in the space, stationary in time and let $\tau = \tau_\xi$ be the scale defined in (1.6). Then

$$\frac{\xi(t_0, x_0, y_0) - \xi(t, x, y)}{h_u} \sim \frac{(t - t_0)^2}{\tau^2} + \frac{(x - x_0)^2}{l_1^2} + \frac{(y - y_0)^2}{l_2^2}, \tag{1.19}$$

in some neighborhood of the point $(t_0,\ x_0,\ y_0)$ of local maximum exceeding the high level u. Thus, the shape of high level excursions as a space-time phenomenon is described by a parabolic ellipsoid. The length τ of its time semi-axis can be interpreted as the duration of the excursion (Fig. 1.7). In applications to real fields, we will refer to a high excursion as a large anomaly and its duration as its lifetime. In the framework of the considered model the lifetime τ_u has the Rayleigh distribution with the mean

$$\langle \tau_u \rangle = \sqrt{2\pi}\tau\sigma/u. \tag{1.20}$$

Note, that the area occupied by the large anomaly and its lifetime are related deterministically

$$S_u = \frac{\pi \tau_u^2 l_1 l_2}{4\tau^2}. \tag{1.22}$$

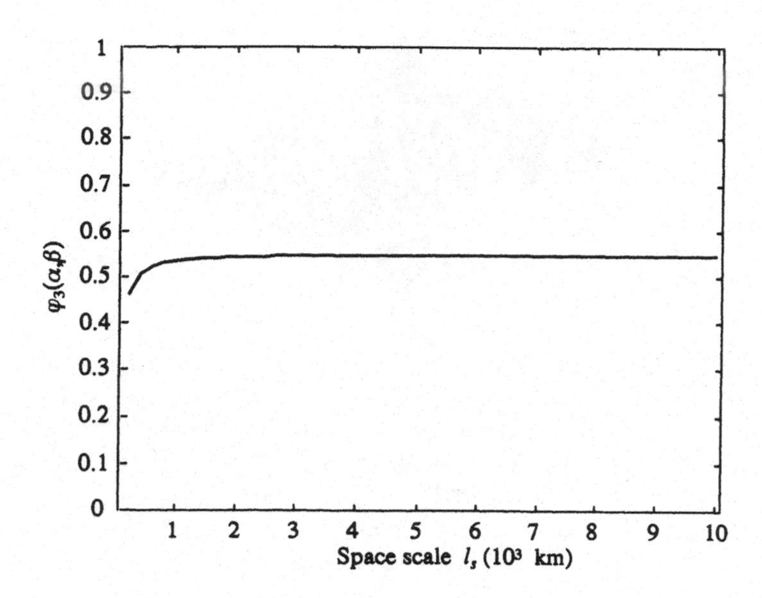

Figure 1.5 The time scale transfer function for the stochastic system with constant diffusivity and feedback and no advection when the input is white noise in time.

Finally, the occurrence time of large anomalies form a Poisson process with the average time between events

$$\langle \theta_u \rangle = \frac{(2\pi)^{3/2}\sigma^2 \tau l_1 l_2}{S_0 u^2 (1 - \Phi(u/\sigma))}, \tag{1.23}$$

where Φ is the standard Gaussian distribution function, and S_0 is the area of the region in question. From the practical viewpoint it is important to know for which u the above asymptotic formulas are valid. The answer to this question has been given in (Belyaev, Nosko, and Philimonova 1972), where these formulas were tested by the Monte-Carlo method for $u = \sigma, \ 1.5\sigma, \ 2\sigma, \ \ldots$. It was shown that for $u = \sigma$ the error of the mentioned formulas is better than 50 percent and for $u = 2\sigma$ the accuracy is practically absolute.

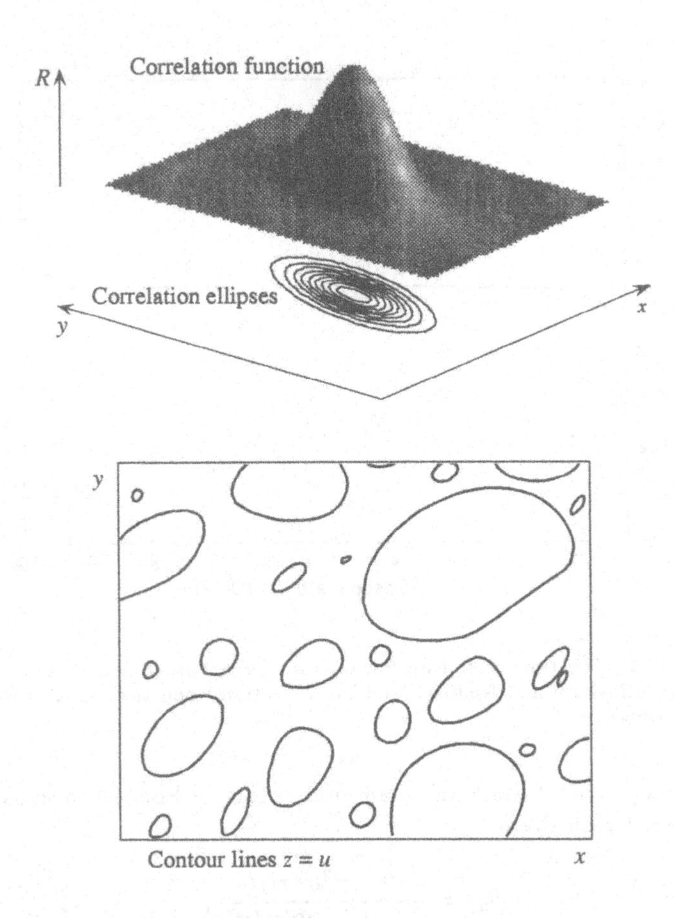

Figure 1.6 The correlation function of a homogeneous Gaussian field and schematic contour lines of high excursions.

1.2 RANDOM FLOW: CLOSURE PROBLEM

In reality, the most interesting physical effects are due to random velocity fields. In this case, quite a different set of mathematical problems arise. When the velocity field is random, it becomes very difficult to obtain even the mean tracer distribution, not to mention higher order statistics. Furthermore, the problem of finding the mean tracer field has not yet been solved for general velocity fields while in the previous setting this was absolutely trivial.

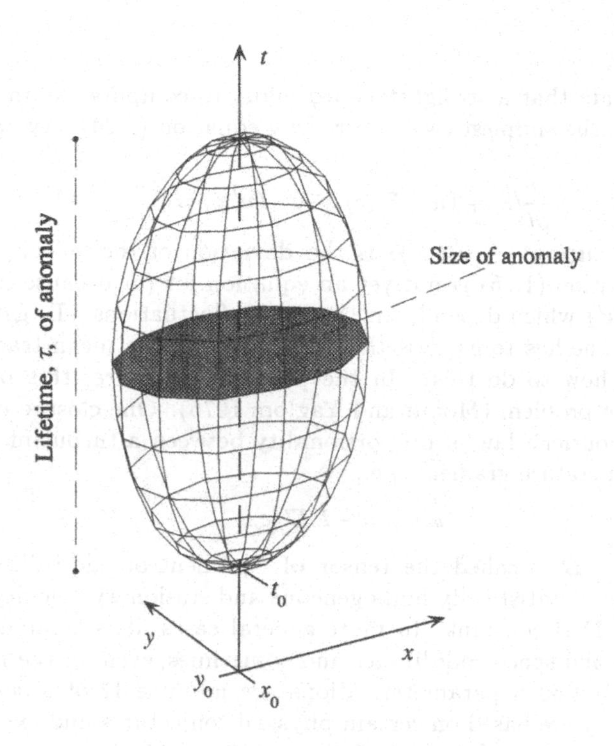

Figure 1.7 The anomaly lifetime and size as defined in high level excursion theory.

Firstly, let us consider the simplest case including only the random advection

$$\frac{\partial c}{\partial t} + \boldsymbol{u} \cdot \nabla c = 0. \tag{1.24}$$

The general problem to be discussed here is how to describe the behavior of the mean field $\langle c(t, \boldsymbol{r})\rangle$ and the statistics of the fluctuating fields $c'(t, \boldsymbol{r}) = c(t, \boldsymbol{r}) - \langle c(t, \boldsymbol{r})\rangle$, $\nabla c'(t, \boldsymbol{r}) = \nabla c'(t, \boldsymbol{r}) - \langle \nabla c'(t, \boldsymbol{r})\rangle$ under given statistics of the velocity field $\boldsymbol{u}(t, \boldsymbol{r})$. It should be stressed that we consider a kinematic problem, i.e. the statistics of $\boldsymbol{u}(\mathrm{t}, \boldsymbol{r})$ are not derived from the equations of motion, but rather are given (for example, by its correlation tensor or spectral tensor in the Gaussian case). On one hand, this approach enables us to deal with the general form of the velocity spectrum or correlation function. On the other hand, the number of physical problems that one can study in this framework is quite restricted and does not include, for example, convection flows.

Let us demonstrate that a straightforward effort to compute the mean tracer field fails even in the simplest case covered by equation (1.24). Averaging this equation yields

$$\frac{\partial}{\partial t}\langle c\rangle + \langle \boldsymbol{u}\rangle \cdot \nabla\langle c\rangle + \nabla \cdot \langle \boldsymbol{u}' c'\rangle = 0, \tag{1.25}$$

where $\boldsymbol{u}'(t,\boldsymbol{r}) = \boldsymbol{u}(t,\boldsymbol{r}) - \langle \boldsymbol{u}(t,\boldsymbol{r})\rangle$ is the deviation of the velocity from its mean value. Relation (1.25) is not yet an equation for $\langle c\rangle$ because it contains the term $\langle \boldsymbol{u}' \cdot \nabla c'\rangle$ which depends on the tracer fluctuations. To get a closed equation for $\langle c\rangle$ one has to express this term through the mean tracer. Then the question is, how to do this? In the physical literature, this problem is called the closure problem (Monin and Yaglom 1975). One closure conjecture corresponds to Fourier's law of proportionality between a turbulent heat flow and a mean temperature gradient, i.e.,

$$\langle \boldsymbol{u}' c'\rangle = -\boldsymbol{D}\nabla\langle c\rangle, \tag{1.26}$$

where the matrix $\boldsymbol{D}$ is called the tensor of turbulent or eddy diffusion. In the simplest case of statistically homogeneous and stationary velocity fields, it is assumed that $\boldsymbol{D}$ is constant. In more general cases, it is assumed that $\boldsymbol{D}$ depends on time and space coordinates and sometimes, even on the mean field $\langle c\rangle$ (Okubo 1980). Such parameterizations are admittedly of a semiempiric character, i.e., they are based on certain physical conjectures and experimental facts. By substituting the closure (1.26) into equation (1.25), one gets a closed equation for the mean tracer field:

$$\frac{\partial}{\partial t}\langle c\rangle + \langle \boldsymbol{u}\rangle \cdot \nabla\langle c\rangle = \nabla \cdot \boldsymbol{D}\nabla\langle c\rangle. \tag{1.27}$$

In a certain sense, equation (1.27) is the canonical description of turbulent diffusion. Its main difficulty consists in parameterization of the tensor $\boldsymbol{D}$; see for discussion and physically reasonable parameterizations (Okubo 1980). However, the procedure of obtaining the equation for the mean tracer field given above is not rigorous, and the first question is whether equation (1.27) is a consequence of (1.24), and if so, then how can the tensor of turbulent diffusion be expressed in terms of statistical characteristics of the velocity field: spectral tensor, correlation tensor and so on?

Now let us show that there is a mathematical reason to expect that the mean tracer can be described by equation (1.27). Let $\boldsymbol{X}_{t,\boldsymbol{r}}(\cdot)$ be the Lagrangian trajectory passing through point $\boldsymbol{r}$ at moment t, i.e. $\boldsymbol{X}_{t,\boldsymbol{r}}(s)$ is the solution of the following integral equation

$$\boldsymbol{X}_{t,\boldsymbol{r}}(s) = \boldsymbol{r} + \int_t^s \boldsymbol{u}(v, \boldsymbol{X}_{t,\boldsymbol{r}}(v))dv. \tag{1.28}$$

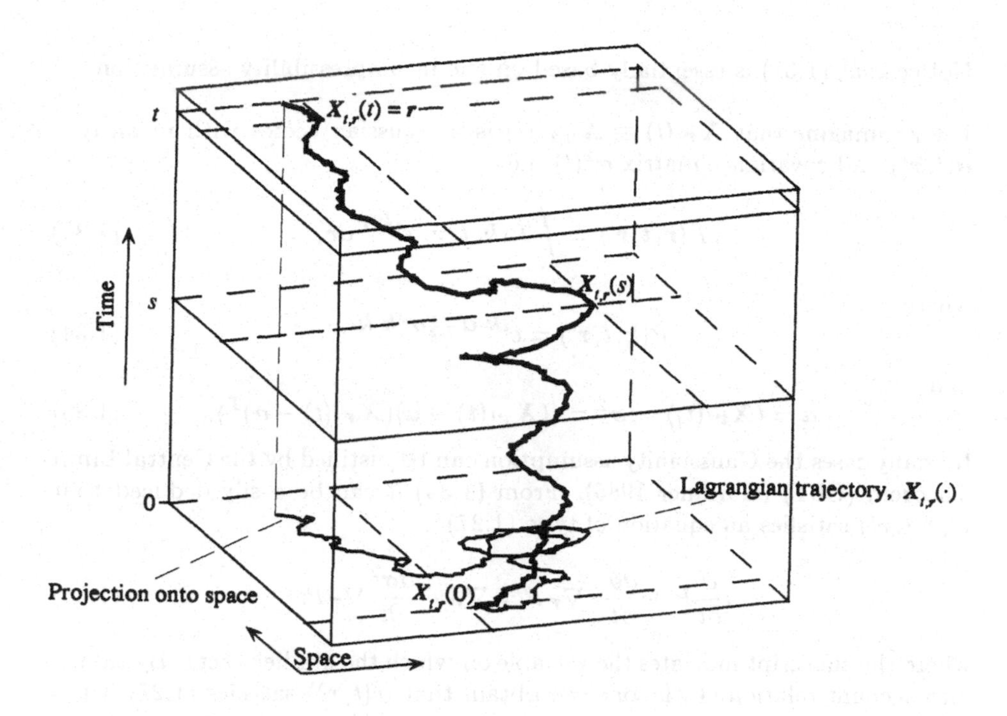

Figure 1.8 The Lagrangian trajectory of an upper ocean drifting buoy passing through point $\boldsymbol{r}$ at time t.

In mathematical language $\boldsymbol{X}_{t,\boldsymbol{r}}(\cdot)$ is the characteristic of equation (1.24) passing through $(t, \boldsymbol{r})$. It can easily be checked that the formula

$$c(t, \boldsymbol{r}) = c_0(\boldsymbol{X}_{t,\boldsymbol{r}}(0)) \tag{1.29}$$

is a solution of the Cauchy problem (1.24) corresponding to the initial condition

$$c|_{t=0} = c_0(\boldsymbol{r}). \tag{1.30}$$

Relation (1.29) expresses the conservation law in an incompressible fluid. From (1.29) it can be deduced (Davis 1982, Bennett 1987) that

$$\langle c(t, \boldsymbol{r}) \rangle = \int c_0(\boldsymbol{r}') P(\boldsymbol{r}; t, \boldsymbol{r}') d\boldsymbol{r}', \tag{1.31}$$

where $P(\boldsymbol{r}; t, \boldsymbol{r}')$ is the probability density of $\boldsymbol{X}_{0,\boldsymbol{r}'}(t)$, i.e.

$$P(\boldsymbol{r}; t, \boldsymbol{r}') d\boldsymbol{r} = Pr\{\boldsymbol{X}_{0,\boldsymbol{r}'}(t) \in (\boldsymbol{r}, \boldsymbol{r} + d\boldsymbol{r})\}. \tag{1.32}$$

Notice that (1.31) is essentially based on the incompressibility assumption.

Let us imagine that $\boldsymbol{X}_{\boldsymbol{r}'}(t) \equiv \boldsymbol{X}_{0,\boldsymbol{r}'}(t)$ is a Gaussian vector with mean $\boldsymbol{a} = \boldsymbol{a}(t, \boldsymbol{r}')$ and covariance matrix $\boldsymbol{\sigma}^2(t)$, i.e.

$$P(\boldsymbol{r}; t, \boldsymbol{r}') = \int \hat{P}(\boldsymbol{k}; t, \boldsymbol{r}') e^{i\boldsymbol{k}\cdot\boldsymbol{r}} d\boldsymbol{r}, \tag{1.33}$$

where

$$\hat{P}(\boldsymbol{k}; t, \boldsymbol{r}') = e^{i\boldsymbol{k}\cdot\boldsymbol{a} - \frac{1}{2}\boldsymbol{\sigma}^2\boldsymbol{k}\cdot\boldsymbol{k}} \tag{1.34}$$

and

$$\boldsymbol{a} = \langle \boldsymbol{X}_{\boldsymbol{r}'}(t) \rangle, \quad \boldsymbol{\sigma}^2 = \langle (\boldsymbol{X}_{\boldsymbol{r}'}(t) - \boldsymbol{a})(\boldsymbol{X}_{\boldsymbol{r}'}(t) - \boldsymbol{a})^T \rangle. \tag{1.35}$$

In many cases the Gaussianity assumption can be justified by the Central Limit Theorem (CLT) (Gardiner 1985). From (1.34) it can be easily deduced that $P(\boldsymbol{r}; t, \boldsymbol{r}')$ satisfies an equation of type (1.27)

$$\frac{\partial}{\partial t} P + \frac{\partial \boldsymbol{a}}{\partial t} \cdot \nabla_{\boldsymbol{r}} P = \nabla_{\boldsymbol{r}} \cdot \frac{\partial \boldsymbol{\sigma}^2}{\partial t} \nabla_{\boldsymbol{r}} P. \tag{1.36}$$

where the subscript indicates the variable on which the gradient acts. By taking into account relation (1.31), one can obtain that $\langle c(t, \boldsymbol{r}) \rangle$ satisfies (1.27) with

$$\boldsymbol{D} = \frac{\partial}{\partial t} \langle (\boldsymbol{X}_{\boldsymbol{r}'}(t) - \boldsymbol{a})(\boldsymbol{X}_{\boldsymbol{r}'}(t) - \boldsymbol{a})^T \rangle. \tag{1.37}$$

This is the famous Taylor's formula (Taylor 1921), which does not solve the problem of expressing the diffusivity in terms of statistics of the Eulerian velocity field, but merely reduces it to the problem of the relationship between the statistics of Lagrangian velocity field and Eulerian velocity field (Davis 1982, Gurbatov, Malakhov, and Saichev 1991).

Thus, we showed that the canonical equation (1.27) holds if the Lagrangian displacement is Gaussian. Gaussianity can be either rigorously derived as an asymptotic case under a scale separation (see next chapter 2) or justified in an heuristic way by using the CLT (Monin and Yaglom 1975, Bennet 1996).

It is worth noting that there is a simple example where the Gaussianity follows directly from the problem statement. Namely, consider the Eulerian velocity field which depends on the time variable only $\boldsymbol{u} = \boldsymbol{u}(t)$. From equation (1.28) it readily follows that

$$\boldsymbol{X}_{\boldsymbol{r}'}(t) = \boldsymbol{r}' + \int_0^t \boldsymbol{u}(s) ds. \tag{1.38}$$

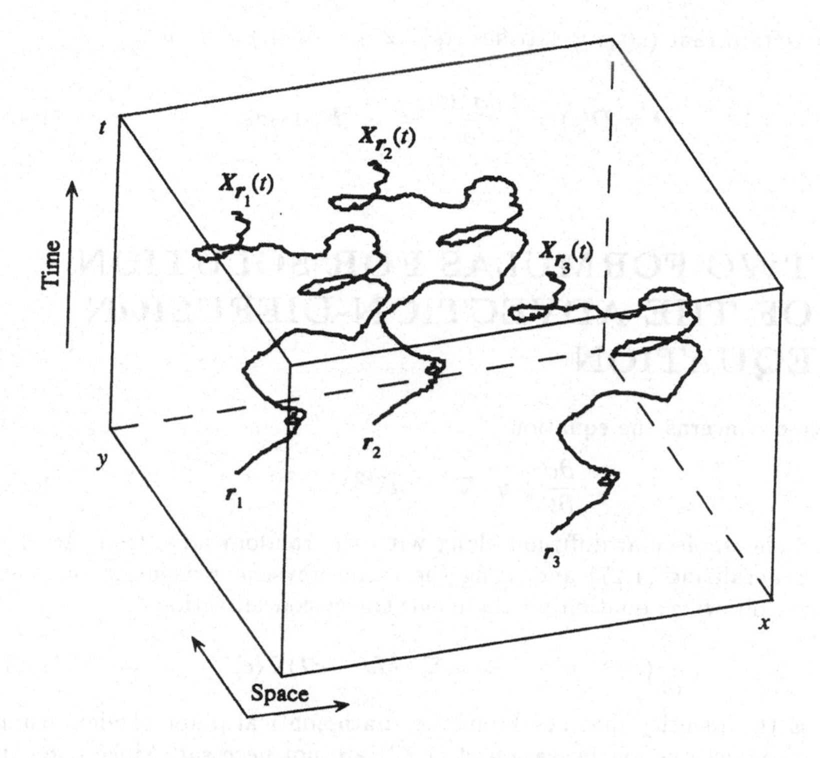

Figure 1.9 Schematic plot of Lagrangian trajectories for a Eulerian velocity field independent of space coordinates.

Thus in this case, all trajectories are parallel (Fig. 1.9). From (1.38) it follows that $\boldsymbol{X}_{\boldsymbol{r}'}(t)$ is Gaussian as an integral of the Gaussian process and its first two moment are trivially computed

$$\langle \boldsymbol{X}_{\boldsymbol{r}'}(t) \rangle = \boldsymbol{r}',$$

$$\langle (\boldsymbol{X}_{\boldsymbol{r}'}(t) - \boldsymbol{a})(\boldsymbol{X}_{\boldsymbol{r}'}(t) - \boldsymbol{a})^T \rangle = \left\langle \int_0^t \boldsymbol{u}(s)ds \left(\int_0^t \boldsymbol{u}(s)ds \right)^T \right\rangle = \tag{1.39}$$

$$= \int_0^t \int_0^t \boldsymbol{R}_{\boldsymbol{u}}(s_1 - s_2)ds_1\, ds_2 = \int_{-t}^t (t - |s|)\boldsymbol{R}_{\boldsymbol{u}}(s)ds,$$

where the correlation function for the velocity is given by

$$\boldsymbol{R}_{\boldsymbol{u}}(s) = \langle \boldsymbol{u}(t)\boldsymbol{u}(t+s)^T \rangle. \tag{1.40}$$

Thus, we obtain that $\langle c(t,\boldsymbol{r})\rangle$ satisfies eq.(1.27) with $\langle \boldsymbol{u}\rangle = 0$ and

$$\boldsymbol{D} = \boldsymbol{D}(t) = \frac{1}{2}\frac{d\sigma^2(t)}{dt} = \int_0^t \boldsymbol{R}_{\boldsymbol{u}}(s)ds. \tag{1.41}$$

1.3 TWO FORMULAS FOR SOLUTION OF THE ADVECTION-DIFFUSION EQUATION

This section concerns the equation

$$\frac{\partial c}{\partial t} + \boldsymbol{u}\cdot\nabla c = \kappa\nabla^2 c, \tag{1.42}$$

which includes molecular diffusion along with the random advection. In this case, by generalizing (1.27) and using the same physical reasoning, one can suggest the following equation for the mean tracer concentration

$$\frac{\partial}{\partial t}\langle c\rangle + \langle \boldsymbol{u}\rangle\cdot\nabla\langle c\rangle = \nabla\cdot(\boldsymbol{D}+\kappa \boldsymbol{I})\nabla\langle c\rangle, \tag{1.43}$$

where $\boldsymbol{I}$ is the identity matrix. From the mathematical point of view using additional conjectures for derivation of (1.43) are not necessary since one can solve (1.42) for given initial conditions and average over the resulting ensemble. More exactly, the solution of (1.42), given the initial condition (1.30), is formally

$$c(t,\boldsymbol{r}) = N_{t,\boldsymbol{r}}(c_0(\boldsymbol{r}'), \boldsymbol{u}(s,\boldsymbol{r}'), \boldsymbol{r}'\in E, 0\le s\le t), \tag{1.44}$$

where the functional in (1.44) is linear in the initial field $c_0(\cdot)$ and nonlinear in the velocity field $\boldsymbol{u}(\cdot,\cdot)$ and depends on $t, \boldsymbol{r}$ and κ. Theoretically speaking, one can average this functional over the ensemble $\{\boldsymbol{u}(\cdot,\cdot)\}$, in order to obtain an expression for $\langle c(t,\boldsymbol{r})\rangle$ and then check the validity of the equation (1.43). Then it is not necessary to state (1.43), because it would be possible to verify its validity. Although in many asymptotic cases this averaging can be done explicitly and, consequently, eq.(1.43) can be derived rigorously, the trouble is that under general conditions on the velocity, this possibility is purely theoretical. Below we give two examples of 'explicit' representation of type (1.44) for the solution of the Cauchy problem for equation (1.42) corresponding to the initial condition (1.30).

Let us forget for while that the velocity field is stochastic and introduce a random Wiener process (Brownian motion) $\boldsymbol{w}(t)$ with values in d-dimensional

Euclidean space E^d, i.e., a Gaussian random process whose first and second moments are given by the formulas

$$\boldsymbol{E}\{\boldsymbol{w}\} = 0, \quad \boldsymbol{E}\{w_i(t)w_j(s)\} = \delta_{ij}\min(t,s), \tag{1.45}$$

where $w_i(t)$ is the i-th component of $\boldsymbol{w}(t)$. Henceforth, the symbol $\boldsymbol{E}\{\cdot\}$ (expectation) will denote averaging over the ensemble of realizations of the Wiener process. We also can view the Wiener process as an integral of Gaussian white noise $\boldsymbol{b}(\cdot)$,

$$\boldsymbol{w}(t) - \boldsymbol{w}(s) = \int_s^t \boldsymbol{b}(v)dv. \tag{1.46}$$

For any pair $(t,\boldsymbol{r})$, we define the generalized Lagrangian trajectory $\boldsymbol{\xi}(\cdot) = \boldsymbol{\xi}_{t,\boldsymbol{r}}(\cdot)$ as the solution of the following stochastic differential equation

$$d\boldsymbol{\xi}(s) = \boldsymbol{u}(s,\boldsymbol{\xi}(s))ds + \sqrt{2\kappa}d\boldsymbol{w}(s), \tag{1.47}$$

satisfying

$$\boldsymbol{\xi}(t) = \boldsymbol{r}, \tag{1.48}$$

which is equivalent to the integral equation

$$\boldsymbol{\xi}_{t,\boldsymbol{r}}(s) = \boldsymbol{r} + \int_t^s \boldsymbol{u}(v,\boldsymbol{\xi}_{t,\boldsymbol{r}}(v))dv + \sqrt{2\kappa}(\boldsymbol{w}(s) - \boldsymbol{w}(t)). \tag{1.49}$$

If $\kappa = 0$, the curve $\boldsymbol{\xi}_{t,\boldsymbol{r}}(s) = \boldsymbol{X}_{t,\boldsymbol{r}}(s)$ is the usual Lagrangian trajectory passing through the point $(t,\boldsymbol{r})$ defined by equation (1.28).

The first representation, we will use further, generalizes (1.29)

$$c(t,\boldsymbol{r}) = \boldsymbol{E}\{c_0(\boldsymbol{\xi}_{t,\boldsymbol{r}}(0))\} \tag{1.50}$$

and the second one is given by

$$\begin{aligned} c(t,\boldsymbol{r}) = \boldsymbol{E}\{\exp[\frac{1}{\sqrt{2\kappa}}\int_0^t \boldsymbol{u}(s,\boldsymbol{y}_s)\cdot d\boldsymbol{w}(s) - \\ -\frac{1}{4\kappa}\int_0^t \boldsymbol{u}^2(s,\boldsymbol{y}_s)ds]c_0(\boldsymbol{y}_t)\}, \end{aligned} \tag{1.51}$$

where

$$\boldsymbol{y}_s = \boldsymbol{r} + \sqrt{2\kappa}\boldsymbol{w}(s) \tag{1.52}$$

and the stochastic integral under the exponent is to be interpreted in the Ito's sense (Gardiner 1985). The rigorous proof of (1.50) can be found in (Friedman 1975, p. 147). The second representation is based on the Girsanov-Cameron-Martin theorem (ibid, p. 156). It should be noted that formulas (1.50), (1.51) hold regardless whether the velocity field is random or not.

The first representation (1.50) is widely used in the physical literature since it has quite clear physical meaning. Namely, the molecular diffusion is interpreted as a result of the Brownian motion of fluid particles. The stochastic equation (1.49), which extends the Langevin equation, describes both macro- and micro-displacement of a particle passing through a given point. In order to get the macro-distribution of the tracer particle we should remove the micro-fluctuations by averaging over the Brownian motion random ensemble what is exactly done in (1.50).

It seems there is no reasonable physical interpretation of the second representation (1.51). For this reason, perhaps, it is not used in applications at all. As we will see later, it nevertheless is helpful for studying turbulent diffusion when the velocity field has a finite correlation time (section 5.1).

Now we give a simple, but not absolutely rigorous derivation of the mentioned representations for the initial conditions given at an arbitrary point

$$c|_{t=s} = f(\boldsymbol{r}). \tag{1.53}$$

In this case formulas (1.50), (1.51) become

$$c(t, \boldsymbol{r}) = \boldsymbol{E}\{f(\boldsymbol{\xi}_{t,\boldsymbol{r}}(s))\} \tag{1.54}$$

and

$$c(t, \boldsymbol{r}) = \boldsymbol{E}\{\exp\left[I_{ts}(\boldsymbol{r})\right] f(\boldsymbol{y}_{ts})\}, \tag{1.55}$$

respectively, where

$$\begin{aligned} \boldsymbol{y}_{ts} &= \boldsymbol{r} + \sqrt{2\kappa}(\boldsymbol{w}(t) - \boldsymbol{w}(s)), \\ I_{ts}(\boldsymbol{r}) &= \frac{1}{\sqrt{2\kappa}} \int_s^t \boldsymbol{u}(v, \boldsymbol{y}_{tv}) \cdot d\boldsymbol{w}(v) - \frac{1}{4\kappa} \int_s^t \boldsymbol{u}^2(v, \boldsymbol{y}_{tv}) dv. \end{aligned} \tag{1.56}$$

To check formulas (1.54), (1.55) we need the following Lemma 1.1, that, perhaps, would be helpful in the search for other representations of the solution to (1.42).

Lemma 1.1. *Consider the Cauchy problem*

$$\frac{\partial c}{\partial t} = L_t c, \qquad t > s, \qquad c|_{t=s} = f(\boldsymbol{r}), \tag{1.57}$$

where L_t is a linear operator acting on the space coordinate $\boldsymbol{r}$ with coefficients depending on t, and let F_{ts} be a family of linear operators depending on the Wiener process $\boldsymbol{w}(\cdot)$ satisfying the following conditions

$$(i) F_{ts} = F_{tv} F_{vs}, \qquad for \quad t \geq v \geq s. \tag{1.58}$$

(ii) F_{ts} depends only on increments $\boldsymbol{w}(t) - \boldsymbol{w}(t')$, $s \leq t' \leq t$.

$$(iii) \frac{\partial}{\partial t} \boldsymbol{E}\{F_{ts}\}\big|_{s=t} = L_t, \tag{1.59}$$

then

$$\frac{\partial}{\partial t} \boldsymbol{E}\{F_{ts}\} = L_t \boldsymbol{E}\{F_{ts}\}, \tag{1.60}$$

i.e. the solution of (1.57) is given by

$$c(t, \boldsymbol{r}) = \boldsymbol{E}\{F_{ts} f(\boldsymbol{r})\}. \tag{1.61}$$

The proof is straightforward. By differentiating both sides of (1.58) with respect to t we have

$$\frac{\partial}{\partial t} F_{ts} f = \frac{\partial}{\partial t} F_{tv} F_{vs} f \tag{1.62}$$

By using the independence of the Wiener process increments on non-overlapping intervals we obtain from (ii)

$$\frac{\partial}{\partial t} \boldsymbol{E}\{F_{ts} f\} = \frac{\partial}{\partial t} \boldsymbol{E}\{F_{tv}\} \boldsymbol{E}\{F_{vs} f\}. \tag{1.63}$$

Setting $v = t$ in (1.63) and using (1.59), we arrive at (1.60). ∘

Now we are ready to check (1.54)-(1.55). As for (1.54), define the transformation A_{ts} of the Euclidean space E by $A_{ts}\boldsymbol{r} = \xi_{t,\boldsymbol{r}}(s)$ where $\xi_{t,\boldsymbol{r}}(s)$ is determined in (1.49). Then set

$$F_{ts} f(\boldsymbol{r}) = f(A_{ts}\boldsymbol{r}). \tag{1.64}$$

Since $A_{ts}\boldsymbol{r}$ is nothing more then the coordinate of the fluid particle at moment s passing through point $\boldsymbol{r}$ at moment t, one can conclude (see Fig. 1.10) that

$$A_{ts} = A_{tv} A_{vs}. \tag{1.65}$$

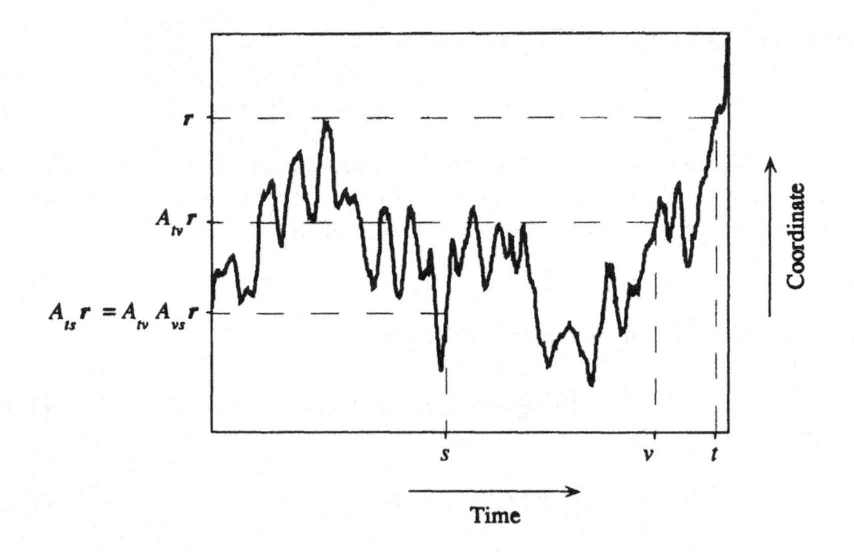

Figure 1.10 To the definition of the semigroup map $A_{ts} : E^d \to E^d$ related to a Lagrangian trajectory.

Thus, condition (*i*) follows from (1.65).

Then, it is easily seen from the integral equation (1.49) that condition (*ii*) is fulfilled.

Setting $t = t + \tau, s = t$ in (1.54) we obtain that condition (*iii*) is equivalent to

$$\lim_{\tau \to 0} \frac{c(t+\tau, \boldsymbol{r}) - c(t, \boldsymbol{r})}{\tau} = -\boldsymbol{u}(t, \boldsymbol{r}) \cdot \nabla c(t, \boldsymbol{r}) + \kappa \nabla^2 c(t, \boldsymbol{r}). \tag{1.66}$$

Finally, (1.66) follows from (1.54). To see this, the integral equation (1.49) gives for small τ

$$\xi_{t+\tau, \boldsymbol{r}}(t) \sim \boldsymbol{r} - \boldsymbol{u}(t, \boldsymbol{r})\tau + \sqrt{2\kappa}(\boldsymbol{w}(t) - \boldsymbol{w}(t+\tau)). \tag{1.67}$$

From Taylor's expansion it follows that

$$c(\boldsymbol{\xi}_{t+\tau, \boldsymbol{r}}(t)) \sim c(t, \boldsymbol{r}) + \Delta\boldsymbol{\xi} \cdot \nabla c(t, \boldsymbol{r}) + \frac{1}{2}(\nabla \cdot \Delta\boldsymbol{\xi} \Delta\boldsymbol{\xi}^T \nabla) c(t, \boldsymbol{r}), \tag{1.68}$$

where

$$\Delta\boldsymbol{\xi} = \boldsymbol{\xi}_{t+\tau, \boldsymbol{r}}(t) - \boldsymbol{r}. \tag{1.69}$$

By combining (1.67) and (1.68) one obtains

$$\begin{aligned} c(t+\tau,\boldsymbol{r}) \sim \boldsymbol{E}\{c(t,\boldsymbol{r}) + (-\tau\boldsymbol{u} + \sqrt{2\kappa}\Delta\boldsymbol{w})\cdot\nabla)c(t,\boldsymbol{r}) \\ +\kappa(\nabla\cdot\Delta\boldsymbol{w}\Delta\boldsymbol{w}^T\nabla)c(t,\boldsymbol{r})\}, \end{aligned} \tag{1.70}$$

where

$$\Delta\boldsymbol{w} = \boldsymbol{w}(t) - \boldsymbol{w}(t+\tau). \tag{1.71}$$

By using (1.45) and passing to the limit $\tau \to 0$ one arrives to (1.66).

In order to prove (1.55) let us define operator F_{ts} as follows

$$F_{ts}f(\boldsymbol{r}) = \exp\left[I_{ts}(\boldsymbol{r})\right]f(\boldsymbol{y}_{ts}). \tag{1.72}$$

First, examine condition (i) of Lemma 1.1.

$$\begin{aligned} &F_{tv}F_{vs}f(\boldsymbol{r}) = F_{tv}\exp\left[I_{vs}(\boldsymbol{r})\right]f(\boldsymbol{r}+\sqrt{2\kappa}(\boldsymbol{w}(v)-\boldsymbol{w}(s))) = \\ &\exp\left[I_{tv}(\boldsymbol{r})\right]\exp\left[I_{vs}(\boldsymbol{r}+\sqrt{2\kappa}(\boldsymbol{w}(t)-\boldsymbol{w}(v)))\right]\times \\ &\times f(\boldsymbol{r}+\sqrt{2\kappa}(\boldsymbol{w}(t)-\boldsymbol{w}(v))+\sqrt{2\kappa}(\boldsymbol{w}(v)-\boldsymbol{w}(s))) = \\ &\exp\left[I_{ts}(\boldsymbol{r})\right]f(\boldsymbol{y}_{ts}). \end{aligned} \tag{1.73}$$

Second, the validity of condition (ii) follows from definition (1.56) of I_{ts}.

Third, check condition (iii) by setting $t = t+\tau$ and $s = t$ in (1.55). One obtains

$$\begin{aligned} &c(t+\tau,\boldsymbol{r}) \sim \boldsymbol{E}\{(-\sqrt{2\kappa}\Delta\boldsymbol{w}\cdot\nabla c(t,\boldsymbol{r}) + \\ &+\kappa(\nabla\cdot\boldsymbol{w}\boldsymbol{w}^T\nabla)c(t,\boldsymbol{r}))(1+\frac{1}{\sqrt{2\kappa}}\boldsymbol{u}\cdot\Delta\boldsymbol{w} + \\ &+\frac{1}{\sqrt{2\kappa}}\nabla\boldsymbol{u}\Delta\boldsymbol{w}\cdot\Delta\boldsymbol{w} + \frac{1}{4\kappa}|\boldsymbol{u}\cdot\Delta\boldsymbol{w}|^2 - \frac{1}{4\kappa}|\boldsymbol{u}|^2\tau)\} = \\ &c(t,\boldsymbol{r}) - \tau\boldsymbol{u}\cdot\nabla c(t,\boldsymbol{r}) + \tau\kappa\nabla^2 c(t,\boldsymbol{r}). \end{aligned} \tag{1.74}$$

By taking into account that $\boldsymbol{E}\{\nabla\boldsymbol{u}\Delta\boldsymbol{w}\cdot\Delta\boldsymbol{w}\} = \tau\nabla\cdot\boldsymbol{u} = 0$ and $\boldsymbol{E}\{(\boldsymbol{u}\cdot\Delta\boldsymbol{w})^2\} = \tau|\boldsymbol{u}|^2$ and by passing to the limit $\tau \to 0$, one can obtain that (1.66) holds and consequently (1.62) is true for any s, t.

Now let us go back to random velocity fields. At first glance, it seems that by using an explicit representation for each realization, it is not difficult to carry out ensemble averaging. Unfortunately, this is not feasible in the general case since the solution of (1.49) cannot be found in explicit form. Moreover, as we have seen before, one can not do this even when $\kappa = 0$. However in the case when the velocity does not depend on the space variable, a closed equation for the mean tracer can be easily derived. Let us do this by using, for example, representation (1.50).

Namely, suppose that $\boldsymbol{u} = \boldsymbol{u}(t)$ is the function of time only. From (1.49) one obtains

$$\boldsymbol{\xi}_{t,\boldsymbol{r}} = \boldsymbol{r} - \int_0^t \boldsymbol{u}(s)ds + \sqrt{2\kappa}\boldsymbol{w}(t), \tag{1.75}$$

where $\boldsymbol{\xi}_{t,\boldsymbol{r}} = \boldsymbol{\xi}_{t,\boldsymbol{r}}(\mathbf{0})$. Hence from (1.50)

$$c(t,\boldsymbol{r}) = \boldsymbol{E}\{c_0(\boldsymbol{r} - \int_0^t \boldsymbol{u}(s)ds + \sqrt{2\kappa}\boldsymbol{w}(t))\} =$$

$$\int_{E^d} c_0(\boldsymbol{r} - \boldsymbol{r}' - \int_0^t \boldsymbol{u}(s)ds) \exp\{-\frac{|\boldsymbol{r}'|^2}{2t\kappa}\} \frac{d\boldsymbol{r}'}{(2\pi t\kappa)^{d/2}}. \tag{1.76}$$

To average this functional with respect to $\boldsymbol{u}$ it is convenient to apply the Fourier transform

$$c_0(\boldsymbol{r}) = \int \hat{c}_0(\boldsymbol{k})\, e^{i\boldsymbol{k}\cdot\boldsymbol{r}} d\boldsymbol{r}. \tag{1.77}$$

Using the well known equality $\langle e^{\xi}\rangle = e^{\frac{1}{2}\langle\xi^2\rangle}$ for the Gaussian mean-zero random variable ξ, one obtains

$$\langle c(t,\boldsymbol{r})\rangle =$$

$$\int_{E^d}\int_{E^d} \hat{c}_0(\boldsymbol{k})e^{i\boldsymbol{k}\cdot(\boldsymbol{r}-\boldsymbol{r}')} \frac{1}{(2\pi t\kappa)^{d/2}} \exp\{-\frac{|\boldsymbol{r}'|^2}{2t\kappa}\}\langle\exp\{-i\boldsymbol{k}\cdot\int_0^t \boldsymbol{u}(s)ds\}\rangle d\kappa d\boldsymbol{r}' =$$

$$\int_{E^d} \hat{c}_0(\boldsymbol{k}) \exp i\boldsymbol{k}\cdot\boldsymbol{r} - \kappa t k^2 - \frac{1}{2}\boldsymbol{\sigma}^2(t)\boldsymbol{k}\cdot\boldsymbol{k} d\boldsymbol{k}, \tag{1.78}$$

where as before

$$\boldsymbol{\sigma}^2(t) = \int_{-t}^t (t-|s|)\boldsymbol{R_u}(s)ds. \tag{1.79}$$

It follows from (1.78) that $\langle c(t,\boldsymbol{r})\rangle$ satisfies eq.(1.43) with

$$\boldsymbol{D} = \boldsymbol{D}(t) = \int_0^t \boldsymbol{R_u}(s)ds. \tag{1.80}$$

This simple example shows that in order to get a closed equation for the mean admixture we do not need any additional closure conjectures, such as $\langle \mathbf{u}'c'\rangle$ is proportional to $\nabla\langle c\rangle$, used in the physical literature (Monin and Yaglom 1975). Difficulties arising when averaging the advection-diffusion equation, have an analytical origin but the problem of computing the average concentration is a well posed mathematical problem.

1.4 SUMMARY

The relationship between the space and time scales of the forcing S and tracer concentration c is studied using the advection-diffusion equation (1.1) which describes passive scalar evolution in a constant flow with constant diffusivity and feedback parameter λ. For the two scale input spectrum (1.9), explicit formulas relating these scales are given (1.10).

Two representations of the solution to the equation governing random advection and molecular diffusion (1.42) are derived. The first, (1.54), is an extension of the common Lagrangian representation and the second, (1.55), is based on the Cameron-Martin-Girsanov formula.

As an example, the exact formula for the mean tracer field is given when the velocity field is Gaussian and independent of the space variable.

2

SCALE CLASSIFICATION OF TURBULENT DIFFUSION MODELS

In this chapter the problem of stirring a passive scalar by a random velocity field is considered in the framework of equation (1.27) which we rewrite for the sake of convenience

$$\frac{\partial c}{\partial t} + \boldsymbol{u} \cdot \nabla c = \kappa \nabla^2 c. \tag{2.1}$$

We assume unless otherwise specified that the non-divergent velocity $\boldsymbol{u}(t, \boldsymbol{r})$ is a Gaussian random field which is homogeneous in the space coordinate $\boldsymbol{r}$, stationary in time t and has zero mean, $\langle \boldsymbol{u} \rangle = \boldsymbol{0}$. Therefore its probability distribution is completely determined by the space-time correlation tensor

$$\boldsymbol{R_u}(s, \boldsymbol{r}) = \langle \boldsymbol{u}(t, \boldsymbol{r}')\boldsymbol{u}(t + s, \boldsymbol{r}' + \boldsymbol{r})^T \rangle \tag{2.2}$$

or by the spectral tensor $\boldsymbol{E}(\omega, \boldsymbol{k})$ given by

$$\boldsymbol{R_u}(s, \boldsymbol{r}) = \int\limits_{E^d} \int\limits_{-\infty}^{\infty} e^{i(\boldsymbol{k} \cdot \boldsymbol{r} - \omega s)} \boldsymbol{E}(\omega, \boldsymbol{k}) d\omega d\boldsymbol{k}, \tag{2.3}$$

where ω is the angular frequency and $\boldsymbol{k}$ is the wave vector.

Under the additional assumption of isotropy the entries of the correlation matrix can be expressed in terms of a scalar function $R_L(t, r)$ called the longitudinal correlation function (Monin and Yaglom 1975)

$$R_{ij}(t, \boldsymbol{r}) = \left(R_L(t, r) + \frac{r}{d-1} \frac{\partial R_L(t, r)}{\partial r} \right) \delta_{ij} - \frac{x_i x_j}{r(d-1)} \frac{\partial R_L(t, r)}{\partial r}, \tag{2.4}$$

where $\boldsymbol{r} = (x_1, \ldots, x_d)$, $r = |\boldsymbol{r}|$, $i, j = 1, \ldots, d$.

The spectral tensor is represented in a similar way

$$E_{ij}(\omega, \boldsymbol{k}) = E_L(\omega, k)\left(\delta_{ij} - \frac{k_i k_j}{k^2}\right), \tag{2.5}$$

where $E_L(\omega, k)$ is the longitudinal spectrum, $k = |\boldsymbol{k}|$.

2.1 APPROACHES TO CLOSURE PROBLEM

Despite the existing 'explicit' representations of the solution for (2.1), discussed in the previous chapter, one cannot extract from them even an equation describing the mean tracer field in the general case because of analytical difficulties. There are different kinds of assumptions which enable us to overcome this obstacle. Roughly speaking these assumptions can be broken down into the three following groups: constraints on scales, constraints on correlations, and others.

For example, the well known weak interaction approach developed by Hasselman (1966) and used by Davis (1982) for the analysis of turbulent diffusion, is based on the assumption that the ratio α of typical particle velocity to typical phase velocity in the Eulerian field, is small, $\alpha \ll 1$. Under this assumption the solution of the equation

$$\boldsymbol{X}(t, \boldsymbol{a}) = \boldsymbol{a} + \int_0^t \boldsymbol{u}(s,\ \boldsymbol{X}(s, \boldsymbol{a}))ds \tag{2.6}$$

for a Lagrangian trajectory starting at point $\boldsymbol{a}$, can be expanded in the perturbation series

$$\boldsymbol{X}(t) = \boldsymbol{X}_0(t\alpha^2) + \alpha \boldsymbol{X}_1(t) + \alpha^2 \boldsymbol{X}_2(t) + \dots. \tag{2.7}$$

As a result, the Lagrangian correlation function can be computed in terms of $\boldsymbol{R_u}(t, \boldsymbol{r})$ with the accuracy $O(\alpha^4)$. Using the Gaussianity of $\boldsymbol{u}$ and relation (1.37) it is not difficult to find the effective diffusivity.

One example of correlation constraints is the classical quasi-normal approximation (Saffman 1969). Assuming that for the following triple correlator

$$\langle \boldsymbol{u}\boldsymbol{u}c' \rangle = 0, \tag{2.8}$$

one arrives at the relation

$$\boldsymbol{R_v}(s) = \langle \boldsymbol{v}(t,\boldsymbol{a})\boldsymbol{v}(t+s,\boldsymbol{a})^T\rangle = \int_{E^d} \boldsymbol{R_u}(s,\boldsymbol{r})\langle c(s,\boldsymbol{r})\rangle d\boldsymbol{r}, \qquad (2.9)$$

where $\boldsymbol{v}(t,\boldsymbol{a}) = \partial \boldsymbol{X}(t,\boldsymbol{a})/\partial t$ is the Lagrangian velocity, $\boldsymbol{R_v}(t)$ is its correlation tensor and $c(t,\boldsymbol{r})$ is the solution of (2.1) with $\kappa = 0$ and initial condition $c_0(\boldsymbol{r}) = \delta(\boldsymbol{r}-\boldsymbol{a})$. In the spectral form (2.9) can be rewritten as

$$\boldsymbol{R_v}(s) = \int_{-\infty}^{\infty}\int_{E} \boldsymbol{E}(\omega,\boldsymbol{k})\; e^{i\omega s}\langle \exp\{i\boldsymbol{k}\cdot(\boldsymbol{X}(s,\boldsymbol{a})-\boldsymbol{a})\}\rangle d\boldsymbol{k}\; d\omega. \qquad (2.10)$$

Notice that under the Gaussian hypothesis for $\boldsymbol{X}(t,\boldsymbol{a})$ relation (2.10) yields the nonlinear integral equation for the Lagrangian correlation function

$$\boldsymbol{R_v}(t) = \int\int e^{i\omega t}\boldsymbol{E}(\omega,\boldsymbol{k})\exp\{-\frac{1}{2}\,(\int_{-t}^{t}(t-|s|)\boldsymbol{R_v}(s)\boldsymbol{k}\cdot\boldsymbol{k}\; ds)\}d\omega d\boldsymbol{k}. \qquad (2.11)$$

It is interesting to check the accuracy of relation (2.11) with simulations. In the successful case equation (2.11) might be viewed as an equation for $\boldsymbol{E}(\omega,\boldsymbol{k})$ and could be used to deduce the Eulerian velocity characteristics from Lagrangian observations. This problem is very important in oceanography, due to the wide use of Lagrangian devices (Davis 1991, Niiller 1995). Equality (2.10) was also derived by Roberts (1961) via the Direct Interaction Approach (DIA) developed by Kraichnan (1959). Basically this approach reduces to correlation constraints as well.

Finally let us mention the method of successive approximation (Phythian 1975), as an example from the group 'others'. Its main idea is to look for the solution of (2.6) by using the recurrence equation

$$\boldsymbol{X}_{n+1}(t) = \boldsymbol{a} + \int_0^t \boldsymbol{u}(s,\boldsymbol{X}_n(s))ds, \quad \boldsymbol{X}_0(t) \equiv \boldsymbol{a}. \qquad (2.12)$$

It appears that the correlation function for the second approximation $\boldsymbol{X}_2(t)$ can be computed in an explicit form and the approximate formula for the turbulent diffusivity

$$\boldsymbol{D} \approx \frac{d}{dt}\langle \boldsymbol{X}_2(t)\boldsymbol{X}_2(t)^T\rangle \qquad (2.13)$$

is highly accurate for some models of random flow.

Among these three types of assumptions, in our opinion the scale constraints approach is most preferable due to two advantages. First, the relation between the scales may be stated in a rigorous mathematical form (later we will demonstrate this). Second, the scale relations can be often verified in practice via experimental data. As for correlation constraints, the essential disadvantage is that of referring to the unknown field $c'(t, \boldsymbol{r})$ which must be found from the original problem. Hence the constraints can be checked only after the problem is solved. For example, in the one dimensional version of the space independent velocity field discussed above we have for the triple correlator,

$$\langle uuc' \rangle = -\frac{\left(\int\limits_0^t R_u(s)ds \right)}{2\pi \left(\int\limits_0^t (t-s) R_u(s)ds \right)^{3/2}} \tag{2.14}$$

at $r = 0$. One can see that this correlation never vanishes, but is small compared to $\langle u^2 \rangle$ and $\langle c'^2 \rangle^{1/2}$ for either big t or small velocity variance $\langle u^2 \rangle$. Thus, in this case conjecture (2.8) is incorrect, however one can expect that (2.9) approximately holds if the observation time is large or the velocity fluctuations are small.

2.2 MAIN TIME SCALES IN THE TURBULENT DIFFUSION PROBLEM

In this section as a basis for unifying different physical and mathematical approaches to turbulent diffusion we suggest 'scale classification'. For the sake of simplicity, we restrict our consideration to homogeneous velocity fields with only one time scale (correlation time) and only one space scale (correlation radius). Therefore in the simplest case there are four independent time scales:

- the observation time (we consider a Cauchy problem),
- the molecular diffusion time,
- the correlation time,
- the turnover time defined as the ratio of the correlation radius and the mean square velocity fluctuation.

If all these scales have the same order of magnitude it is hopeless to attempt any analytical solution of the problem. Explicit formulas for the effective diffusivity or explicit equations for the statistical characteristics of the tracer can be obtained only if one or more of these scales is much smaller or much bigger than the others.

We shall show that many well known exactly solvable models can be formulated in terms of the orders of these time scales. Our focus will be on the so called 'short-correlation approximation', which means that the correlation time of velocity is much smaller than both the observation time and the molecular diffusivity time. In this case an explicit expression for the turbulent diffusivity can be found under a wide range of conditions. However this expression drastically depends on the relationship between the correlation time and the turnover time. If they are of the same order then the turbulent diffusivity depends on the molecular diffusivity which can essentially intensify the mixing of the fluid. In contrast, if the turnover time is much bigger then the correlation time but still much less than the observation time we arrive at the classical Fokker-Planck equation where the effective diffusivity depends on the statistics of the velocity field only.

Let us introduce accurate definitions for the correlation radius and the correlation time of the underlying velocity field. There are different ways to define these scales. Our definitions reveal the purpose for studying velocity fields which are smooth in space and quickly decorrelating in time. Both of these features are reasonable from a hydrodynamics point of view, due to the dissipation of energy through small space scales and the short memory of developed turbulence (Batchelor 1982).

Restricting ourselves to the isotropic case we define the correlation radius and the correlation time by using the longitudinal spectrum

$$l_u = \left(\frac{\int\int E_L(\omega, k) k^{d-1} d\omega dk}{\int\int E_L(\omega, k) k^{d+1} d\omega dk} \right)^{1/2} \tag{2.15}$$

and

$$\tau_u = \frac{\int E_L(o, k) k^{d-1} dk}{\int\int E_L(\omega, k) k^{d-1} d\omega dk}. \tag{2.16}$$

The integration with respect to ω is over the entire real line and with respect to k from zero to infinity unless otherwise specified. Let us note that the variance

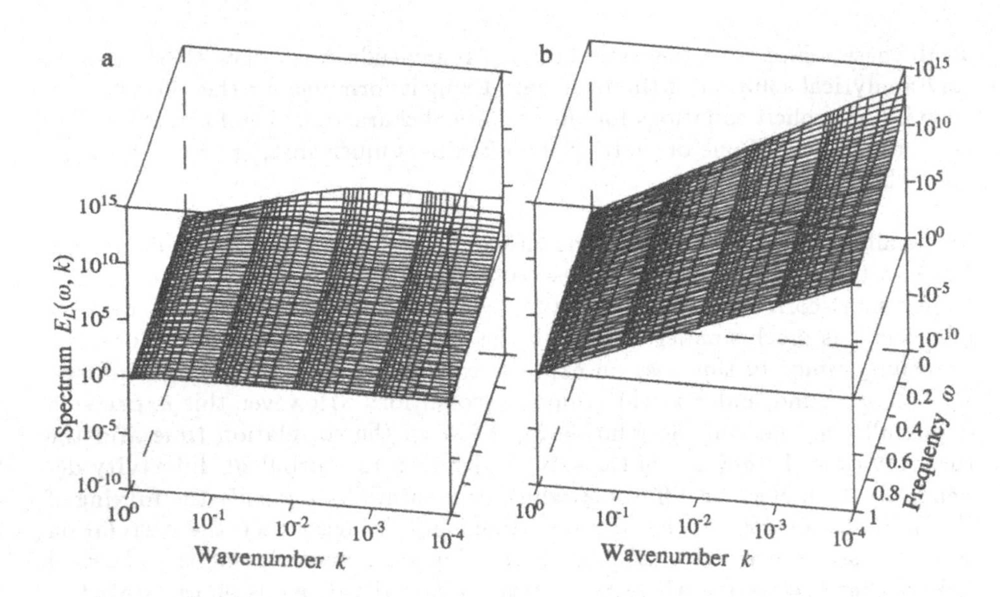

Figure 2.1 The wavenumber-frequency spectra given by (2.18) for the parameters defined as follows: *a*) $\alpha = 1$, $z = 1$, *b*) $\alpha = 3$, $z = 0.5$

$\sigma_u^2 = \langle |\boldsymbol{u}|^2 \rangle$ is given by

$$\sigma_u^2 = \frac{\vartheta_d(d-1)}{d} \int\int E_L(\omega, k) k^{d-1} d\omega dk, \tag{2.17}$$

where ϑ_d as before is the volume of a unit ball.

To illustrate different regimes of turbulent diffusion, we will consider the spectrum (Avellaneda and Majda 1990, 1992)

$$E_L(\omega, k) = \left\{ \begin{array}{ll} \dfrac{Cak^{2-d-\alpha+z}}{\omega^2 + a^2 k^{2z}} & C_0 \bar{k} < k < C_1 \bar{k} \\ 0 & \text{otherwise} \end{array} \right\}. \tag{2.18}$$

Here $\bar{k}$ is the wave length scale. The constants C_0 and C_1 determine the upper and lower cutoffs, and C, a, α and z are positive parameters. For the wave spectrum $E_L(k) = \int E_L(\omega, k) d\omega$ we have

$$E_L(k) = C\pi k^{2-d-\alpha} \quad , \quad C_0 \bar{k} < k < C_1 \bar{k} \tag{2.19}$$

and hence α determines the spectral slope for the above wave number range. In fig. 2.1 we have plotted $E_L(\omega, k)$ for different values of α and z.

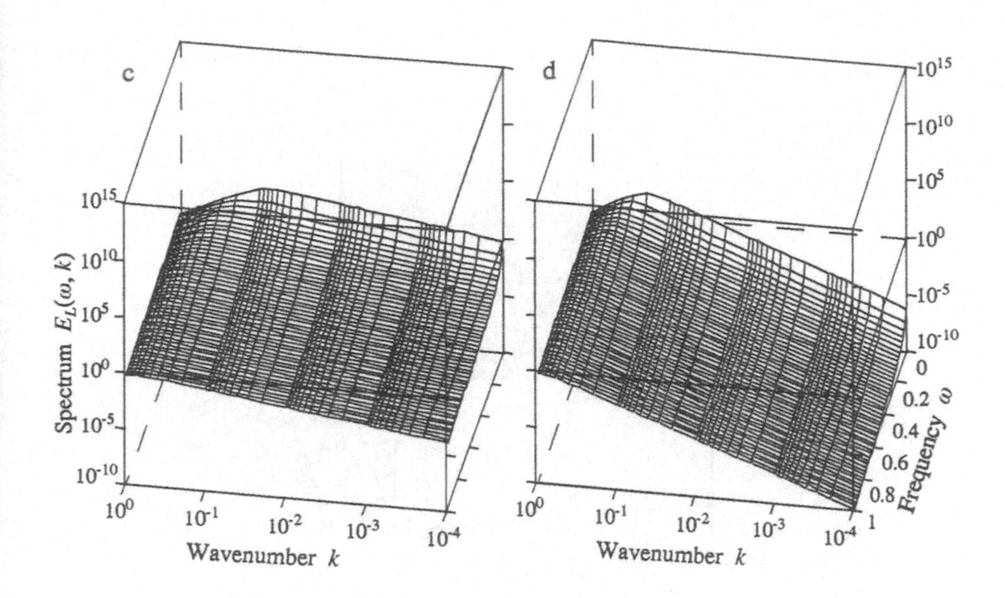

Figure 2.1 *(continued)* c) $\alpha = 1$, $z = 2$, d) $\alpha = 0.5$, $z = 3$.

The time correlation function of the harmonic due to the wavenumber k decays as $exp\{-k^z t\}$. Thus the parameter z characterizes the rate of decorrelation. Via straightforward calculations, one can obtain from (2.15-2.17)

$$l_u = \frac{1}{\bar{k}} \sqrt{\frac{(C_1^{-\alpha+2} - C_0^{-\alpha+2})(4-\alpha)}{(2-\alpha)(C_1^{4-\alpha} - C_0^{4-\alpha})}}, \tag{2.20}$$

$$\tau_u = \frac{1}{a\bar{k}^z} \frac{(2-\alpha)(C_1^{2-\alpha-z} - C_0^{2-\alpha-z})}{\pi(2-\alpha-z)(C_1^{2-\alpha} - C_0^{2-\alpha})}, \tag{2.21}$$

$$\sigma_u^2 = \frac{\vartheta_d(d-1)}{d} \frac{C\pi\bar{k}^{2-\alpha}}{a(2-\alpha)} (C_1^{2-\alpha} - C_0^{2-\alpha}). \tag{2.22}$$

These quantities are plotted in Fig. 2.2 and Fig. 2.3.

Now we proceed to the scale analysis of equation (2.1) with a general velocity field. Namely, let us consider the Cauchy problem with an isotropic Gaussian velocity field whose statistics are fully determined by parameters σ_u, τ_u, and l_u and the initial field $c_0(\boldsymbol{r})$ is determined by the single length scale, l_0. In doing so we assume that $c_0(\boldsymbol{r})$ is either random or deterministic. If for example $c_0(\boldsymbol{r})$ is assumed to be an isotropic random field then l_0 can be defined as in chapter

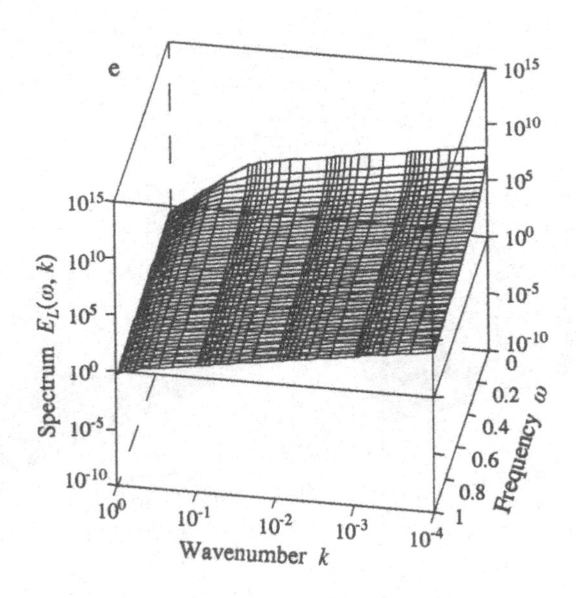

Figure 2.1 *(continued)* e) $\alpha = 3$, $z = 2$.

1 (1.5). Therefore we have 6 parameters

$$\{t,\ \sigma_u,\ \tau_u,\ l_u,\ l_0,\ \kappa\}, \tag{2.23}$$

which fully determine the mean field $\langle c(t, \boldsymbol{r})\rangle$ and the rest of the tracer statistics.

There are four independent time scales that are expressed in terms of these parameters. Namely, the observation time, t, the Eulerian correlation time,

$$\tau_E \equiv \tau_u, \tag{2.24}$$

where τ_u is given by (2.16), the turnover time,

$$\tau_T = \frac{l_u}{\sigma_u}, \tag{2.25}$$

and, finally, the molecular diffusion time,

$$\tau_D = \frac{l_0^2}{\kappa}. \tag{2.26}$$

Thus τ_E characterizes the correlation time of the time series $\boldsymbol{u_r}(t) = \boldsymbol{u}(t, \boldsymbol{r})$ obtained at the arbitrary fixed point $\boldsymbol{r}$, whereas τ_T characterizes how fast a

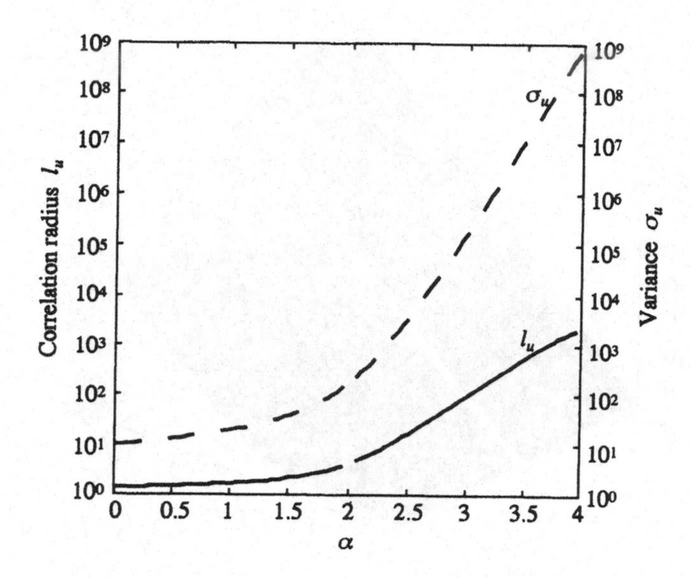

Figure 2.2 Dependence of the variance σ_u and space scale l_u on the parameter α for the spectrum (2.18). The blanks indicate that σ_u and l_u are not defined.

Lagrangian particle trapped by an eddy performs one revolution. In 'eddy' terminology the scale τ_E also can be interpreted as the eddy lifetime.

We call the introduced scales independent since each of them is determined by a distinct subset of the full set (2.23) of the problem parameters. Notice that there is another molecular diffusion time scale $\widetilde{\tau_D} = l_u^2/\kappa$ in this problem, but we cannot consider $\widetilde{\tau_D}$ as independent of τ_T because both include the same quantity, l_u.

If all scales have the same order

$$t \sim \tau_E \sim \tau_T \sim \tau_D, \tag{2.27}$$

then any effort to find an explicit formula for $\langle c(t, \boldsymbol{r}) \rangle$ fails. To make progress in computing the tracer statistics it is necessary to have a separation of scales. For instance the separation assumption

$$t \sim \tau_E \sim \tau_D \ll \tau_T \tag{2.28}$$

may be interpreted mathematically as a velocity field $\boldsymbol{u}(t, \boldsymbol{r}) = \boldsymbol{u}(t)$ independent of the space coordinate. This case was discussed in detail above. Thus

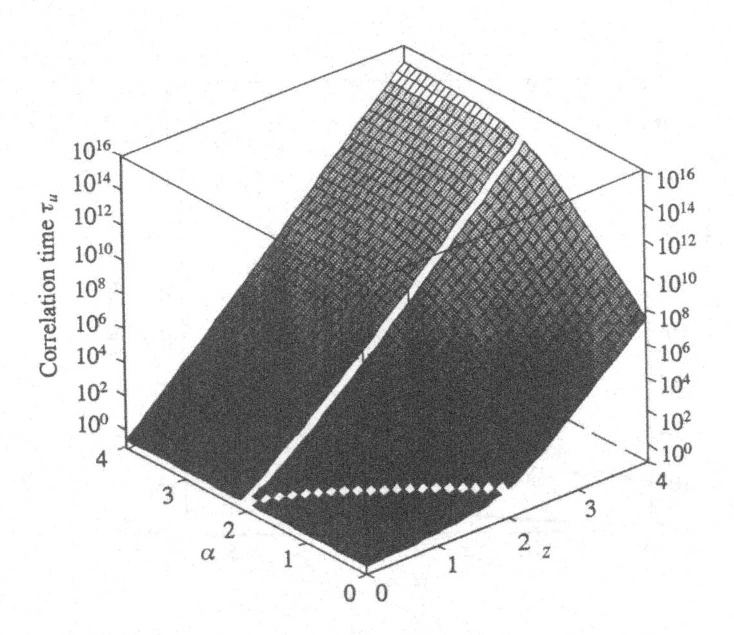

Figure 2.3 Dependence of the correlation time τ_u on the parameters α and z. The blanks indicate where τ_u is not defined.

Figure 2.4 Classification of scale separation.

under (2.28) all statistics of the passive scalar can be found immediately. Before we discuss other separation relationships which imply solvability of the turbulent diffusion problem, let us enumerate all possible separations. In other words let us find out how many different ways one can place $t, \tau_E, \tau_D, \tau_T$ in the circles on Fig. 2.4, where the arrow means either $\sim$ or $\ll$.

In doing so, we should take into account that some combinations are equivalent. For instance, $\tau_E \sim t \sim \tau_D \ll \tau_T$ is equivalent to (2.28). There are 74 provided the configuration (2.27) is excluded. Let us exhibit some of these order relationships under which an exact asymptotic can be found.

In view of the proposed classification the mentioned weak interaction approach (section 2.1) corresponds to the following relation

$$\tau_T \ll \tau_E \sim t \ll \tau_D. \tag{2.29}$$

The assumption $t \ll \tau_D$ means that the molecular diffusion is ignored.

In the absence of molecular diffusion Batchelor (1952) and Roberts (1961) computed the short time asymptotics. In our classification this case can be written as

$$t \ll \tau_T \sim \tau_E \ll \tau_D. \tag{2.30}$$

Under (2.30) we have

$$P(\boldsymbol{r}; t, \boldsymbol{a}) \sim \frac{1}{(2\pi t^2 \sigma_u^2)^{d/2}} \exp\{-\frac{(\boldsymbol{r}-\boldsymbol{a})^2}{2t^2\sigma_u^2}\}, \tag{2.31}$$

where the Lagrangian pdf $P(\boldsymbol{r}; t, \boldsymbol{a})$ is defined in chapter 1 (1.32) and therefore $\boldsymbol{D}(t) = D(t)\boldsymbol{I}$, where $\boldsymbol{I}$ is the identity matrix and

$$D(t) \sim \sigma_u^2 t. \tag{2.32}$$

Adding the molecular diffusion is obvious. If

$$t \ll \tau_T \sim \tau_E \sim \tau_D, \tag{2.33}$$

then

$$P(\boldsymbol{r}; t, \boldsymbol{a}) \sim \frac{1}{2\pi\left[(t^2\sigma_u^2 + \kappa t)\right]^{d/2}} \exp\{-\frac{(\boldsymbol{r}-\boldsymbol{a})^2}{2(t^2\sigma_u^2 + \kappa t)}\} \tag{2.34}$$

and hence

$$D(t) \sim \sigma_u^2 t + \kappa. \tag{2.35}$$

Notice that since formulas (2.34, 2.35) do not include in explicit form τ_E and τ_T, they cover more than one limiting case in our classification. More exactly they are valid for example if $t \ll \tau_D \sim \tau_E \ll \tau_T$ or $t \ll \tau_D \sim \tau_T \ll \tau_E$ etc. The same remark is true for some cases discussed below.

The opposite case where t is large, namely

$$\tau_D \sim \tau_E \sim \tau_T \ll t \tag{2.36}$$

will be discussed in chapter 5.

A reasonable mathematical model for large Eulerian correlation

$$t \sim \tau_D \sim \tau_T \ll \tau_E \tag{2.37}$$

is a steady , i.e. time independent velocity field

$$\boldsymbol{u}(t, \boldsymbol{r}) = \boldsymbol{u}(\boldsymbol{r}). \tag{2.38}$$

In general, assumption (2.38) does not simplify the problem at all. Thus, the scale separation (2.37) is not sufficient for solvability. But the additional order constraint

$$\tau_T \ll t \sim \tau_D \ll \tau_E \tag{2.39}$$

leads to a substantial mathematical theory which will be commented on below. Notice that if a scale separation includes two or more symbols $\ll$, then the problem becomes undetermined in the sense that there can be a variety of asymptotical regimes for $\langle c \rangle$. Let us give a simple example concerning the multiscale problem as whole. Imagine that there are three scales m_1, m_2, m_3 determining the behavour of a variable $z = z(m_1, m_2, m_3)$. The relation

$$m_1 \ll m_2 \ll m_3 \tag{2.40}$$

means that $m_1/m_2 \to 0, \quad m_2/m_3 \to 0$. If the variable z is a function of $m_1 m_3/m_2^2$ then assumption (2.40) is not enough to compute the asymptotic of z.

Further, we restrict our consideration to the case where all time scales are are defined in such a way that there will not be room for such ambiguities. Namely, let $\varepsilon > 0$ be a formal dimensionless small parameter and let us make the change coordinates

$$t' = \alpha(\varepsilon)t, \quad \boldsymbol{r}' = \beta(\varepsilon)\boldsymbol{r}, \tag{2.41}$$

where $\alpha(\varepsilon)$, $\beta(\varepsilon)$ are some dimensionless functions. As a result equation (2.1) becomes

$$\frac{\partial c}{\partial t'} + \boldsymbol{u}_\varepsilon(t', \boldsymbol{r}') \cdot \nabla c = \kappa_\varepsilon \nabla^2 c. \tag{2.42}$$

where

$$\boldsymbol{u}_\varepsilon(t', \boldsymbol{r}') = \frac{\beta(\varepsilon)}{\alpha(\varepsilon)} \boldsymbol{u}\left(\frac{t'}{\alpha(\varepsilon)}, \frac{\boldsymbol{r}'}{\beta(\varepsilon)}\right), \quad \kappa_\varepsilon = \frac{\beta(\varepsilon)^2}{\alpha(\varepsilon)}\kappa. \tag{2.43}$$

Such a transformation of variables is called the rescaling or renormalization. It is clear that in this situation all the time scales are functions of ε only and hence any scale separation completely determines the asymptotical behavior of the mean tracer. For example, (2.39) can be obtained by choosing $\alpha(\varepsilon) = \varepsilon^2$,

$\beta(\varepsilon) = \varepsilon$. The latter is equivalent to the following renormalization of the steady velocity field

$$\boldsymbol{u}_\varepsilon(\boldsymbol{r}) = \frac{1}{\varepsilon}\boldsymbol{u}\left(\frac{\boldsymbol{r}}{\varepsilon}\right). \tag{2.44}$$

The most important and well-known result concerning (2.44) is the following statement on homogenization. Roughly, if $c_\varepsilon(t, \boldsymbol{r})$ is the solution of

$$\frac{\partial c_\varepsilon}{\partial t} + \boldsymbol{u}_\varepsilon \cdot \nabla c_\varepsilon = \kappa c_\varepsilon \tag{2.45}$$

due to a square integrable initial condition, then the limiting tracer field

$$\bar{c}(t, \boldsymbol{r}) = \lim_{\varepsilon \to 0} c_\varepsilon(t, \boldsymbol{r}) \tag{2.46}$$

is deterministic and satisfies an equation of the canonical form (1.27). This assertion was proved in a rigorous mathematical way by different authors (Kozlov 1983, Varadhan and Papanickolaou 1982, Avellaneda and Majda 1990). Later, in chapter 4 we will discuss the homogenization for time-dependent velocity fields.

2.3 APPROXIMATION OF SHORT-CORRELATED VELOCITIES

We now proceed to the main subject of our consideration, the short-correlation approximation. The basis of this approximation is the following assumption

$$\tau_E \ll t \sim \tau_D, \tag{2.47}$$

i.e. the Eulerian correlation time is much less than both the observation time and the diffusion time which is allowed to be much bigger than t as well.

Let us note that thus far τ_T is not fixed in our definition of short-correlation velocity. One can see that there are 5 options for the turnover time (see Fig. 2.5) corresponding to the relations

$$\begin{array}{ll} \text{(A)} & \tau_T \gg 1, \\ \text{(B)} & \tau_T \sim 1, \\ \text{(C)} & \tau_E \ll \tau_T \ll 1, \\ \text{(D)} & \tau_T \sim \tau_E \ll 1, \\ \text{(E)} & \tau_T \ll \tau_E \ll 1. \end{array} \tag{2.48}$$

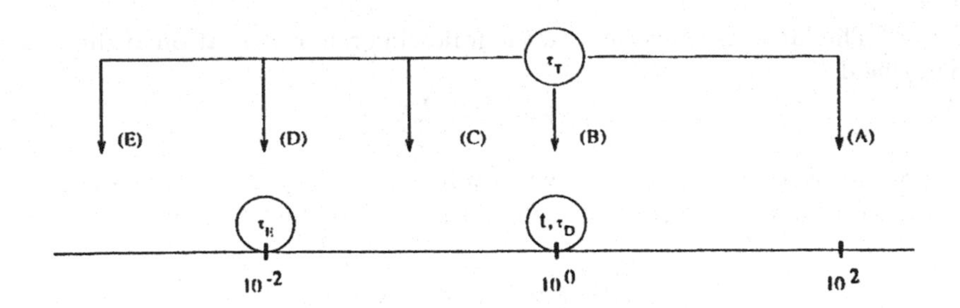

Figure 2.5 Five versions of the short correlation approximation ($\varepsilon = 10^{-1}$).

The assumptions of scale separation (A),(B),(C),(D) can be satisfied for a velocity field with a finite low-frequency energy

$$\int_0^\infty E_L(0,k)k^{d-1}dk < \infty \tag{2.49}$$

under the following rescaling

$$\begin{array}{ll} \text{(A)} & \boldsymbol{u}_\varepsilon(t,\boldsymbol{r}) = \boldsymbol{u}(t/\varepsilon^2), \\ \text{(B)} & \boldsymbol{u}_\varepsilon(t,\boldsymbol{r}) = \boldsymbol{u}(t/\varepsilon^2,\boldsymbol{r}), \\ \text{(C)} & \boldsymbol{u}_\varepsilon(t,\boldsymbol{r}) = \varepsilon^{-1}\boldsymbol{u}(t/\varepsilon^2,\boldsymbol{r}), \\ \text{(D)} & \boldsymbol{u}_\varepsilon(t,\boldsymbol{r}) = \varepsilon^{-1}\boldsymbol{u}(t/\varepsilon^2,\boldsymbol{r}/\varepsilon), \end{array} \tag{2.50}$$

and $\kappa_\varepsilon = \kappa$.

Indeed, from definitions (2.15), (2.16), (2.25) and (2.26) it follows that for the renormalized velocity field $\boldsymbol{u}_\varepsilon(t,\boldsymbol{r})$

$$\tau_E^{(\varepsilon)} \sim \varepsilon^2 \ , \ \tau_D^{(\varepsilon)} \sim 1 \tag{2.51}$$

and

$$\begin{array}{ll} \text{(A)} & \tau_T^{(\varepsilon)} = \infty, \\ \text{(B)} & \tau_T^{(\varepsilon)} \sim 1, \\ \text{(C)} & \tau_T^{(\varepsilon)} \sim \varepsilon, \\ \text{(D)} & \tau_T^{(\varepsilon)} \sim \varepsilon^2. \end{array} \tag{2.52}$$

The situation with separation (E) from (2.48) is much more complicated. Nevertheless, this case is very important for applications in hydrodynamics and oceanography. Here we only give a particular example of separation (E) and a detailed discussion is given in chapter 4.

Lets consider the spectrum (2.18) with the following restrictions on the parameters

$$\alpha + z > 2, \quad \alpha + 2z < 4, \tag{2.53}$$

and assume that the lower boundary $C_0 \equiv C_0\varepsilon$ is small, e.g. in the classical turbulence theory $\varepsilon = (Re)^{-3/4}$ (Batchelor 1982). Let $c_\varepsilon(t, \boldsymbol{r})$ be the solution of the Cauchy problem

$$\frac{\partial c_\varepsilon}{\partial t} + \boldsymbol{u}_\varepsilon(t, \boldsymbol{r}) \cdot \nabla c_\varepsilon = \kappa_\varepsilon \nabla^2 c_\varepsilon, \quad c_0(\boldsymbol{r}) \in L_2(E^d) \tag{2.54}$$

with the following renormalization

$$\boldsymbol{u}_\varepsilon(t, \boldsymbol{r}) = \frac{\varepsilon}{\rho^2(\varepsilon)} \boldsymbol{u}\left(\frac{t}{\rho^2(\varepsilon)}, \frac{\boldsymbol{r}}{\varepsilon}\right), \quad \kappa_\varepsilon = \frac{\varepsilon^2}{\rho^2(\varepsilon)}\kappa, \tag{2.55}$$

It was shown in (Avellaneda and Majda 1992) that if

$$\rho = \varepsilon^{(4-\alpha-z)/2}, \tag{2.56}$$

then limit $\lim_{\varepsilon \to 0} c_\varepsilon(t, \boldsymbol{r}) = \bar{c}(t, \boldsymbol{r})$ is nondegenerate and the limiting function satisfies an equation of the form (1.27).

The renormalized velocity field (2.55) is determined by the same type of spectrum as (2.18), but its parameters are

$$C^{(\varepsilon)} = \frac{C\varepsilon^{4-\alpha}}{\rho^4}, \quad C_1^{(\varepsilon)} = C_1/\varepsilon, \quad a_\varepsilon = \frac{\varepsilon^z}{\rho^2}a. \tag{2.57}$$

Taking into account conditions (2.53) we obtain from the general formulas (2.20-2.22) and definition (2.26) of the diffusion time

$$\tau_E^{(\varepsilon)} \sim \left\{ \begin{array}{ll} \varepsilon^{6-2\alpha-2z} & \alpha < 2 \\ \varepsilon^{4-\alpha-2z} & \alpha > 2 \end{array} \right\}. \tag{2.58}$$

$$\tau_T^{(\varepsilon)} \sim \varepsilon^{4-\alpha-z}, \tag{2.59}$$

$$\tau_D^{(\varepsilon)} \sim \varepsilon^{2-\alpha-z}, \tag{2.60}$$

From (2.58-2.60) it follows that in the region (2.53) we have the following separation of scales

$$\tau_T \ll \tau_E \ll t \ll \tau_D. \tag{2.61}$$

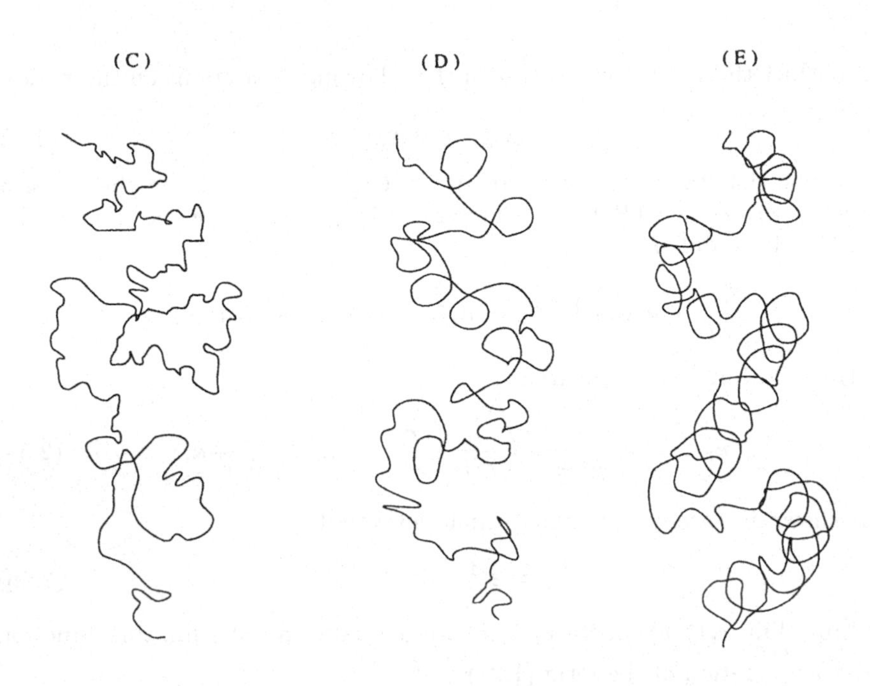

Figure 2.6 Typical Lagrangian trajectories $\boldsymbol{X}(t)$ in 2D random flow (C): $\tau_T \gg \tau_E$ (D): $\tau_T \sim \tau_E$ (E): $\tau_T \ll \tau_E$.

To illustrate the difference between cases (C), (D) and (E), Fig. 2.6 shows sketches of typical Lagrangian trajectories for $d = 2$. In the first case the trajectory does not contain loops, because the eddy is broken earlier than a particle trapped by it can complete a turnover. Of course, this case does not preclude self-intersection. In case (D) the particle trapped by an eddy completes a number of full revolutions before the eddy dies. Finally in case (E), the number of complete revolutions can be very large.

From an analytical viewpoint cases (A),(B) are not difficult. For model (A) we have Fickian equation for the mean concentration with effective diffusivity (Kubo 1963)

$$\boldsymbol{D} \sim \varepsilon^2 \int_0^\infty \boldsymbol{R_u}(t)dt. \tag{2.62}$$

In case (B) we have a similar result

$$D \sim \varepsilon^2 \int_0^\infty \boldsymbol{R_u}(t,0)dt. \tag{2.63}$$

In both cases the effective diffusivity is of the order ε^2 and hence is much smaller than molecular diffusivity ($\tau_D \sim 1$). For this reason we do not focus on these cases. The second reason is that (A) and (B) can be deduced from case (C) considered in detail below, by simply multiplying the velocity by ε. Sometimes assumptions (A) and (B) are called the quasilinear approximation.

2.4 SUMMARY

Four independent time scales related to the advection-diffusion equation are introduced in a rigorous way and their physical meaning is explained. They are the correlation time of the Eulerian velocity field $\tau_E \equiv \tau_u$ defined by (2.16), the turnover time τ_T given by (2.25),(2.15), the molecular diffusion time (2.26) and, finally, the observation time t. It is shown that many standard approximations in the advection-diffusion equation can be expressed in the form of a separation of the introduced scales.

A classification of different approaches based on the short-correlation approximation $\tau_E \ll t$ is given. Examples of rigorous interpretations of some of the scale separations in terms of a small parameter are shown.

3

DELTA-CORRELATION APPROXIMATION

This chapter addresses the velocity field renormalized as follows

$$\boldsymbol{u}_\varepsilon(t,\boldsymbol{r}) = \frac{1}{\varepsilon}\boldsymbol{u}\left(\frac{t}{\varepsilon^2},\boldsymbol{r}\right), \tag{3.1}$$

where ε is small.

From (2.51) and (2.52) it follows that (3.1) implies the following separation of scales

$$\tau_E \ll \tau_T \ll t. \tag{3.2}$$

The results of this chapter are also valid for a more general rescaling $\boldsymbol{u}_\varepsilon(t,\boldsymbol{r}) = \varepsilon^{-1}\boldsymbol{u}(t\varepsilon^{-2},\boldsymbol{r}\varepsilon^{-\alpha})$, where $0 \le \alpha < 1$ (Komorowski 1996). Obviously, the relation (3.2) holds in this case since $\tau_T \sim \varepsilon^{1+\alpha}$.

Under certain general conditions the random field $\boldsymbol{u}_\varepsilon(t,\boldsymbol{r})$ defined by (3.1) converges to a white noise process in the sense of distributions, i.e.

$$\lim_{\varepsilon\to 0} \boldsymbol{R}_{\boldsymbol{u}}^{(\varepsilon)}(s,\boldsymbol{r}) = \delta(s)\boldsymbol{B}_{\boldsymbol{u}}(\boldsymbol{r}), \tag{3.3}$$

where $\boldsymbol{R}_{\boldsymbol{u}}^{(\varepsilon)}(s,\boldsymbol{r}) = \varepsilon^{-2}\boldsymbol{R}_{\boldsymbol{u}}(s\varepsilon^{-2},\boldsymbol{r})$ is the correlation function of the field $\boldsymbol{u}_\varepsilon(t,\boldsymbol{r})$ and

$$\boldsymbol{B}_{\boldsymbol{u}}(\boldsymbol{r}) = \int_{-\infty}^{\infty} \boldsymbol{R}_{\boldsymbol{u}}(s,\boldsymbol{r})ds. \tag{3.4}$$

As a result one can readily obtain the following equation for the mean concentration

$$\frac{\partial\langle c\rangle}{\partial t} = \nabla\cdot\boldsymbol{D}\nabla\langle c\rangle + \kappa\nabla^2\langle c\rangle \tag{3.5}$$

with

$$D = \frac{1}{2}B_{u}(0) = \int_{0}^{\infty} R_{u}(t,0)dt = \pi \int E_{u}(0,k)dk. \tag{3.6}$$

This is well known Fokker-Planck equation for incompressible flow and we do not give its derivation. Instead, we derive the mean tracer field equation in the general situation when a source and potential are present

$$\frac{\partial c}{\partial t} + u \cdot \nabla c + \lambda c = \kappa \nabla^2 c + S(t, r), \tag{3.7}$$

in addition, the fields u, λ, and S are not assumed to be homogeneous. Revoking the homogeneity assumption is of a great importance for oceanographic applications. Moreover, we exhibit two derivations. The first one, based on the Furutsu-Novikov formula (Furutsu 1963, Novikov 1964), is quite short and simple but is valid only when u, λ, and S are Gaussian fields. The second method is more straightforward and general. It is based on the Lagrangian representation of the solution to (3.7) and the assumption of innovation of the fields u, λ, and S in time. This assumption is restrictive, but it allows us to analyze the situation when ε is small but finite.

For the sake of simplicity our consideration is restricted to divergence free velocity fields. A rigorous derivation of the mean tracer equation for compressible flows can be found e.g. in (Semenov 1989). In absence of source and potential this equation is nothing more than the backward Kolmogorov equation.

3.1 DERIVING THE MEAN TRACER EQUATION: FUNCTIONAL AND LAGRANGIAN APPROACHES

Let us start with the functional approach developed in (Klyatskin 1994 and Klyatskin, Woyczynski, and Gurarie 1996). Let

$$u = \langle u \rangle + u', \quad \lambda = \langle \lambda \rangle + \lambda', \quad S = \langle S \rangle + S', \tag{3.8}$$

be a decomposition into the mean fields (expectations) and deviations from them (fluctuations) denoted by the primes. Substituting (3.8) into (3.7) and averaging yields

$$\partial \langle c \rangle / \partial t + \langle u \rangle \cdot \nabla \langle c \rangle + \nabla \langle u'c' \rangle + \langle \lambda \rangle \langle c \rangle + \langle \lambda' c' \rangle = \langle S \rangle. \tag{3.9}$$

Now we suppose that the fluctuation fields are Gaussian and stationary in time and set

$$\begin{aligned}
\langle u_i'(t,\boldsymbol{r}_1)u_j'(s,\boldsymbol{r}_2)\rangle &= R_{ij}(t-s,\boldsymbol{r}_1,\boldsymbol{r}_2),\\
\langle u_i'(t,\boldsymbol{r}_1)\lambda'(s,\boldsymbol{r}_2)\rangle &= R_{i,\lambda}(t-s,\boldsymbol{r}_1,\boldsymbol{r}_2),\\
\langle u_i'(t,\boldsymbol{r}_1)S'(s,\boldsymbol{r}_2)\rangle &= R_{i,S}(t-s,\boldsymbol{r}_1,\boldsymbol{r}_2),\\
\langle \lambda'(t,\boldsymbol{r}_1)S'(s,\boldsymbol{r}_2)\rangle &= R_{\lambda,S}(t-s,\boldsymbol{r}_1,\boldsymbol{r}_2),\\
\langle \lambda'(t,\boldsymbol{r}_1)\lambda'(s,\boldsymbol{r}_2)\rangle &= R_{\lambda,\lambda}(t-s,\boldsymbol{r}_1,\boldsymbol{r}_2),
\end{aligned} \tag{3.10}$$

where $\boldsymbol{r}_1$ and $\boldsymbol{r}_2$ are arbitrary points in space and $i,j=1,\dots,d$.

Let us renormalize the fields $\boldsymbol{u}',\lambda',S'$ in the fashion (3.1). Then by passing to the limit $\varepsilon\to 0$ we obtain similar to (3.3)

$$\begin{aligned}
\langle u_i'(t,\boldsymbol{r}_1)u_j'(s,\boldsymbol{r}_2)\rangle &= \delta(t-s)B_{ij}(\boldsymbol{r}_1,\boldsymbol{r}_2),\\
\langle u_i'(t,\boldsymbol{r}_1)\lambda'(s,\boldsymbol{r}_2)\rangle &= \delta(t-s)B_{i,\lambda}(\boldsymbol{r}_1,\boldsymbol{r}_2),\\
\langle u_i'(t,\boldsymbol{r}_1)S'(s,\boldsymbol{r}_2)\rangle &= \delta(t-s)B_{i,S}(\boldsymbol{r}_1,\boldsymbol{r}_2),\\
\langle \lambda'(t,\boldsymbol{r}_1)S'(s,\boldsymbol{r}_2)\rangle &= \delta(t-s)B_{\lambda,S}(\boldsymbol{r}_1,\boldsymbol{r}_2),\\
\langle \lambda'(t,\boldsymbol{r}_1)\lambda'(s,\boldsymbol{r}_2)\rangle &= \delta(t-s)B_{\lambda,\lambda}(\boldsymbol{r}_1,\boldsymbol{r}_2),
\end{aligned} \tag{3.11}$$

where

$$B_{ij}(\boldsymbol{r}_1,\boldsymbol{r}_2)=\int_{-\infty}^{\infty} R_{ij}(t,\boldsymbol{r}_1,\boldsymbol{r}_2)dt \tag{3.12}$$

and others B's in (3.11) are expressed through corresponding R's in the same way.

Let $\zeta_1(t,\boldsymbol{r})$,
$\dots,\zeta_N(t,\boldsymbol{r})$ be arbitrary Gaussian random fields and $F[\zeta_1(\cdot,\cdot),\dots,\zeta_N(\cdot,\cdot)]$ be a smooth functional of these fields. Then Furutsu-Novikov formula yields

$$\langle F[\zeta_1(\cdot,\cdot),\dots,\zeta_N(\cdot,\cdot)]\zeta_i(t,\boldsymbol{r})\rangle =$$

$$\sum_j\int\int\langle\frac{\delta F}{\delta\zeta_j(s,\hat{\boldsymbol{r}})}\rangle\langle\zeta_j(s,\hat{\boldsymbol{r}})\zeta_i(t,\boldsymbol{r})\rangle ds d\hat{\boldsymbol{r}} \quad . \tag{3.13}$$

Applying this formula in the case $N = d+2$ to $c(t,\boldsymbol{r})$ as a functional of $u_i'(\cdot,\cdot)$, $\lambda'(\cdot,\cdot)$, $S'(\cdot,\cdot)$ we obtain

$$\langle c'(t,\boldsymbol{r})u_i'(t,\boldsymbol{r})\rangle = \langle c(t,\boldsymbol{r})u_i'(t,\boldsymbol{r})\rangle =$$

$$\int [\sum_j \langle \frac{\delta c(t,\boldsymbol{r})}{\delta u_j'(t,\hat{\boldsymbol{r}})}\rangle B_{ij}(\boldsymbol{r},\hat{\boldsymbol{r}}) + \langle \frac{\delta c(t,\boldsymbol{r})}{\delta \lambda'(t,\hat{\boldsymbol{r}})}\rangle B_{i,\lambda}(\boldsymbol{r},\hat{\boldsymbol{r}}) \tag{3.14}$$

$$+\langle \frac{\delta c(t,\boldsymbol{r})}{\delta S'(t,\hat{\boldsymbol{r}})}\rangle B_{i,S}(\boldsymbol{r},\hat{\boldsymbol{r}})]d\hat{\boldsymbol{r}},$$

$$\langle c'(t,\boldsymbol{r})\lambda'(t,\boldsymbol{r})\rangle = \langle c(t,\boldsymbol{r})\lambda'(t,\boldsymbol{r})\rangle =$$

$$= \int [\sum_j \langle \frac{\delta c(t,\boldsymbol{r})}{\delta u_j'(t,\hat{\boldsymbol{r}})}\rangle B_{j,\lambda}(\hat{\boldsymbol{r}},\boldsymbol{r}) + \langle \frac{\delta c(t,\boldsymbol{r})}{\delta \lambda'(t,\hat{\boldsymbol{r}})}\rangle B_{\lambda,\lambda}(\hat{\boldsymbol{r}},\boldsymbol{r}) + \tag{3.15}$$

$$+\langle \frac{\delta c(t,\boldsymbol{r})}{\delta S'(t,\hat{\boldsymbol{r}})}\rangle B_{\lambda,S}(\boldsymbol{r},\hat{\boldsymbol{r}})]d\hat{\boldsymbol{r}},$$

Let us compute the variational derivatives on both sides of the original equation (3.7)

$$\frac{\partial}{\partial t}\frac{\delta c(t,\boldsymbol{r})}{\delta u_i'(s,\hat{\boldsymbol{r}})} + \delta(t-s,\hat{\boldsymbol{r}}-\boldsymbol{r})\frac{\partial c(t,\boldsymbol{r})}{\partial x_i}$$
$$+\boldsymbol{u}\cdot\nabla \frac{\delta c(t,\boldsymbol{r})}{\delta u_i'(s,\hat{\boldsymbol{r}})} + \lambda \frac{\delta c(t,\boldsymbol{r})}{\delta u_i'(s,\hat{\boldsymbol{r}})} = 0, \tag{3.16}$$

$$\frac{\partial}{\partial t}\frac{\delta c(t,\boldsymbol{r})}{\delta \lambda'(s,\hat{\boldsymbol{r}})} + \delta(t-s,\hat{\boldsymbol{r}}-\boldsymbol{r})c(t,\boldsymbol{r})$$
$$+\boldsymbol{u}\cdot\nabla \frac{\delta c(t,\boldsymbol{r})}{\delta \lambda'(s,\hat{\boldsymbol{r}})} + \lambda \frac{\delta c(t,\boldsymbol{r})}{\delta \lambda'(s,\hat{\boldsymbol{r}})} = 0, \tag{3.17}$$

$$\frac{\partial}{\partial t}\frac{\delta c(t,\boldsymbol{r})}{\delta S'(s,\hat{\boldsymbol{r}})}$$
$$+\boldsymbol{u}\cdot\nabla \frac{\delta c(t,\boldsymbol{r})}{\delta S'(s,\hat{\boldsymbol{r}})} + \lambda \frac{\delta c(t,\boldsymbol{r})}{\delta S'(s,\hat{\boldsymbol{r}})} = \delta(t-s,\hat{\boldsymbol{r}}-\boldsymbol{r}). \tag{3.18}$$

These equations are equivalent to the following Cauchy problems respectively

$$\frac{\partial}{\partial t}\frac{\delta c(t,\boldsymbol{r})}{\delta u_i'(s,\hat{\boldsymbol{r}})} + \boldsymbol{u}\cdot\nabla \frac{\delta c(t,\boldsymbol{r})}{\delta u_i'(s,\hat{\boldsymbol{r}})} + \lambda \frac{\delta c(t,\boldsymbol{r})}{\delta u_i'(s,\hat{\boldsymbol{r}})} = 0,$$
$$\frac{\delta c(t,\boldsymbol{r})}{\delta u_i'(s,\hat{\boldsymbol{r}})}\Big|_{t=s} = -\frac{1}{2}\delta(\hat{\boldsymbol{r}}-\boldsymbol{r})\frac{\partial c(s,\boldsymbol{r})}{\partial x_i}, \tag{3.19}$$

$$\frac{\partial}{\partial t}\frac{\delta c(t,\boldsymbol{r})}{\delta\lambda'(s,\hat{\boldsymbol{r}})}+\boldsymbol{u}\cdot\nabla\frac{\delta c(t,\boldsymbol{r})}{\delta\lambda'(s,\hat{\boldsymbol{r}})}+\lambda\frac{\delta c(t,\boldsymbol{r})}{\delta\lambda'(s,\hat{\boldsymbol{r}})}=0,$$
$$\frac{\delta c(t,\boldsymbol{r})}{\delta\lambda'(s,\hat{\boldsymbol{r}})}\Big|_{t=s}=-\frac{1}{2}\delta(\hat{\boldsymbol{r}}-\boldsymbol{r})c(s,\boldsymbol{r}),\tag{3.20}$$

$$\frac{\partial}{\partial t}\frac{\delta c(t,\boldsymbol{r})}{\delta S'(s,\hat{\boldsymbol{r}})}+\boldsymbol{u}\cdot\nabla\frac{\delta c(t,\boldsymbol{r})}{\delta S'(s,\hat{\boldsymbol{r}})}+\lambda\frac{\delta c(t,\boldsymbol{r})}{\delta S'(s,\hat{\boldsymbol{r}})}=0,$$
$$\frac{\delta c(t,\boldsymbol{r})}{\delta S'(s,\hat{\boldsymbol{r}})}\Big|_{t=s}=\frac{1}{2}\delta(\hat{\boldsymbol{r}}-\boldsymbol{r})\tag{3.21}$$

The factors 1/2 on the right-hand sides are due to the agreement $\int_0^\infty \delta(t)dt=\frac{1}{2}$. Let us substitute the expressions for the variational derivatives of c which are given in (3.19-3.21) into (3.14, 3.15). As a result we have

$$\langle c'(t,\boldsymbol{r})u_i'(t,\boldsymbol{r})\rangle=-\frac{1}{2}\sum_j B_{ij}(\boldsymbol{r},\boldsymbol{r})\frac{\partial c(t,\boldsymbol{r})}{\partial x_j}$$
$$-\frac{1}{2}c(t,\boldsymbol{r})B_{i,\lambda}(\boldsymbol{r},\boldsymbol{r})+\frac{1}{2}B_{i,S}(\boldsymbol{r},\boldsymbol{r}),\tag{3.22}$$

$$\langle c'(t,\boldsymbol{r})\lambda'(t,\boldsymbol{r})\rangle=-\frac{1}{2}\sum_j B_{j,\lambda}(\boldsymbol{r},\boldsymbol{r})\frac{\partial c(t,\boldsymbol{r})}{\partial x_j}$$
$$-\frac{1}{2}c(t,\boldsymbol{r})B_{\lambda,\lambda}(\boldsymbol{r},\boldsymbol{r})+\frac{1}{2}B_{\lambda,S}(\boldsymbol{r},\boldsymbol{r}).\tag{3.23}$$

By substituting the obtained correlators into the non-closed equation (3.9) we find the following equation for the mean tracer

$$\frac{\partial}{\partial t}\langle c\rangle+(\langle\boldsymbol{u}\rangle-\boldsymbol{B_{u,\lambda}})\cdot\nabla\langle c\rangle=$$
$$\nabla\cdot\boldsymbol{D}\nabla\langle c\rangle-(\langle\lambda\rangle-\frac{1}{2}\nabla\cdot\boldsymbol{B_{u\lambda}}-\frac{1}{2}B_{\lambda,\lambda})\langle c\rangle\tag{3.24}$$
$$+\langle S\rangle-\tfrac{1}{2}(B_{\lambda,S}+\nabla\cdot\boldsymbol{B_{u,S}}),$$

where $\boldsymbol{D}=\kappa\boldsymbol{I}+\frac{1}{2}\boldsymbol{B}$, $\boldsymbol{B}$ is the matrix with entries $B_{ij}(\boldsymbol{r},\boldsymbol{r})$ and $\boldsymbol{B_{u,\lambda}}$, $\boldsymbol{B_{u,S}}$ are vectors with coordinates $B_{i,\lambda}(\boldsymbol{r},\boldsymbol{r})$ and $B_{i,S}(\boldsymbol{r},\boldsymbol{r})$ respectively.

Now we show another derivation of this equation. It is based on the following formula for solution of equation (3.7) which is a generalization of (1.54) to the case when the source and potential are nonzero,

$$
\begin{aligned}
c(t,\boldsymbol{r}) = \boldsymbol{E}\{c(s,\boldsymbol{\xi}(s))\exp\{-\int_s^t \lambda(\sigma,\boldsymbol{\xi}(\sigma))d\sigma\} \\
+\int_s^t S(\sigma,\boldsymbol{\xi}(\sigma))\exp\{-\int_\sigma^t \lambda(v,\boldsymbol{\xi}(v))dv\}d\sigma\},
\end{aligned}
\tag{3.25}
$$

where the generalized Lagrangian trajectory $\boldsymbol{\xi}(s)$ is defined by (1.47) and (1.48). This representation can be found in (Friedman 1985, p.148) or deduced similarly to (1.54), using the expansion (1.67). For derivation of the equation for the mean tracer we need to extend this expansion up to the order τ^2. Namely, when τ is small,

$$
\begin{aligned}
\Delta\xi_i(\tau) = -u_i(t,\boldsymbol{r})\tau + \sqrt{2\kappa}\Delta w_i(\tau) + \frac{\tau^2}{2}u_j(t,\boldsymbol{r})\frac{\partial}{\partial x_j}u_i(t,\boldsymbol{r}) \\
-\sqrt{2\kappa}\frac{\partial u_i(t,\boldsymbol{r})}{\partial x_j}\int_0^\tau \Delta w_j(\sigma)d\sigma + o(\tau^2),
\end{aligned}
\tag{3.26}
$$

where as before $\Delta\boldsymbol{\xi}(\tau) = \boldsymbol{\xi}_{t+\tau,\boldsymbol{r}}(t) - \boldsymbol{r}$, $\Delta\boldsymbol{w}(\tau) = \boldsymbol{w}(t+\tau) - \boldsymbol{w}(t)$ and the variable subscripts refer to the corresponding components of the vectors. Note that the second term of the expansion (3.26) is of order $\tau^{1/2}$, the order of the fourth term is $\tau^{3/2}$.

For the sake of simplicity let us suppose that the mean fields $\langle\boldsymbol{u}\rangle, \langle\lambda\rangle, \langle S\rangle$ are zero. For this reason we omit the primes on the variables representing the deviation from mean. The main assumption is that

$$
\boldsymbol{u}(t,\boldsymbol{r}) = \boldsymbol{u}_{[t/\tau_0]}(\boldsymbol{r}), \quad \lambda(t,\boldsymbol{r}) = \lambda_{[t/\tau_0]}(\boldsymbol{r}), \quad S(t,\boldsymbol{r}) = S_{[t/\tau_0]}(\boldsymbol{r}), \tag{3.27}
$$

where $[a]$ denotes the integer part of a, τ_0 is a fixed time scale, the vector sequence $\{\boldsymbol{u}_n(\boldsymbol{r}), \lambda_n(\boldsymbol{r}), S_n(\boldsymbol{r})\}$ of dimension $d+2$ is a sequence of independent identically distributed random fields with zero means. Dependence between the components with the same n is admissible. So, we suppose that all random fields appearing in the problem are innovated in the same time interval τ_0 and within this interval they are independent of t (Fig. 3.1). From this assumption it follows that the time correlation functions of the underlying fields do not depend on time if the two moments belong to the same interval $I_n \equiv ((n-1)\tau_0, n\tau_0)$, $n = 1, 2, \ldots$. More exactly, if $t, s \in I_n$, then

$$
\langle u_i(t,\boldsymbol{r}_1)u_j(s,\boldsymbol{r}_2)\rangle = R_{ij}(\boldsymbol{r}_1,\boldsymbol{r}_2),
$$

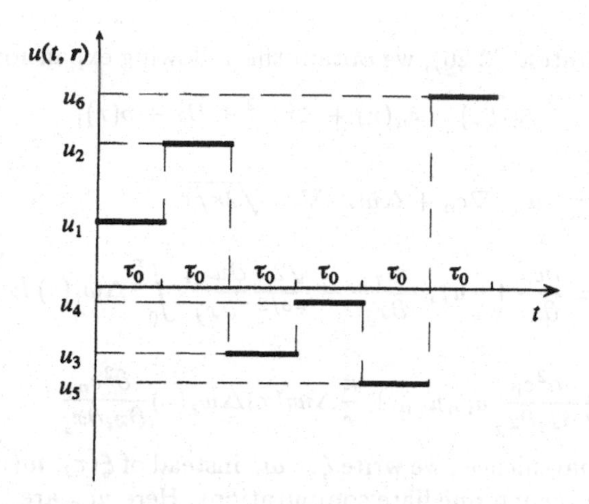

Figure 3.1 Innovating velocity component at a fixed spatial location $\boldsymbol{r}$. The velocity time series $\boldsymbol{u}_1, \boldsymbol{u}_2, \boldsymbol{u}_3, \ldots$ is a sequence of independent random values.

$$\begin{aligned}
\langle u_i(t,\boldsymbol{r}_1)\lambda(s,\boldsymbol{r}_2)\rangle &= R_{i,\lambda}(\boldsymbol{r}_1,\boldsymbol{r}_2), \\
\langle u_i(t,\boldsymbol{r}_1)S(s,\boldsymbol{r}_2)\rangle &= R_{i,S}(\boldsymbol{r}_1,\boldsymbol{r}_2), \\
\langle \lambda(t,\boldsymbol{r}_1)S(s,\boldsymbol{r}_2)\rangle &= R_{\lambda,S}(\boldsymbol{r}_1,\boldsymbol{r}_2), \\
\langle \lambda(t,\boldsymbol{r}_1)\lambda(s,\boldsymbol{r}_2)\rangle &= R_{\lambda,\lambda}(\boldsymbol{r}_1,\boldsymbol{r}_2),
\end{aligned} \tag{3.28}$$

with the corresponding space correlation functions on the right-hand side.

After rescaling via (3.1) we obtain

$$\begin{aligned}
\boldsymbol{u}_\varepsilon(t,\boldsymbol{r}) &= \varepsilon^{-1}\boldsymbol{u}_n(\boldsymbol{r}), \quad t \in I_{n,\varepsilon}, \\
\lambda_\varepsilon(t,\boldsymbol{r}) &= \varepsilon^{-1}\lambda_n(\boldsymbol{r}), \quad t \in I_{n,\varepsilon}, \\
S_\varepsilon(t,\boldsymbol{r}) &= \varepsilon^{-1}S_n(\boldsymbol{r}), \quad t \in I_{n,\varepsilon},
\end{aligned} \tag{3.29}$$

where $I_{n,\varepsilon} = ((n-1)\varepsilon^2\tau_0, n\varepsilon^2\tau_0)$.

Now we go back to the representation (3.25). Let us set $\tau = \tau_0\varepsilon^2$ and $c_n(\boldsymbol{r}) = c(n\tau,\boldsymbol{r})$.

Taking into account Taylor's expansion

$$c(t+\tau,\ \boldsymbol{\xi}_{t+\tau,\boldsymbol{r}}(t)) = c(t,\boldsymbol{r}) + \Delta\boldsymbol{\xi}\cdot\nabla c + \frac{1}{2}\nabla\cdot\Delta\boldsymbol{\xi}\Delta\boldsymbol{\xi}^T\nabla c\ , \tag{3.30}$$

and the representation (3.26), we obtain the following expansion

$$c_n(\boldsymbol{\xi}_\tau) = c_n(\boldsymbol{r}) + A\tau^{1/2} + B\tau + o(\tau), \tag{3.31}$$

where

$$A = -\boldsymbol{u}_n \cdot \nabla c_n + \Delta \boldsymbol{w}_\tau \cdot \nabla c_n \sqrt{2\kappa/\tau},$$

$$B = \frac{\partial c_n}{\partial x_i}\left(\frac{1}{2}u_{j,n}\frac{\partial u_{j,n}}{\partial x_j} - \frac{\sqrt{2\kappa}}{\tau^{3/2}}\frac{\partial u_{n,i}}{\partial x_j}\int_0^\tau \Delta w_j(s)ds\right) \tag{3.32}$$

$$+\frac{1}{2}\frac{\partial^2 c_n}{\partial x_i \partial x_j}u_{i,n}u_{j,n} + \frac{\kappa}{\tau}\Delta w_i(\tau)\Delta w_j(\tau)\frac{\partial^2 c_n}{\partial x_i \partial x_j}.$$

For the sake of convenience , we write $\boldsymbol{\xi}_\tau$, $\boldsymbol{w}_\tau$ instead of $\boldsymbol{\xi}(\tau)$, $\boldsymbol{w}(\tau)$, respectively, and set $\tau_0 = 1$ in the intermediate computations. Here $u_{i,n}$ are the components of the vector $\boldsymbol{u}_n(\boldsymbol{r})$, $i = 1, \ldots, d$. The values of the fields under consideration are for fixed $\boldsymbol{r}$.

To obtain the expansion on the right-hand side of (3.25) for small τ it is sufficient to expand the functions $\lambda_n(\boldsymbol{\xi}_s)$ and $S_n(\boldsymbol{\xi}_s)$ in a series up to the terms of order $s^{1/2}$:

$$\lambda_n(\boldsymbol{\xi}_s) = \lambda_n(\boldsymbol{r}) - s\nabla \cdot \boldsymbol{u}_n\lambda_n(\boldsymbol{r})/\sqrt{\tau} + \sqrt{2\kappa}\Delta\boldsymbol{w}_s \cdot \nabla\lambda_n(\boldsymbol{r}),$$
$$S_n(\boldsymbol{\xi}_s) = S_n(\boldsymbol{r}) - s\boldsymbol{u}_n \cdot \nabla S_n(\boldsymbol{r})/\sqrt{\tau} + \sqrt{2\kappa}\Delta\boldsymbol{w}_s \cdot \nabla S_n(\boldsymbol{r}), \tag{3.33}$$

so that

$$\exp\left\{-\frac{1}{\sqrt{\tau}}\int_0^\tau \lambda_n(\boldsymbol{\xi}_s)ds\right\} = 1 - \lambda_n(\boldsymbol{r})\sqrt{\tau} + C\tau + o(\tau), \tag{3.34}$$

where

$$C = \frac{1}{2}\boldsymbol{u}_n \cdot \nabla\lambda_n(\boldsymbol{r}) - \frac{1}{2}\lambda_n^2(\boldsymbol{r}) - \sqrt{\frac{2\kappa}{\tau}}\int_0^\tau \Delta\boldsymbol{w}_s ds \cdot \nabla\lambda_n(\boldsymbol{r}). \tag{3.35}$$

When estimating the term containing the source function, it is sufficient to restrict our consideration to the expansion

$$\exp\left\{-\frac{1}{\sqrt{\tau}}\int_0^s \lambda_n(\boldsymbol{\xi}_\sigma)d\sigma\right\} = 1 - s\tau^{-1/2}\lambda_n(\boldsymbol{r}), \tag{3.36}$$

whereby it follows that the last term in (3.25) is equal to

$$S_n(\boldsymbol{r})\sqrt{\tau} + D\tau + o(\tau), \tag{3.37}$$

where

$$D = -\frac{1}{2}\left(\lambda_n(\boldsymbol{r})S_n(\boldsymbol{r}) + \boldsymbol{u}_n \cdot \nabla S_n(\boldsymbol{r}) + \frac{\sqrt{2\kappa}}{\tau^{3/2}}\int_0^\tau \Delta \boldsymbol{w}_s ds \cdot \nabla S_n(\boldsymbol{r})\right). \quad (3.38)$$

Let us set $s = n\tau$, $t = (n+1)\tau$ in (3.25) and then substitute (3.31), (3.34) and (3.37) into the right-hand side. Consecutively averaging the result over the Wiener process and over the ensemble corresponding to the random fields $\boldsymbol{u}_n$, S_n and λ_n, we obtain

$$\langle c_{n+1}\rangle = \langle c_n\rangle + (\langle Cc_n\rangle + \langle B\rangle + \langle A\lambda_n\rangle - \langle D\rangle)\tau + o(\tau). \quad (3.39)$$

Since the values of the fields $\boldsymbol{u}_n$, S_n, λ_n do not depend on c_n (because the latter is defined only by values of these fields up to and including the moment $(n-1)\tau$), we have $\langle Cc_n\rangle = \langle C\rangle\langle c_n\rangle$, and the averaging of $A\lambda_n$ and B leads to splitting of the correlations. Taking all this into account, we get

$$\begin{aligned} \langle B\rangle &= \frac{1}{2}\frac{\partial\langle c_n\rangle}{\partial x_i}\left\langle u_{j,n}\frac{\partial u_{i,n}}{\partial x_j}\right\rangle + \frac{1}{2}\frac{\partial^2\langle c_n\rangle}{\partial x_i \partial x_j}\langle u_{i,n}u_{j,n}\rangle + \kappa\nabla^2\langle c_n\rangle, \\ \langle C\rangle &= \frac{1}{2}(\langle \boldsymbol{u}_n \cdot \nabla\lambda_n\rangle + \langle\lambda_n^2\rangle), \\ \langle A\lambda_n\rangle &= \langle\lambda_n \boldsymbol{u}_n\rangle \cdot \nabla c_n, \\ \langle D\rangle &= \frac{1}{2}(\langle\lambda_n S_n\rangle + \langle \boldsymbol{u}_n \cdot \nabla S_n\rangle). \end{aligned} \quad (3.40)$$

Transferring $\langle c_n\rangle$ to the left-hand side in (3.39) and dividing both sides by τ, we get (as $\tau \to 0$) the equation for the mean field:

$$\begin{aligned} &\frac{\partial}{\partial t}\langle c\rangle - \boldsymbol{B}_{\boldsymbol{u},\lambda} \cdot \nabla\langle c\rangle = \\ &\nabla \cdot \boldsymbol{D}\nabla\langle c\rangle + +(\frac{1}{2}\nabla \cdot \boldsymbol{B}_{\boldsymbol{u}\lambda} + \frac{1}{2}B_{\lambda,\lambda})\langle c\rangle - \frac{1}{2}(B_{\lambda,S} + \nabla \cdot \boldsymbol{B}_{\boldsymbol{u},S}), \end{aligned} \quad (3.41)$$

where as before $\boldsymbol{D} = \kappa\boldsymbol{I} + \frac{1}{2}\boldsymbol{B}$, $\boldsymbol{B}$ is the matrix with entries $B_{ij}(\boldsymbol{r},\boldsymbol{r})$, $\boldsymbol{B}_{\boldsymbol{u},\lambda}$, $\boldsymbol{B}_{\boldsymbol{u},S}$ are vectors with coordinates $B_{i,\lambda}(\boldsymbol{r},\boldsymbol{r})$ and $B_{i,S}(\boldsymbol{r},\boldsymbol{r})$ respectively. Here $B's$ are defined in a different way, namely, $B_{ij}(\boldsymbol{r}_1,\boldsymbol{r}_2) = \tau_0 R_{ij}(t,\boldsymbol{r}_1,\boldsymbol{r}_2)$ and others B's in (3.11) are expressed through corresponding R's in the same way.

Obviously, this is the same equation as (3.24). To make sure one should take into account (3.28) and recall that when deriving (3.41) the mean fields are assumed to be zero. Note that if $S \equiv 0, \lambda \equiv 0$ then (3.41) gives the well known Fokker-Planck equation

$$\frac{\partial}{\partial t}\langle c\rangle = \nabla \cdot (\kappa \boldsymbol{I} + \frac{1}{2}\boldsymbol{B}(\boldsymbol{r},\boldsymbol{r}))\nabla\langle c\rangle \ . \tag{3.42}$$

3.2 EQUATIONS FOR THE HIGHER STATISTICAL MOMENTS

This section is aimed at obtaining equations for the higher statistical moments of the tracer and its gradient in the framework of the model (3.27). First, we focus on the general inhomogeneous case and assume non-zero mean velocity, source and λ. Then the homogeneous and isotropic cases will be considered.

While obtaining the equation for the correlation function of the tracer fluctuations, we shall use the following expansions:

$$\begin{aligned}
c_{n+1}(\boldsymbol{r}) &= \boldsymbol{E}_{\boldsymbol{w}_1}\{(c_n(\boldsymbol{r}) + A(\boldsymbol{r})\tau^{1/2} + B(\boldsymbol{r})\tau) \\
&\times(1 - \lambda_n(\boldsymbol{r})\tau^{1/2} + C(\boldsymbol{r})\tau) + S_n(\boldsymbol{r})\tau^{1/2} + D(\boldsymbol{r})\tau\} + o(\tau), \\
c_{n+1}(\tilde{\boldsymbol{r}}) &= \boldsymbol{E}\boldsymbol{w}_2\{(c_n(\tilde{\boldsymbol{r}}) + A(\tilde{\boldsymbol{r}})\tau^{1/2} + B(\tilde{\boldsymbol{r}})\tau) \\
&\times(1 - \lambda_n(\tilde{\boldsymbol{r}})\tau^{1/2} + C(\tilde{\boldsymbol{r}})\tau) + S_n(\tilde{\boldsymbol{r}})\tau^{1/2} + D(\tilde{\boldsymbol{r}})\tau\} + o(\tau),
\end{aligned} \tag{3.43}$$

where $\boldsymbol{r}, \tilde{\boldsymbol{r}}$ are arbitrary and A, B, C, D are given by (3.32), (3.35), (3.38). By multiplying equalities (3.43), averaging over the Wiener processes $\boldsymbol{w}_1$ and $\boldsymbol{w}_2$, and taking into account their independence, we obtain the following equation in the limit as $\tau \to 0$:

$$\begin{aligned}
&\frac{\partial R}{\partial t} + (\langle\lambda(t,\boldsymbol{r})\rangle + \langle\lambda(t,\tilde{\boldsymbol{r}})\rangle - B_{\lambda,\lambda}(\boldsymbol{r},\tilde{\boldsymbol{r}}))R \\
&+\langle\boldsymbol{u}(t,\boldsymbol{r})\rangle \cdot \nabla_{\boldsymbol{r}} R + \langle\boldsymbol{u}(t,\tilde{\boldsymbol{r}})\rangle \cdot \nabla_{\tilde{\boldsymbol{r}}} R \\
&+\boldsymbol{B}_{\boldsymbol{u},\lambda}(\boldsymbol{r},\tilde{\boldsymbol{r}}) \cdot \nabla_{\boldsymbol{r}} R + \boldsymbol{B}_{\boldsymbol{u},\lambda}(\tilde{\boldsymbol{r}},\boldsymbol{r}) \cdot \nabla_{\tilde{\boldsymbol{r}}} R = \\
&\nabla_{\boldsymbol{r}} \cdot \boldsymbol{B}(\boldsymbol{r},\tilde{\boldsymbol{r}})\nabla_{\tilde{\boldsymbol{r}}} R + \frac{1}{2}\nabla_{\boldsymbol{r}} \cdot \boldsymbol{B}(\boldsymbol{r},\boldsymbol{r})\nabla_{\boldsymbol{r}} R \\
&+\frac{1}{2}\nabla_{\tilde{\boldsymbol{r}}} \cdot \boldsymbol{B}(\tilde{\boldsymbol{r}},\tilde{\boldsymbol{r}})\nabla_{\tilde{\boldsymbol{r}}} R + \kappa\nabla^2_{\boldsymbol{r}} R + \kappa\nabla^2_{\tilde{\boldsymbol{r}}} R + B_{S,S}(\boldsymbol{r},\tilde{\boldsymbol{r}}),
\end{aligned} \tag{3.44}$$

where

$$R \equiv R(t,\boldsymbol{r},\tilde{\boldsymbol{r}}) = \langle c'(t,\boldsymbol{r})c'(t,\tilde{\boldsymbol{r}})\rangle, \quad c'(t,\boldsymbol{r}) = c(t,\boldsymbol{r}) - \langle c(t,\boldsymbol{r})\rangle,$$
$$B_{S,S}(\boldsymbol{r},\tilde{\boldsymbol{r}}) = \tau_0\langle \tilde{S}_n(\boldsymbol{r})\tilde{S}_n(\tilde{\boldsymbol{r}})\rangle, \quad \tilde{S} = S' - \boldsymbol{u}'\cdot\nabla\langle c\rangle. \tag{3.45}$$

Note that, since the velocity field is nondivergent, the first term on the right-hand side of the equation for the correlation function can be represented as $B_{ij}(\boldsymbol{r},\tilde{\boldsymbol{r}})\partial^2 R/\partial x_i \partial\tilde{x}_j$, where the $B_{ij}(\boldsymbol{r},\tilde{\boldsymbol{r}})$ are the components of the matrix $\boldsymbol{B}(\boldsymbol{r},\tilde{\boldsymbol{r}})$.

While deriving the equations for the higher moments $m_k(t,\boldsymbol{r}_1,\ldots,\boldsymbol{r}_k) = \langle c'(t,\boldsymbol{r}_1)$ $c'(t,\boldsymbol{r}_2)\ldots c'(t,\boldsymbol{r}_n)\rangle$, we restrict ourselves to the case when the mean current, $\langle \boldsymbol{u}\rangle$, and the feedback parameter λ are constant. In this case one can ignore λ in the intermediate computations and add the term $-k\lambda m_k$ to the right-hand side of the final equation for k-th moment. Thus by setting $\lambda_n \equiv 0$ we obtain $C_n \equiv 0$, where C is given by (3.35) and hence the following expansion is true:

$$c_{n+1}(\boldsymbol{r}) = \boldsymbol{E}\{c_n(\boldsymbol{r}) + G_n(\boldsymbol{r})\tau^{1/2} + H_n(\boldsymbol{r})\tau\} + o(\tau), \tag{3.46}$$

where $G_n = A_n + S_n$,

$$H_n = B_n - \frac{1}{2}\left[\boldsymbol{u}_n\cdot\nabla S_n + \frac{\sqrt{2\kappa}}{\tau^{3/2}}\int_0^\tau \Delta\boldsymbol{w}_s ds\cdot\nabla S_n\right], \tag{3.47}$$

A, B are given by (3.32). The cross multiplication of relations (3.47) taken at the k points $\boldsymbol{r}_1,\ldots,\boldsymbol{r}_k$ yields

$$\begin{aligned}
&c_{n+1}(\boldsymbol{r}_1)\ldots c_{n+1}(\boldsymbol{r}_k) = \\
&= \boldsymbol{E}_{\boldsymbol{w}_1\ldots\boldsymbol{w}_k}\{c_n(\boldsymbol{r}_1)\ldots c_n(\boldsymbol{r}_k) + \tau^{1/2}G_n(\boldsymbol{r}_\alpha)c_n(\boldsymbol{r}_1)\ldots c_n(\boldsymbol{r}_{\alpha-1}) \\
&\times c_n(\boldsymbol{r}_{\alpha+1})\ldots c_n(\boldsymbol{r}_k) + \tau(G_n(\boldsymbol{r}_\alpha)G_n(\boldsymbol{r}_\beta)c_n(\boldsymbol{r}_1)\ldots c_n(\boldsymbol{r}_{\alpha-1}) \\
&\times c_n(\boldsymbol{r}_{\alpha+1})\ldots c_n(\boldsymbol{r}_{\beta-1})c_n(\boldsymbol{r}_{\beta+1})\ldots c_n(\boldsymbol{r}_k) + H_n(\boldsymbol{r}_\alpha)c_n(\boldsymbol{r}_1)\ldots \\
&\times c_n(\boldsymbol{r}_{\alpha-1})c_n(\boldsymbol{r}_{\alpha+1})\ldots c_n(\boldsymbol{r}_k))\},
\end{aligned} \tag{3.48}$$

where $\boldsymbol{w}_1,\ldots,\boldsymbol{w}_k$ are independent Wiener processes. Let us average both sides of the last relation and pass to the limit as $\tau\to 0$. Because of the splitting of correlations, the second summand is equal to zero, the third summand yields a term containing the order $k-2$ moment and a diffusion term with coefficients depending on various space coordinates, and the fourth term yields the traditional diffusion term. Finally we obtain

$$\frac{\partial}{\partial t}m_k = B_{ij}(\boldsymbol{r}_\alpha,\boldsymbol{r}_\beta)\frac{\partial^2 m_k}{\partial x_{\alpha,i}\partial x_{\beta,j}} + \frac{1}{2}\frac{\partial}{\partial x_{\alpha,i}}(2\kappa\delta_{ij} + B_{ij}(\boldsymbol{r}_\alpha,\boldsymbol{r}_\alpha))\frac{\partial m_k}{\partial x_{\alpha,j}} + m_{k-2,\alpha,\beta}B_S(\boldsymbol{r}_\alpha,\boldsymbol{r}_\beta) - k\lambda m_k, \tag{3.49}$$

where $m_{k-2,\alpha,\beta}$ is the $(k-2)$-th point moment for the set of points $\{\boldsymbol{r}_1,\ldots,\boldsymbol{r}_{\alpha-1}, \boldsymbol{r}_{\alpha+1},\ldots,\boldsymbol{r}_{\beta-1},\boldsymbol{r}_{\beta+1},\ldots,\boldsymbol{r}_k\}$ (here $\boldsymbol{r}_\alpha$ and $\boldsymbol{r}_\beta$ are excluded from the initial set and we summed over the indices i, j, $\alpha \neq \beta$).

Let

$$\mu_{\boldsymbol{i}}(t,\boldsymbol{r}_1,\ldots,\boldsymbol{r}_k) = \left\langle \frac{\partial c'(\boldsymbol{r}_1)}{\partial x_{1,i_1}}\cdots\frac{\partial c'(\boldsymbol{r}_k)}{\partial x_{1,i_k}}\right\rangle = \frac{\partial^k m_k}{\partial x_{1,i_1}\partial x_{2,i_2}\ldots\partial x_{k,i_k}} \tag{3.50}$$

be a joint k-th order moment of tracer fluctuation gradients for the set of indices $\boldsymbol{i} = (i_1,\ldots,i_k)$, $1 \leq i_\alpha \leq d$, $\alpha = 1,\ldots,k$, where $x_{p,q}$ is the q-th coordinate of the vector $\boldsymbol{r}_p$. By differentiating both sides of equation (3.49), we obtain the following system of equations for the moments if mean current and source are absent:

$$\frac{\partial}{\partial t}\mu_{\boldsymbol{i}} = \frac{\partial^2 B_{pj}(\boldsymbol{r}_\alpha,\boldsymbol{r}_\beta)}{\partial x_{\alpha,i_\alpha}\partial x_{\beta,i_\beta}}\mu_{\boldsymbol{i}(\alpha,p;\beta,j)} + \frac{\partial B_{pj}}{\partial x_{\beta,i_\beta}}\frac{\partial \mu_{\boldsymbol{i}(\beta,j)}}{\partial x_{\alpha,p}} + \frac{\partial B_{pj}}{\partial x_{\alpha,i_\alpha}}\frac{\partial \mu_{\boldsymbol{i}(\alpha,p)}}{\partial x_{\beta,j}} + B_{pj}\frac{\partial^2 \mu_{\boldsymbol{i}}}{\partial x_{\alpha,p}\partial x_{\beta,j}} + \kappa\nabla^2_{\boldsymbol{r}_\alpha}\mu_{\boldsymbol{i}} + \frac{1}{2}\frac{\partial}{\partial x_{\alpha,p}}\left(B_{pj}(\boldsymbol{r}_\alpha,\boldsymbol{r}_\alpha)\frac{\partial \mu_{\boldsymbol{i}}}{\partial x_{\beta,j}}\right) + \frac{\partial}{\partial x_{\alpha,p}}\left(\mu_{\boldsymbol{i}(\alpha,p)}\frac{\partial B_{pj}(\boldsymbol{r}_\alpha,\boldsymbol{r}_\alpha)}{\partial x_{\alpha,i_\alpha}}\right) - k\lambda\mu_{\boldsymbol{i}}, \tag{3.51}$$

where the symbol $\boldsymbol{i}(\alpha,p;\beta,j)$ denotes the vector $\boldsymbol{i}$ with the components i_α and i_β replaced by p and j, respectively. The meaning of the notation $\boldsymbol{i}(\alpha,p)$ and $\boldsymbol{i}(\beta,j)$ is the same. In (3.51) we sum over all p,j and $\alpha \neq \beta$.

Equation (3.51) can also be obtained in another way. Applying the operator ∇ to both sides of (3.7), we get the following equation for the gradient $\boldsymbol{g} = \nabla c$:

$$\frac{\partial \boldsymbol{g}}{\partial t} + \boldsymbol{u}\cdot\nabla\boldsymbol{g} + \boldsymbol{g}\cdot\nabla\boldsymbol{u} = \kappa\nabla^2\boldsymbol{g} - \lambda\boldsymbol{g} + \nabla S. \tag{3.52}$$

Repeating the procedure used when deriving (3.49) from (3.7), we obtain (3.51).

3.3 HOMOGENEOUS AND ISOTROPIC FLOW

Let us suppose that the velocity field and the source field are homogeneous, i.e., their means are constant and their correlation functions depend only on the difference of their arguments $\boldsymbol{B}(\boldsymbol{r},\tilde{\boldsymbol{r}}) = \boldsymbol{B}(\boldsymbol{r}-\tilde{\boldsymbol{r}})$, $B_{S,S}(\boldsymbol{r},\tilde{\boldsymbol{r}}) = B_S(\boldsymbol{r}-\tilde{\boldsymbol{r}})$. In this case the equations for the mean and the higher moments can be reduced to the following form (λ is still assumed constant):

$$\frac{\partial}{\partial t}\langle c\rangle = \nabla\cdot\boldsymbol{D}\nabla\langle c\rangle - \lambda\langle c\rangle + \langle S\rangle - \tau_0\langle \boldsymbol{u}'\cdot\nabla S'\rangle,$$

$$\frac{\partial}{\partial t}m_n = (2\kappa\delta_{ij} + B_{ij}(\boldsymbol{0}) - B_{ij}(\boldsymbol{r}_\alpha - \boldsymbol{r}_\beta))\frac{\partial^2}{\partial x_{\alpha,i}\partial x_{\beta,j}} \tag{3.53}$$

$$+m_{n-2,\alpha,\beta}B_S(\boldsymbol{r}_\alpha - \boldsymbol{r}_\beta) - n\lambda m_n.$$

If the initial tracer field is also homogeneous, we obtain for the tracer correlation function $R(t,\boldsymbol{r}) \equiv R(t,\tilde{\boldsymbol{r}},\tilde{\boldsymbol{r}}+\boldsymbol{r})$,

$$\frac{\partial}{\partial t}R = (2\kappa\delta_{ij} + B_{ij}(\boldsymbol{0}) - B_{ij}(\boldsymbol{r}))\frac{\partial^2}{\partial x_i\partial x_j}R - 2\lambda R + B_S(\boldsymbol{r}). \tag{3.54}$$

Under the additional assumption of isotropy, the covariation tensor of the velocity field has the form (2.4). Substituting that representation of the covariation tensor into equation (3.54), we get

$$\frac{\partial}{\partial t}R(t,r) = HR(t,r) - 2\lambda R(t,r) + B_S(r), \tag{3.55}$$

where the correlation function of the tracer $R(t,r)$ and source $B_S(r)$ now are functions of $r = |\boldsymbol{r}|$ and the operator H is defined by

$$H = \frac{1}{r^{d-1}}\frac{\partial}{\partial r}r^{d-1}K(r)\frac{\partial}{\partial r}, \tag{3.56}$$

where $K(r) = 2\kappa + F_L(r)$, $F_L(r) = B_L(0) - B_L(r)$, $B_L(r) = \tau_0 R_L(r)$ and $R_L(r)$ is the longitudinal correlation function defined in (2.4).

Equation (3.55) in the case $B_s(r) \equiv 0$, $\lambda = 0$ is well known. Its predecessor had appeared in (Richardson 1926) before the notion of correlation function became widely used. Then equations of form (3.55) were repeatedly recovered in a number of papers (Roberts 1961, Kraichnan 1974, Lindgren 1981, and Bennet 1987). Those authors proposed different expressions for the 'two-particle' diffusion coefficient $K(r)$ and different assumptions were used. In a rigorous way, only for a δ-correlation velocity field can such an equation be obtained (Semenov 1990, Molchanov and Piterbarg 1994).

It follows from (3.53) that the equation for the mean tracer gradient in the homogeneous case has the same form as that for the mean temperature:

$$\frac{\partial}{\partial t}\langle \boldsymbol{g} \rangle = \nabla \cdot \boldsymbol{D} \nabla \langle \boldsymbol{g} \rangle. \tag{3.57}$$

In the general isotropic case, no closed equation for the trace of the gradient's correlation matrix is available, but the behavior of the square of the gradient can be analyzed by solving an equation for the auxiliary function $G(t,r) = -K^{-1}HR(t,r)$. The form of the operator H implies that the value $G(t,0)$ coincides with the mean square fluctuation of the gradient magnitude. Applying the operator H to both sides of (3.55) and dividing by K, we obtain the following equation for the function G:

$$\frac{\partial}{\partial t}G(t,r) = \widetilde{H}G(t,r) + U(r)G(t,r) - 2\lambda G(t,r) - \frac{1}{K}HB_S(r), \tag{3.58}$$

where

$$\widetilde{H} = H + 2\frac{dK}{dr}\frac{\partial}{\partial r}, \qquad U = \frac{d^2K}{dr^2} + \frac{d-1}{r}\frac{dK}{dr} + \frac{1}{K}\left(\frac{dK}{dr}\right)^2. \tag{3.59}$$

For the longitudinal correlation function $B_L(r) = \exp(-r^2/l_u^2)$ the drift $\dfrac{dK}{dr}$ and the potential $U(\boldsymbol{r})$ look like those in Figs. 3.2 and 3.3.

Equation (3.58) differs in principle from equation (3.55) due to the potential term UG. It is this term that causes possible growth of the gradient. In the next section this fact will be discussed in detail.

3.4 THE LIFETIME OF TRACER FLUCTUATIONS, STIRRING, AND MIXING

Here we shall discuss the process by which tracer fluctuations decay due to the combined effects of stochastic velocity and molecular diffusion in the framework

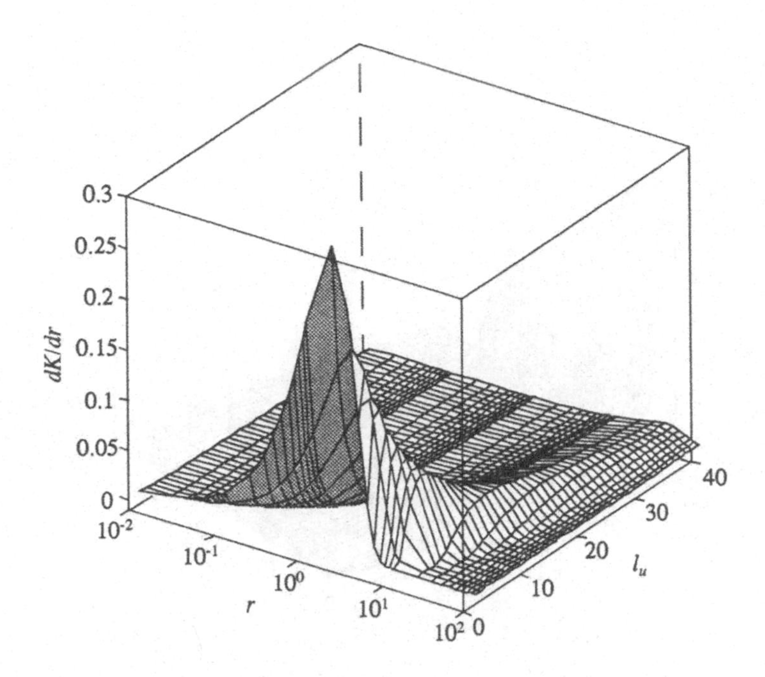

Figure 3.2 The typical behavior of the drift $\frac{dK}{dr}$ in the diffusion equation (3.55) for the tracer correlation function.

of equations (3.55) and (3.58). To clarify these effects it is worth excluding the source from these equations and setting λ=0. Thus, we will discuss the time behavior of the solutions to the equations

$$\frac{\partial}{\partial t}R(t,r) = HR(t,r) \tag{3.60}$$

and

$$\frac{\partial}{\partial t}G(t,r) = \widetilde{H}G(t,r) + U(r)G(t,r), \tag{3.61}$$

where the operators H and $\widetilde{H}$ are given by (3.56) and (3.59), respectively.

We assume that the initial tracer field $c_0(\boldsymbol{r})$ is an isotropic random field with zero mean and correlation function $R_0(r)$. We will now consider the Cauchy problem for (3.60) and (3.61) due to the initial conditions $R_0(r)$ and $G_0(r) = -K^{-1}HR_0(r)$ respectively.

It is clear that if $t \to 0$, the solutions of (3.60) and (3.61) decay due to molecular diffusion, whereas if $\kappa = 0$, $R(t,0)$ remains constant for all t and $G(t,0)$ grows

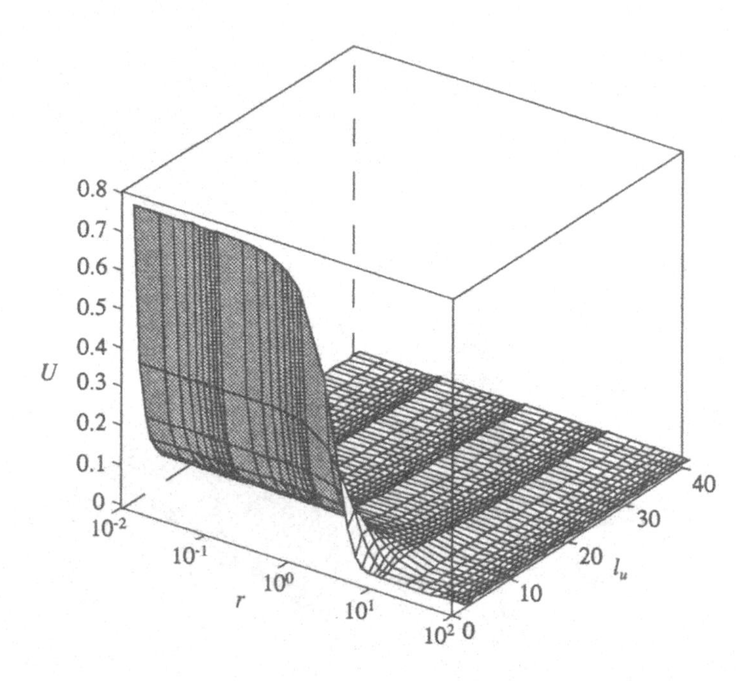

Figure 3.3 The typical behavior of the potential $U(\boldsymbol{r})$ in the diffusion equation (3.58) for the mean square gradient.

exponentially. A detailed consideration of the last case is given in (Saichev and Woyczynski 1996).

The first question to focus on is how fast $R(t,0)$ decays when κ is small. In applications, the decay time can be thought of as the lifetime of passive tracer anomalies. Indeed, as we mentioned in chapter 1 the shape of high peaks of a homogeneous random field is similar to that of the correlation function of this field. So, the evolution of the correlation function $R(t,r)$ described by (3.60) can be interpreted as an evolution of a 'typical' individual excursion over a high level for the tracer field itself. Even if this interpretation appears to be doubtful, the characteristic decay time of the tracer variance $\sigma_c^2(t) = R(t,0)$ provides an appropriate estimator for the lifetime of the tracer fluctuations.

The second question is addressed to equation (3.61) which describes the behavior of the tracer gradient. It can be shown that its solution grows at least at the initial stage due to the second term on the right-hand side. We shall uncover when it really occurs and how long it lasts.

Thus, we consider the Cauchy problem for eq.(3.60) where

$$H = a(r)\frac{d}{dr} + \mathfrak{k}(r)\frac{d^2}{dr^2}, \tag{3.62}$$

with

$$\begin{aligned} a(r) &= -B_L'(r) + \frac{d-1}{r}(2\kappa + B_L(0) - B_L(r)), \\ \mathfrak{k}(r) &= 2\kappa + B_L(0) - B_L(r). \end{aligned} \tag{3.63}$$

Here the prime stands for derivative with respect to r.

It is well known (Gardiner 1985) that equation (3.62) is the inverse Fokker-Plank equation for a Lagrangian particle whose coordinate satisfies the Ito equation

$$d\xi = adt + \sqrt{2\mathfrak{k}}dw(t). \tag{3.64}$$

Consequently the solution of (3.60), due to the initial condition $R_0(r)$, is represented in the form

$$R(t,r) = \boldsymbol{E}\{R_0(\xi_r(t))\}, \tag{3.65}$$

where $\xi_r(t)$ is the solution of (3.64) with $\xi_r(0) = r$ and $\boldsymbol{E}$ as before is the averaging operator over the ensemble of the Wiener process $w(\cdot)$.

Let l_0 be the correlation scale of $c_0(\boldsymbol{r})$ defined for example by (1.5) and $\tau(l_0)$ be the first time when the particle, starting from zero, leaves the interval $(-l_0, l_0)$. Obviously, the value $m(l_0) = \boldsymbol{E}\{\tau(l_0)\}$ can be viewed as a decay time for $\sigma_c^2(t)$ (Fig. 3.4). The following statement provides an asymptotic for $\tau(l_0)$ and shows that under small molecular diffusion the value of $\tau(l_0)$ is deterministic.

Proposition 3.1. *The ratio $\tau(l)/m(l)$ converges in probability to* 1 *as $\kappa \to 0$ for any $l > 0$, and the leading term of the asymptotic for* m(*l*) *has the following form*

$$m(l) \sim -\frac{1}{dB_L''(0)} \log\left(-\frac{l^2 B_L''(0)}{\kappa}\right). \tag{3.66}$$

Using definitions (2.4),(2.5),(2.15) and the relation $B_L(r) = \tau_0 R_L(r)$ it is not difficult to express the lifetime in terms of the Peclet number

$$Pe = \frac{\tau_0 \sigma_u^2}{\kappa}. \tag{3.67}$$

Namely, first we obtain

$$R_L''(0) = -\frac{\sigma_u^2}{c_0(d) l_u^2}, \tag{3.68}$$

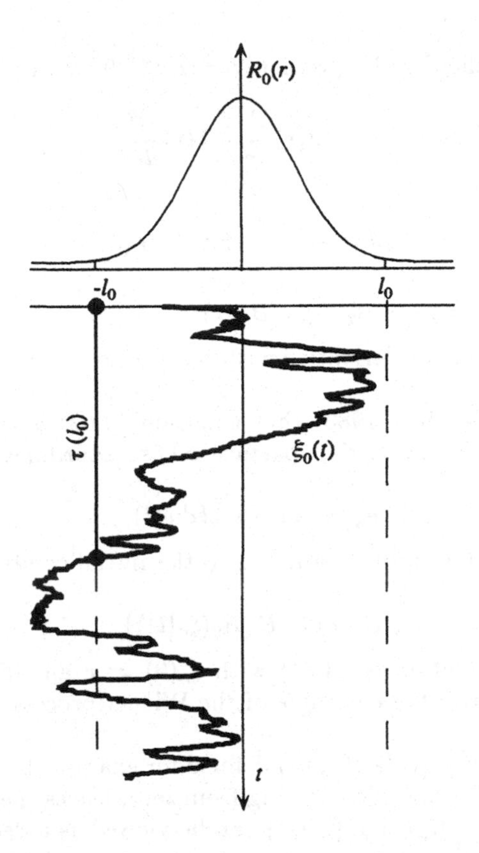

Figure 3.4 Interpretation of the lifetime of tracer fluctuations as a time when the Lagrangian particle leaves the correlation cell for the first time.

where $c_0(d) = 2(d+2)/d$ and then

$$m(l_0) \sim c_1(d) \frac{l_u^2}{\tau_0 \sigma_u^2} \log \left(\frac{l_0^2 Pe}{l_u^2} \right), \tag{3.69}$$

where $c_1(d) = c_0(d)/d$.

Thus, Proposition 3.1 claims that if the Peclet number is sufficiently large, the lifetime of tracer fluctuations is given by (3.69). This estimate drastically differs from the traditional one, $l_0^2/\sigma_u^2 \tau_0$, which is based on the analogy between turbulent and molecular diffusion. It can be shown that the traditional estimate is valid when $l_0 \gg l_u$. In the case of small and moderate l_0 (compared with the

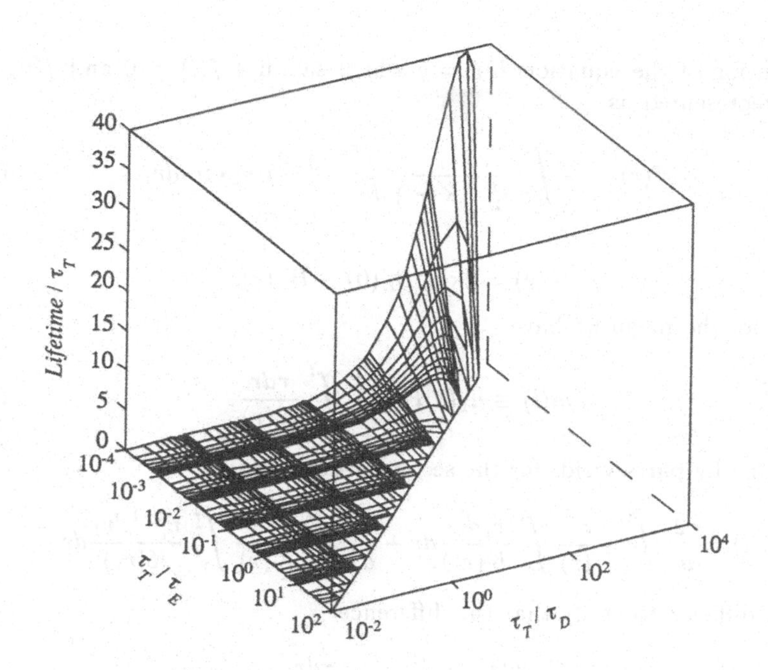

Figure 3.5 Dependence of the lifetime on the Eulerian time scale τ_E, the turnover time τ_T, and the diffusion time τ_D.

correlation scale of the velocity field's fluctuations), the estimate (3.69) should be used.

Finally note that (3.69) can be written in terms of the time scales introduced in the previous section

$$\textit{Lifetime} \sim c_1(d)\frac{\tau_T^2}{\tau_E}\log\left(\frac{\tau_E \tau_D}{\tau_T^2}\right), \tag{3.70}$$

where the innovation time τ_0 is interpreted as the Eulerian correlation time. Fig. 3.5 shows a plot of (3.70).

Let us prove Proposition 3.1.

Proof. Let $\tau(r,l)$ be the first time that a particle which starts from the point r, leaves the interval $(-l,l)$. It is well-known (Gardiner 1985, p.185) that the statistical moments $\mu_n = \boldsymbol{E}\{\tau(r,l)^n\}$ satisfy the following equation

$$H\mu_n = -n\mu_{n-1}, \qquad \mu_n\big|_{|r|=l} = 0. \tag{3.71}$$

The solution of the equation $Hf = g$ which satisfies $f(l) = 0$ and $f(0) < \infty$ can be represented as

$$f(r) = -\int_r^l \frac{1}{r_1^{d-1} K(r_1)} \int_0^{r_1} r_2^{d-1} g(r_2) dr_2 dr_1, \tag{3.72}$$

where

$$K(r) = 2\kappa + B_L(0) - B_L(r). \tag{3.73}$$

In part, for the mean we have

$$m(l) \equiv \mu_1(0, l) = \frac{1}{d} \int_0^l \frac{r dr}{K(r)}. \tag{3.74}$$

Integrating by parts yields for the second order moment

$$\mu_2(0, l) = \frac{2}{d^2} \int_0^l \frac{r}{K(r)} \int_r^l \frac{r_1 dr_1}{K(r_1)} dr + \frac{2}{d^2} \int_0^l \frac{r^{1-d}}{K(r)} \int_0^r \frac{r_1^{d+1} dr_1}{K(r_1)} dr. \tag{3.75}$$

It is not difficult to verify that the difference

$$\int_0^l \frac{r dr}{K(r)} - \int_0^l \frac{r dr}{2\kappa - \frac{1}{2} r^2 B_L''(0)} \tag{3.76}$$

is finite as $\kappa \to 0$, whereby we obtain formula (3.66). Then, the first term on the right-hand side of (3.75) is equal to $\mu_1(0, l)^2$. The inner integral in the second term is finite when $\kappa \to 0$ and hence the second term is of order $\mu_1(0, \delta)$. Therefore

$$m_2(l) \equiv \mu_2(0, l) = \mu_1(0, l)^2 + O(\mu_1(0, l)). \tag{3.77}$$

From the Chebyshev's inequality it follows that for any $\delta > 0$

$$\begin{aligned} Pr\{|\frac{\tau(l)}{m(l)} - 1| \geq \delta\} &\leq \frac{E\{\tau(l) - m(l)\}^2}{m(l)^2 \delta^2} = \\ \frac{m_2(l) - m(l)^2}{m(l)^2 \delta^2} &= \frac{O(m)}{m^2 \delta^2} \to 0. \end{aligned} \tag{3.78}$$

Proposition 3.1 is proven. ∘

The investigation of equation (3.53) in the general case $\lambda > 0$, $B_S \neq 0$ can be carried out similarly. Piterbarg (1987) considered numerical solutions of this equation illustrating an interaction between generation and dissipation of temperature anomalies.

The fundamental difference between turbulent and molecular diffusion is also illustrated by the behavior of the gradient. A very important result can be obtained by very simple means directly from the equation (3.61). Namely, let us set $r = 0$ and $t = 0$ in (3.61). Taking into account the relations

$$U(0) = -dB_L''(0), \quad \widetilde{H}G\big|_{r=0} = HG\big|_{r=0} = 2\kappa d G_{rr}''(t,0) \tag{3.79}$$

we can conclude that at $t = 0$ the derivative $(\partial/\partial t)G(t,0)$ is positive (i.e., the mean square gradient increases) if and only if

$$-B_L''(0)G_0(0) + 2\kappa G_0''(0) > 0. \tag{3.80}$$

One can easily check that $G_0(0) = -2l_g^2 G_0''(0)$, where the gradient space scale, l_g^2 is defined by (1.5). Taking into account (3.67) and (3.68) one can rewrite (3.80) as follows

$$l_g^2 > c_0(d)(Pe)^{-1}l_u^2, \tag{3.81}$$

where $c_0(d)$ is the same as in (3.68).

Thus, the initial tracer gradient must be sufficiently large to produce an increasing gradient (although it can be much smaller than that of the velocity field if the Peclet numbers are large). We note that the fact of gradient's growth was established long ago in numerous experiments, the simplest one was described by Eckart (1948): "It is useful to consider a trivial experiment by way of introduction: the mixing of coffee and cream.Three more or less distinct stages can be observed: 1. The initial stage, in which rather large volumes of cream and coffee are distinctly visible; there are sharp gradients at the interfaces between the volumes, but elsewhere the gradient is practically zero. Averaged over the entire volume, the gradient is small. If motion of the liquids is avoided, this state persists for a considerable time. 2. The intermediate stage, after motion has been induced by stirring the liquids; the masses of cream and coffee are distorted, with a rapid increase in the extent of the interfacial regions having high concentration gradients. The average value of the gradient is correspondingly increased. 3. The final stage, in which the gradients disappear, apparently quite suddenly and spontaneously, with the liquid becoming homogeneous." The second stage of this process was called 'stirring', and the third one - 'mixing'. In the physical oceanography literature, the stirring is often referred to as transport without mixing (Haidvogel et. al. 1983) and it results in the streakiness of the tracer field. Though in the ocean both processes occur simultaneously and the stirring generally enhances mixing, for some time and space scales mixing can be negligible. During the onset of heating the horizontal stirring becomes more pronounced at geostrophic turbulence scales while the surface cooling in the autumn favors the enhancement of three dimensional

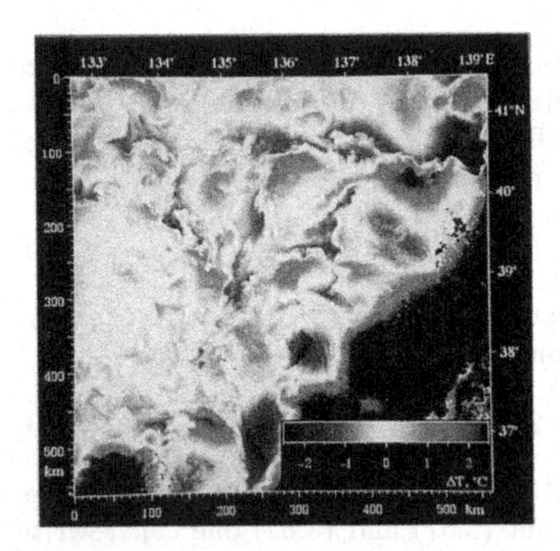

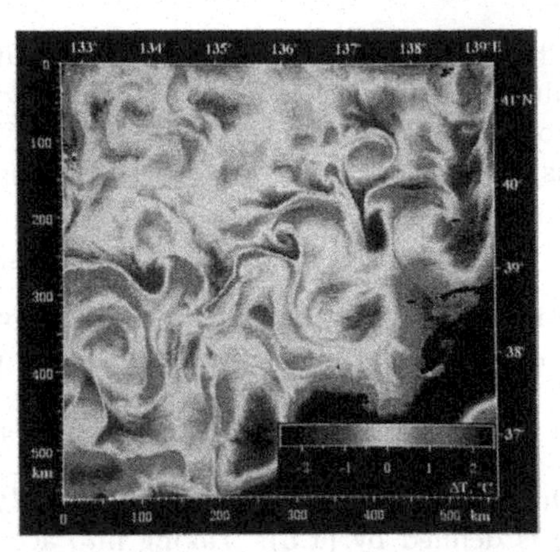

A B

Figure 3.6 The differences between the NOAA-10 SST images and the wavelet transform reconstructions for large scales (> 140 km). Left plate, a, is for the image of 08:58 UT, October 27, 1993. Right plate, b, is for the image of 08:51 UT, May 19, 1994. The dark areas in the bottom right corners indicate land and clouds. Reprinted with permission from original color figures (Ostrovskii 1995).

mixing (Ostrovskii 1995). This was found from the results of wavelet analysis of the NOAA satellite radiometry data of the Japan Sea in autumn 1993 and spring 1994. By using this observation, let us briefly demonstrate the typical patterns of stirring and mixing.

The original sea surface temperature (SST) fields of the Japan Sea had almost the same temperature range in May 1994 as in October 1993. But SST patterns appeared fuzzy in October and sharp in May. We visualized the SST variability at individual spatial scales by subtracting wavelet transform reconstructions for the scales larger than scale of interest from original image. Fig. 3.6a and 3.6b show such differences between original images and reconstructions for the large scales (> 140 km). Fig. 3.7a and 3.7b show the differences between the same original images and the reconstructions for the meso- and submesoscales (> 9 km). It is important that the signals on Fig. 3.6 and 3.7 have essentially the same energy as the original SST variations at scales smaller than $\approx$ 140 km and $\approx$ 9 km, respectively. Thus the decoding-encoding wavelet transform allows a decomposition of the SST variability into different spatial modes which conserve the energy of the original signals at certain wavenumber ranges.

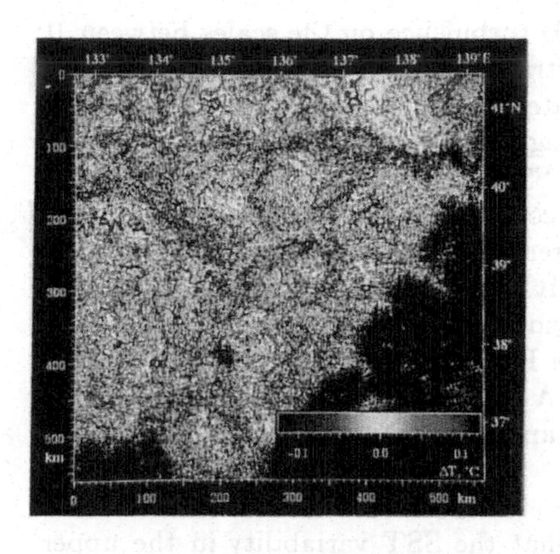

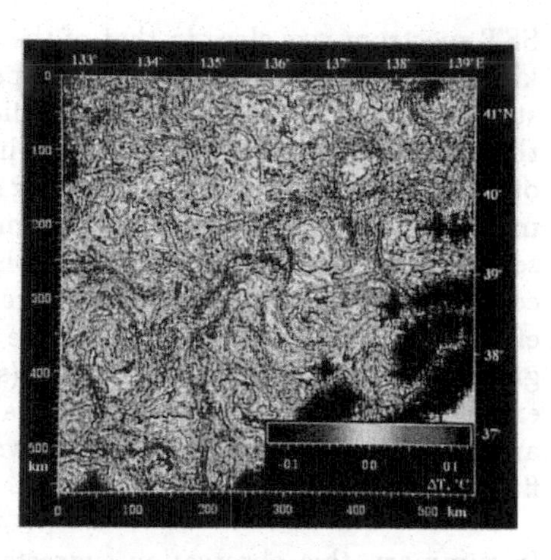

A B

Figure 3.7 **The differences between the NOAA-10 SST images and the wavelet transform reconstructions for meso- and submesoscales (> 9 km). Left plate, *a*, is for the image of 08:58 UT, October 27, 1993. Right plate, *b*, is for the image of 08:51 UT, May 19, 1994. The dark areas in the bottom right corners indicate land and clouds. Reprinted with permission from original color figures (Ostrovskii 1995).**

It is easy to see the difference between the autumn and spring SST patterns by comparing Fig. 3.6a and 3.7a with Fig. 3.6b and 3.7b, respectively. In October (Fig. 3.6a and 3.7a) the small scale variability was characterized by plumes with irregular shapes. These features were usually less than 30 km in size though some extended for up to 50 km. The irregularity of the small scale features had a noise-like appearance in Fig. 3.7a. In May (Fig. 3.6b and 3.7b) SST patterns were characterized by elongated and narrow streaks; the streaks could be longer than 100 km while their widths did not exceed a few kilometers. They wrapped around mesoscale vortices so SST changed gradually along them. The streak boundaries appeared as thin curves in Fig. 3.7b. The small scale features were anisotropic in spring and the isotherms followed streamlines.

Noticeably, the power spectra of SST had different slopes in October-November of the spectra was about $k^{-2.0}$ at the wavenumbers $0.04 < k < 1.2$, cpk. The spectrum steepness began to grow in April. In May the spectral density distribution varied as $k^{-2.8}$ in the wavenumber band $0.06 < k < 0.8$, cpk. Existing theory states that the slope $k^{-5/3}$ corresponds to the Kolmogorov spectrum of three-dimensional (3D) turbulence. The power law k^{-3} is predicted by the statistical theory of two-dimensional (2D) turbulence. Thus in May the

SST spectrum was close to that of the 2D turbulence on the scales between 10 km and 100 km. The fact that the spectra of the SST variability in May are steeper than in October-November indicates a development of 3D turbulence in the upper sea in the autumn. This is in agreement with the above observation of the increase of the horizontal mixing in SST patterns of October visualized by means of wavelet transform. Under some conditions the turbulence in the upper sea can resemble 2D one with its coherent structures, while under different conditions it appears to be 3D in nature. It is worth to notice that the combined effect of perfect geostrophic turbulence and a system of horizontal temperature gradients was suggested by J. Woods (see Fedorov 1986) for explanation of the exponent 2.2 found in the SST spectra. A superposition of the 2D turbulence and the 3D turbulence is a well known approach in numerical simulations of fluid dynamics.

In summary, this observation suggests that the SST variability in the upper ocean appears to behave differently at the scales attributed to geostrophic turbulence at the beginning of the cooling season (October-November) and at the beginning of the heating season (April-May). Observations of the Japan Meteorological Agency buoy (37.9oN, that in October the wind variations were of the same amplitude as in May. In October the average air-sea net heat flux becomes directed upward from the Japan Sea to the air. While stability of the upper mixed layer decreases, the rate of turbulence generation is increased by convection. The increase of the turbulent energy leads not only to vertical entrainment of cold water from below but also to enhancement of three dimensional mixing. Mixing tends to increase the eddy viscosity and consequently the horizontal SST gradients decrease. In May the upper sea becomes hydrostatically stable due to the surface heating. The upper mixed layer gains buoyancy; the heating strengthens stratification and leads to dampening of vertical mixing. The horizontal stirring by mesoscale eddies becomes more pronounced. The eddy viscosity is then weaker so the horizontal SST gradients are sharp. While there is inflow of the Tsushima Warm Current into the southern Japan Sea horizontal advection and stirring contribute to the upper layer stability further to the north.

After the above brief detour to oceanic stirring and mixing, let us resume our theoretical considerations. The simple formula (3.81) answers the well known question on the feasibility of stirring but the question of duration of the stage of the gradient's growth is less trivial.

Let $\tilde{\xi}_r(\cdot)$ be the trajectory of a Lagrange particle starting from the point r, whose distribution is described by the following inverse Fokker-Planck equation

$$\frac{\partial}{\partial t}G(t,r) = \widetilde{H}G(t,r). \tag{3.82}$$

Then the solution of equation (3.61) with the initial condition $G_0(r)$ can be represented by the Feynman-Kac formula which follows from (3.25)

$$G(t,r) = \boldsymbol{E}\{G_0(\tilde{\xi}_r(t))\exp[\int_0^t U(\tilde{\xi}_r(s))ds]\}, \tag{3.83}$$

Let $\tilde{\tau}(r,l)$, be the time that the particle leaves the interval of radius l.

Proposition 3.2. *The value $\tilde{\tau}(l)/\widetilde{m}(l)$, where $\tilde{\tau}(l) = \tilde{\tau}(0,l)$ and $\widetilde{m}(l) = \boldsymbol{E}\{\tilde{\tau}(l)\}$, converges in probability to 1 as $Pe \to \infty$, and the leading term of the asymptotics of $\widetilde{m}(l)$ has the following form:*

$$\widetilde{m}(l) \sim \frac{c_2(d)l_u^2}{\tau_0\sigma_u^2}\log(\frac{l^2 Pe}{l_u^2}), \tag{3.84}$$

where $c_2(d) = 2(d+2)/d(d+1)$.

The proof of this proposition is analogous to that of Proposition 3.1. We only give an exact expression for the mean exit time

$$\widetilde{m}(r,l) = \int_r^l \frac{r_1^{1-d}}{K(r_1)^3}\int_0^{r_1} K(r_2)^2 r_2^{d-1} dr_2 dr_1. \tag{3.85}$$

The representation (3.83) and Proposition 3.2 enable us to derive a rough estimate for the stirring time t_{str}, i.e. the duration of gradient growth.

Note that $U(0) > 0$ and assume that $U(r) > 0$ for all $r < l_u$. This assumption is reasonable in view of the explicit expression (3.59) for U. Consider an initial tracer field such that $G_0(r)$ is almost constant when $r < l_u$. Of course, this implies that $l_g > l_u$ and hence (3.80) readily holds. Under these assumptions it follows from (3.83) that $G(t,0) \equiv \langle|g|^2\rangle$ grows at least during the time $\widetilde{m}(l)$. Hence for large Pe

$$t_{str} > \frac{c_2(d)l_u^2}{\tau_0\sigma_u^2}\log Pe. \tag{3.86}$$

Let us estimate t_{str} in the case when $l_g < l_u$ but (3.81) holds.

We can neglect molecular diffusion for the stirring stage. Setting $\kappa = 0$ and $r = 0$ in equation (3.61) we find that the mean square gradient increases exponentially at this stage:

$$G(t,0) = G_0(0)\exp\{-dB_L''(0)t\}. \tag{3.87}$$

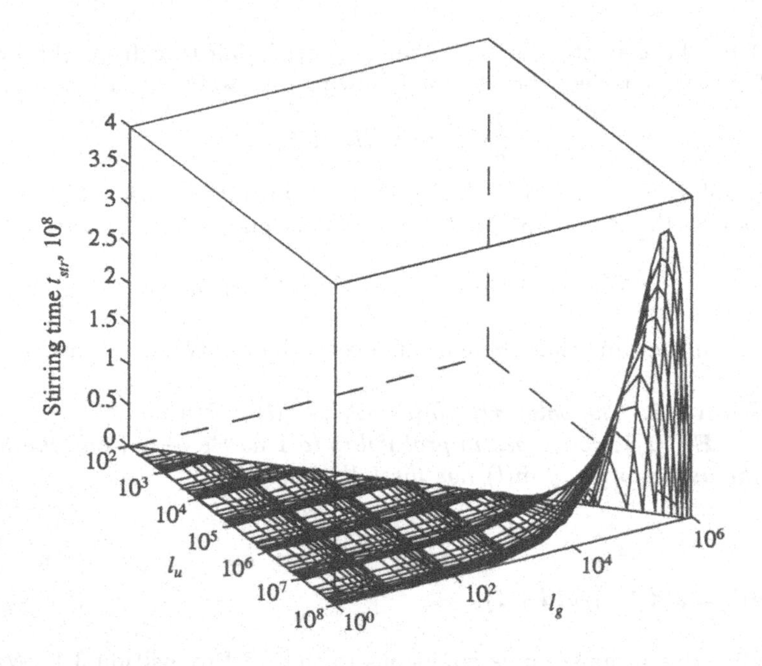

Figure 3.8 Dependence of the stirring duration on the space scales of the velocity fluctuation and the velocity correlation radius.

By the way, the exponential growth of the gradient can also be predicted in the general case using the Feynman-Kac formula (3.83). By differentiating twice both parts of equation (3.61) with respect to r and setting once again $r = 0$, $\kappa = 0$, we obtain

$$\frac{\partial}{\partial t} G''_{rr}(t,0) = -d(2d+5)B''_L(0)G''_{rr}(t,0) + U''(0)G(t,0), \tag{3.88}$$

i.e.,

$$G''_{rr}(t,0) \sim G''_{rr}(0,0)\exp\{-d(2d+5)B''_L(0)t\}. \tag{3.89}$$

In accordance with (3.80) the gradient will increases so long as

$$-B''_L(0)G(t,0) + 2\kappa G''_{rr}(t,0) > 0. \tag{3.90}$$

Thus we can estimate t_{str} from the equality

$$-B''_L(0)G(t_{str},0) + 2\kappa G''_{rr}(t_{str},0) = 0. \tag{3.91}$$

Substituting expressions (3.87) and (3.89) into (3.91) yields

$$t_{str} \sim \frac{l_u^2}{d^2 \tau_0 \sigma_u^2} \log \left(\frac{l_g^2 Pe}{l_u^2} \right). \tag{3.92}$$

Certainly, formula (3.92) is hypothetical and needs a rigorous proof. t_{str} is plotted in Fig. 3.8.

In Fig.3.9 from (Piterbarg 1989), typical graphs of $\sigma_g^2(t)$ are shown for different values of l_g.

3.5 SUMMARY

The well-known δ-correlation approximation leading to the Fokker-Planck equation is revisited. The scale separation behind this approximation is given, (3.2). The mean tracer equation (3.24) is derived for the general advection-diffusion equation which includes inhomogeneous random velocities, together with source and potential terms. Two independent derivations are shown. The first is based on the Furutsu-Novikov formula and the second exploits the Lagrangian representation (3.25) of the solution to (3.7). The second method is used to get equations for the tracer correlation function (3.44), higher statistical moments (3.49) and higher statistical moments of the tracer gradient (3.51).

In the homogeneous case the equations for the mean tracer and its higher moments are simplified to (3.53), (3.54). Also equations for the tracer correlation function and an auxiliary function closely related to the correlation function of the gradient are given in the isotropic case, (3.55), (3.58).

The isotropic equations are used for studying tracer anomalies and stirring-mixing processes when only the effects of random advection and molecular diffusion are taken into account. For small molecular diffusivities the fluctuation lifetime is expressed in terms of the introduced time scales (3.70). A necessary and sufficient condition for the gradient's growth (stirring) is given in terms of the space scales of the velocity field, initial tracer field and Peclet number, (3.81). Estimates are also obtained for the stirring duration (3.86), (3.92).

We investigated the behavior of the first two moments of a tracer and its gradient. Higher moments and the probabilistic distribution of a tracer were studied in (Sinai and Yakhot 1992, Chen and Chen 1989, Majda 1993, Saichev and

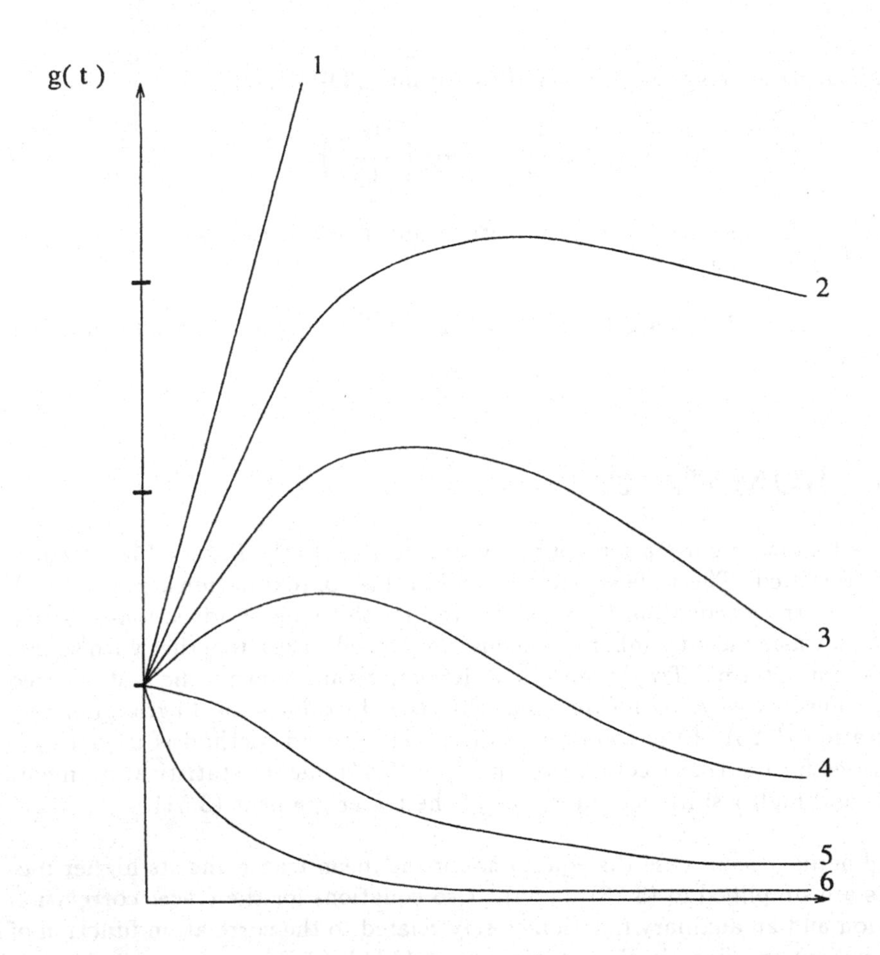

Figure 3.9 Mean square tracer gradient verses time for $Pe = 20$ and different values of $s \equiv l_0^2 Pe/l_u^2$: (1) $s = 2000$, (2) $s = 125$, (3) $s = 70$, (4) $s = 31$, (5) $s = 8$, (6) $s = 0.5$. The curves are obtained by numerically solving (3.61) with $R_L(r) = \exp(-r^2/l_u^2)$ and $R_c^{(0)}(r) = \exp(-r^2/l_0^2)$.

Woyczynski 1996). Non-Gaussian velocity field was considered in (Chechetkin, Lutovinov, and Samokhin 1991).

4

HOMOGENIZATION AND SUPERDIFFUSION

In this chapter we first discuss the behavior of the solution of the advection-diffusion equation

$$\frac{\partial c_\varepsilon}{\partial t} + \boldsymbol{u}_\varepsilon \cdot \nabla c_\varepsilon = \kappa \nabla^2 c_\varepsilon \tag{4.1}$$

with the renormalization

$$\boldsymbol{u}_\varepsilon(t, \boldsymbol{r}) = \frac{1}{\varepsilon} \boldsymbol{u}\left(\frac{t}{\varepsilon^2}, \frac{\boldsymbol{r}}{\varepsilon}\right), \tag{4.2}$$

where the subscript ε for c_ε indicates the dependence of the solution of (4.1) on ε. Here it is assumed that the low frequency energy $|| \int \boldsymbol{E}(0, \boldsymbol{k}) d\boldsymbol{k} ||$ is finite. In other words we consider the case when $\tau_T \sim \tau_E \ll t \sim \tau_D$. Under certain additional constraints with respect to the velocity field and the condition $c_0(\boldsymbol{r}) \in L^2(E)$, the following homogenization statement is true

$$\lim_{\varepsilon \to 0} c_\varepsilon(t, \boldsymbol{r}) = \bar{c}(t, \boldsymbol{r}), \tag{4.3}$$

where $\bar{c}(t, \boldsymbol{r})$ is the deterministic field satisfying the equation

$$\frac{\partial \bar{c}}{\partial t} = \nabla \cdot \boldsymbol{D} \nabla \bar{c} \tag{4.4}$$

with the effective diffusivity given by

$$\boldsymbol{D} = \int\int \boldsymbol{E}(\omega, \boldsymbol{k}) \frac{\boldsymbol{D}_0 \boldsymbol{k} \cdot \boldsymbol{k}}{(\boldsymbol{D}_0 \boldsymbol{k} \cdot \boldsymbol{k})^2 + \omega^2} d\omega d\boldsymbol{k} + \kappa \boldsymbol{I}, \tag{4.5}$$

where

$$\boldsymbol{D}_0 = \kappa \boldsymbol{I} + \pi \int \boldsymbol{E}(0, \boldsymbol{k}) d\boldsymbol{k}. \tag{4.6}$$

We will refer to this case as 'homogenization'.

Then we consider an isotropic velocity field with spectrum (2.18) in which the infrared cutoff C_0 is proportional to ε. The focus is on the region $\alpha + z > 2$, where the superdiffusion condition $\tau_T \ll \tau_E$ holds. Different regimes of superdiffusion corresponding to different regions of (α, z)-variability will be described. For 2D-shear flow such a classification was done by Avellaneda and Majda (1990). For this flow, the solution of (4.1) can be represented in an explicit form.

To study both, homogenization and superdiffusion, we use the parametrix method presented below.

4.1 PARAMETRIX METHOD: FINITE LOW-FREQUENCY KINETIC ENERGY

As before lets assume that $\boldsymbol{u}(t, \boldsymbol{r})$ is a centered random field, stationary in t, homogeneous in $\boldsymbol{r}$, $\nabla \cdot \boldsymbol{u} = 0$. We denote by $P(t, \boldsymbol{r}; t_0, \boldsymbol{r}_0)$, $t \geq t_0$, $\boldsymbol{r}, \boldsymbol{r}_0 \in E$ the fundamental solution of (2.1), i.e., P satisfies

$$\frac{\partial P}{\partial t} + \boldsymbol{u} \cdot \nabla_{\boldsymbol{r}} P = \kappa \nabla_{\boldsymbol{r}}^2 P, \tag{4.7}$$

with the initial condition $P\big|_{t=t_0} = \delta(\boldsymbol{r} - \boldsymbol{r}_0)$.

Along with (2.1), let us consider the equation with 'frozen' space coordinate:

$$\frac{\partial c}{\partial t} + \boldsymbol{u}(t, \boldsymbol{r}_0) \cdot \nabla_{\boldsymbol{r}} c = \kappa \nabla_{\boldsymbol{r}}^2 c, \tag{4.8}$$

and let $P^0(t, \boldsymbol{r}; t_0, \boldsymbol{r}_0)$ satisfy equation (4.8) with the same initial condition, $P^0\big|_{t=t_0} = \delta(\boldsymbol{r} - \boldsymbol{r}_0)$. Existence of solutions to these problems is guaranteed by the smoothness of the coefficients in (2.1) (see e.g. (Friedman 1964)).

Using well-known properties of fundamental solutions, we get the following integral equation for $P(t, \boldsymbol{r}; t_0, \boldsymbol{r}_0)$:

$$P(t, \boldsymbol{r}; t_0, \boldsymbol{r}_0) = P^0(t, \boldsymbol{r}; t_0, \boldsymbol{r}_0) +$$

$$+ \int_{t_0}^{t} \int_E P^0(t, \boldsymbol{r}; s, \boldsymbol{z})(\boldsymbol{u}(s, \boldsymbol{r}_0) - \boldsymbol{u}(s, \boldsymbol{z})) \cdot \nabla_{\boldsymbol{z}} P(s, \boldsymbol{z}; t_0, \boldsymbol{r}_0) d\boldsymbol{z} ds. \tag{4.9}$$

Symbolically we can write,

$$P = P^0 + RP, \tag{4.10}$$

where R is an integral operator with the kernel

$$R(t, \boldsymbol{r}; s, \boldsymbol{z}) = -\nabla_{\boldsymbol{z}} P^0(t, \boldsymbol{r}; s, \boldsymbol{z}) \cdot (\boldsymbol{u}(s, \boldsymbol{r}_0) - \boldsymbol{u}(s, \boldsymbol{z}))\chi_{[t_0, t]}(s), \tag{4.11}$$

and $\chi_A(\cdot)$ is the indicator function of the set A. The solution of (4.10) can be represented in the form of a formal series:

$$P = P^0 + RP^0 + R^2 P^0 + \ldots \tag{4.12}$$

Let us establish its convergence. Let $V_{\boldsymbol{a}}(\boldsymbol{r}) = \{\boldsymbol{z} \in E : a_j \le z_j \le x_j, \quad j = 1, \ldots, d\}$ be a parallelepiped with sides of lengths $x_j - a_j > 0$, where $\boldsymbol{r} = (x_1, \ldots, x_d)$ and $\boldsymbol{a} = (a_1, \ldots, a_d)$ are points in the d-dimensional Euclidean space E. Suppose that the following generalized ergodic property holds. For almost each random $\boldsymbol{a}$

$$\lim_{x_1, \ldots, x_d \to \infty} \frac{1}{|V_{\boldsymbol{a}}(\boldsymbol{r})|} \int_{V_{\boldsymbol{a}}(\boldsymbol{r})} F(\boldsymbol{u}(t, \boldsymbol{z})) d\boldsymbol{z} = \langle F(\boldsymbol{u}(t, \boldsymbol{z})) \rangle \tag{4.13}$$

with probability one for any fixed t and function F for which the right-hand side in (4.13) is finite. $|V|$ denotes the volume of V. Thus (4.13) extends the classical ergodic property to averaging over a random box with increasing volume. Our conjecture is that (4.13) is equivalent to classical ergodicity under wide conditions.

Lemma 4.1 *If (4.13) is satisfied, then the series (4.12) is convergent in the metric of the space* $L_1 = L_1(E)$ *and the function* $P(t, \boldsymbol{r}) \equiv P(t, \boldsymbol{r}; t_0, \boldsymbol{r}_0)$ *defined by this series satisfies the following conditions:*

$P(t, \boldsymbol{r}) \in L_1, \quad \boldsymbol{u}(t, \boldsymbol{r}) \cdot \nabla_{\boldsymbol{r}} P(t, \boldsymbol{r}) \in L_1$ *for any fixed* t.

Proof. It is easy to establish that

$$P^0(t, \boldsymbol{r}; s, \boldsymbol{r}_0) = \frac{1}{[4\pi\kappa(t - s)]^{d/2}} \exp\left\{ -\frac{(\boldsymbol{r} - \boldsymbol{r}_0 - \int_s^t \boldsymbol{u}(v, \boldsymbol{r}_0) dv)^2}{4\kappa(t - s)} \right\}. \tag{4.14}$$

Further, let $t_0 = 0$ and

$$f(t,\boldsymbol{r}) \equiv Rg = \int_0^t \int P^0(t,\boldsymbol{r};s,\boldsymbol{z})(\boldsymbol{u}(s,\boldsymbol{r}_0) - \boldsymbol{u}(s,\boldsymbol{z})) \cdot \nabla_{\boldsymbol{z}} g(s,\boldsymbol{z}) d\boldsymbol{z} ds, \quad (4.15)$$

where $g(\cdot,\cdot)$ is a function such that

$$\int |(\boldsymbol{u}(s,\boldsymbol{r}_0) - \boldsymbol{u}(s,\boldsymbol{z})) \cdot \nabla_{\boldsymbol{z}} g(s,\boldsymbol{z})| d\boldsymbol{z} \le M(s), \quad s \in (0,t), \quad (4.16)$$

and $M(s)$ is a random function for which the integral $\int_0^t (t-s)^{-1/2} M(s) ds$ is finite. Here and below the spatial domain of integration is not indicated if it coincides with E.

Then $\int |f(t,\boldsymbol{r})| d\boldsymbol{r} \le \int_0^t M(s) ds$ and

$$\begin{aligned} &\int |(\boldsymbol{u}(t,\boldsymbol{r}_0) - \boldsymbol{u}(t,\boldsymbol{r})) \cdot \nabla_{\boldsymbol{r}} f(t,\boldsymbol{r})| d\boldsymbol{r} \le \\ &\int_0^t \int \int |(\boldsymbol{u}(t,\boldsymbol{r}_0) - \boldsymbol{u}(t,\boldsymbol{r})) \cdot \nabla_{\boldsymbol{r}} P^0(t,\boldsymbol{r};s,\boldsymbol{z})| d\boldsymbol{r} \\ &\times |(\boldsymbol{u}(s,\boldsymbol{r}_0) - \boldsymbol{u}(s,\boldsymbol{z})) \cdot \nabla_{\boldsymbol{z}} g(s,\boldsymbol{z})| d\boldsymbol{z} ds. \end{aligned} \quad (4.17)$$

Let us estimate the inner integral

$$\begin{aligned} J = &\int |(\boldsymbol{u}(t,\boldsymbol{r}_0) - \boldsymbol{u}(t,\boldsymbol{r})) \cdot \nabla_{\boldsymbol{r}} P^0(t,\boldsymbol{r};s,\boldsymbol{z})| d\boldsymbol{r} \le \\ \le &\int |\boldsymbol{u}(t,\boldsymbol{r}_0) - \boldsymbol{u}(t,\boldsymbol{r})| |\nabla_{\boldsymbol{r}} P^0(t,\boldsymbol{r};s,\boldsymbol{z})| d\boldsymbol{r}. \end{aligned} \quad (4.18)$$

Let

$$h(\boldsymbol{r}) = \frac{1}{|V_{\boldsymbol{a}}(\boldsymbol{r})|} \int_{V_{\boldsymbol{a}}(\boldsymbol{r})} |\boldsymbol{u}(t,\boldsymbol{r}_0) - \boldsymbol{u}(t,\boldsymbol{v})| d\boldsymbol{v}, \quad (4.19)$$

where $\boldsymbol{a} = \boldsymbol{z} + \int_s^t \boldsymbol{u}(v,\boldsymbol{z}) dv$; then

$$|\boldsymbol{u}(t,\boldsymbol{r}_0) - \boldsymbol{u}(t,\boldsymbol{r})| = D(\boldsymbol{a},\boldsymbol{r}) h(\boldsymbol{r}), \quad (4.20)$$

where operator $D(\boldsymbol{a},\boldsymbol{r})$ is given by

$$D(\boldsymbol{a},\boldsymbol{r}) = \frac{\partial^d}{\partial x_1 \dots \partial x_d} (x_1 - a_1) \dots (x_d - a_d). \quad (4.21)$$

Substituting (4.20) into (4.18) and integrating by parts, we obtain

$$J \leq \int h(\boldsymbol{r}) D^*(\boldsymbol{a}, \boldsymbol{r}) |\nabla_{\boldsymbol{r}} P^0(t, \boldsymbol{r}; s, \boldsymbol{z})| d\boldsymbol{r}. \tag{4.22}$$

According to condition (4.13), when t and $\boldsymbol{r}_0$ are fixed, the absolute value of the function $h(\boldsymbol{r})$ is bounded by a random variable H, and hence

$$J \leq H \int |D^*(\boldsymbol{a}, \boldsymbol{r}) |\nabla_{\boldsymbol{r}} P^0(t, \boldsymbol{r}; s, \boldsymbol{z})|| \, d\boldsymbol{r}. \tag{4.23}$$

The integral on the right-hand side can easily be calculated. By denoting

$$\boldsymbol{v} = (\boldsymbol{r} - \boldsymbol{a}) / \sqrt{2\kappa(t-s)}, \tag{4.24}$$

we see that it equals $J_0(2\kappa(t-s))^{-1/2}$, where

$$J_0 = \frac{1}{(2\pi)^{d/2}} \int_E D^*(0, \boldsymbol{v}) |\nabla_{\boldsymbol{v}} \exp\left\{-\frac{\boldsymbol{v}^2}{2}\right\} | d\boldsymbol{v}. \tag{4.25}$$

Thus, $J \leq H J_0 (2\kappa(t-s))^{-1/2}$, and hence

$$\int |(\boldsymbol{u}(t, \boldsymbol{r}_0) - \boldsymbol{u}(t, \boldsymbol{r})) \cdot \nabla_{\boldsymbol{r}} f(t, \boldsymbol{r})| d\boldsymbol{r} \leq \frac{H J_0}{\sqrt{2\kappa}} \int_0^t \frac{M(s) ds}{\sqrt{t-s}}. \tag{4.26}$$

Let $g(s, \boldsymbol{z}) = P^0(s, \boldsymbol{z}; 0, \boldsymbol{r}_0)$, where $\boldsymbol{r}_0$ is fixed, and set $M(s) = H J_0 / \sqrt{2\kappa s}$. Then, as in (4.26),

$$\int |(\boldsymbol{u}(t, \boldsymbol{r}_0) - \boldsymbol{u}(t, \boldsymbol{r})) \cdot \nabla_{\boldsymbol{r}} g(t, \boldsymbol{r})| d\boldsymbol{r} \leq \frac{H J_0}{\sqrt{2\kappa t}}. \tag{4.27}$$

Further, applying (4.26) again we get:

$$\int |(\boldsymbol{u}(t, \boldsymbol{r}_0) - \boldsymbol{u}(t, \boldsymbol{r})) \cdot \nabla_{\boldsymbol{r}} R P^0| d\boldsymbol{r} \leq \pi \left(\frac{H J_0}{\sqrt{2\kappa}}\right)^2. \tag{4.28}$$

Reiterating this procedure $n-2$ times, we obtain

$$\int |(\boldsymbol{u}(t, \boldsymbol{r}_0) - \boldsymbol{u}(t, \boldsymbol{r})) \cdot \nabla_{\boldsymbol{r}} R^{n-1} P^0| d\boldsymbol{r} \leq \left(\frac{H J_0}{\sqrt{2\kappa}}\right)^n t^{(n-2)/2} \frac{\pi^{n/2}}{\Gamma((n+2)/2)}, \tag{4.29}$$

and hence

$$\int |R^n P^0| d\boldsymbol{r} \leq \left(\frac{H J_0}{\sqrt{2\kappa}}\right)^n \frac{2 t^{n/2} \pi^{n/2}}{n \Gamma\left(\frac{n+2}{2}\right)}, \tag{4.30}$$

whereby we obtain the lemma. ∘

Let us consider the Cauchy problem for equation (4.1) with the initial condition $c_0(\boldsymbol{r}) \in L_2 = L_2(E)$. Let $c_\varepsilon(t,\boldsymbol{r})$ be the solution of this problem. Note that $c_\varepsilon(t,\boldsymbol{r}) \equiv c(t\varepsilon^{-2}, \boldsymbol{r}\varepsilon^{-1})$, where $c(t,\boldsymbol{r})$ is the solution of (2.1) with the initial condition $c_0(\varepsilon\boldsymbol{r})$. Thus, $c_\varepsilon(t,\boldsymbol{r})$ can also be interpreted as the tracer field for large values of time with a large-scale initial condition.

We assume that the velocity field satisfies the following additional conditions:

(1) $\boldsymbol{u}(t,\boldsymbol{r})$ is Gaussian,

(2) $\boldsymbol{u}(t,\boldsymbol{r})$ is regular with respect to $\boldsymbol{r}$, i.e., the following condition is satisfied:

$$\bigcap_{\delta>0} L_\delta(\boldsymbol{u}) = \{0\}, \tag{4.31}$$

where $L_\delta(\boldsymbol{u})$ is the linear space of random variables generated by the values of the field $\boldsymbol{u}(t,\boldsymbol{r})$, while $\boldsymbol{r}$ varies outside of a cube with center at zero and the rib length δ^{-1}, and t varies in $M_+ = (0,\infty)$,

(3) the function

$$\boldsymbol{S}_{\boldsymbol{r}}(\omega) = \boldsymbol{S}(\omega,\boldsymbol{r}) = (2\pi)^{-1}\int_{-\infty}^{\infty} \boldsymbol{R}_{\boldsymbol{u}}(s,\boldsymbol{r})e^{i\omega s}\,ds, \tag{4.32}$$

is N times continuously differentiable in ω ($N \geq 1$), i.e.

$$\boldsymbol{S}_{\boldsymbol{r}}(\omega) \in C^N. \tag{4.33}$$

and $\boldsymbol{R}_{\boldsymbol{u}}(s,\boldsymbol{r}) = \langle \boldsymbol{u}(t,\boldsymbol{r}_0)\; \boldsymbol{u}(t+s,\boldsymbol{r}_0+\boldsymbol{r})^T\rangle$ is the correlation tensor of the velocity field.

Proposition 4.1. *Under conditions* (1), (2), (3) *the field* $c_\varepsilon(t,\boldsymbol{r})$ *converges with probability* 1 *as* $\varepsilon \to 0$ *to a nonrandom field* $\bar{c}(t,\boldsymbol{r})$ *satisfying the equation*

$$\frac{\partial \bar{c}}{\partial t} = \nabla\cdot(\kappa\boldsymbol{I} + \boldsymbol{D})\nabla\bar{c}, \tag{4.34}$$

where $\boldsymbol{D}$ *is given by (4.5).*

Proof. The regularity condition (4.31) implies the ergodicity condition (4.13), and hence, the solution of equation (4.1) can be expanded in a series of the form (4.12) whose terms depend on ε.

The m-th term of this expansion has the following form:

$$\begin{aligned}
&\xi_{\varepsilon,m} \equiv R_\varepsilon^m P_\varepsilon^0 = \int_{G_m}\int_{E^m} P_\varepsilon^0(t,\boldsymbol{r};s_m,\boldsymbol{z}_m)(\boldsymbol{u}_\varepsilon(s_m,\boldsymbol{z}_m)\\
&-\boldsymbol{u}_\varepsilon(s_m,\boldsymbol{z}_{m-1}))\cdot\nabla_m P_\varepsilon^0(s_m,\boldsymbol{z}_m;s_{m-1},\boldsymbol{z}_{m-1})\\
&\times(\boldsymbol{u}_\varepsilon(s_{m-1},\boldsymbol{z}_{m-1})\\
&-\boldsymbol{u}_\varepsilon(s_{m-1},\boldsymbol{z}_{m-2}))\cdot\nabla_{m-1}\dots(\boldsymbol{u}_\varepsilon(s_1,\boldsymbol{z}_1)\\
&-\boldsymbol{u}_\varepsilon(s_1,\boldsymbol{r}_0))\cdot\nabla_1 P_\varepsilon^0(s_1,\boldsymbol{z}_1;0,\boldsymbol{r}_0)(d\boldsymbol{z})^m(ds)^m,
\end{aligned} \tag{4.35}$$

where

$$E^m = \underbrace{E\times\cdots\times E}_{m\text{ times}},\quad \nabla_j = \nabla_{\boldsymbol{z}_j},\quad j=1,\dots,m;$$

$$\begin{aligned}
&(d\boldsymbol{z})^m = d\boldsymbol{z}_1\dots d\boldsymbol{z}_m,\quad (ds)^m = ds_1\dots ds_m;\\
&G_m = \{s_1,\cdots,s_m:\ t\ge s_m\ge s_{m-1}\ge\cdots\ge s_1\},
\end{aligned} \tag{4.36}$$

and

$$\begin{aligned}
&P_\varepsilon^0(t,\boldsymbol{r};s,\boldsymbol{z}) = \frac{1}{(4\pi\kappa(t-s))^{d/2}}\exp\left\{-\frac{(\boldsymbol{r}-\boldsymbol{z}-\int_s^t \boldsymbol{u}_\varepsilon(v,\boldsymbol{z})dv)^2}{4\kappa(t-s)}\right\} =\\
&\frac{1}{(2\pi)^d}\int_E \exp\{i\boldsymbol{k}\cdot(\boldsymbol{r}-\boldsymbol{z}-\int_s^t \boldsymbol{u}_\varepsilon(v,\boldsymbol{z})dv)-\kappa\boldsymbol{k}^2(t-s)\}d\boldsymbol{k}.
\end{aligned} \tag{4.37}$$

It follows from (4.35) that the random field $\xi_{\varepsilon,m} = \xi_{\varepsilon,m}(\boldsymbol{r}_0)$ also satisfies the regularity condition (2) as a random function with respect to $\boldsymbol{r}_0$ while t, $\boldsymbol{r}$ are fixed, because $L_\delta(\xi_{\varepsilon,m}) \subset L_\delta(\boldsymbol{u})$ for any $\delta > 0$.

Let $c_0(\boldsymbol{r}_0) = c\chi_V(\boldsymbol{r}_0)$, where c is a nonrandom value, V is a cube with the center at an arbitrary point and $\chi_V(\cdot)$ is the indicator of the set V. Then

$$\int_E R_\varepsilon^m P_\varepsilon^0 c_0(\boldsymbol{r}_0)d\boldsymbol{r}_0 = \frac{c\varepsilon^d}{|V|}\int_{\varepsilon^{-1}V}\xi_{\varepsilon,m}(\varepsilon\boldsymbol{r}_0)d\boldsymbol{r}_0, \tag{4.38}$$

where $|V|$ is the volume of the cube V. The regularity of the field $\xi_{\varepsilon,m}(\varepsilon\boldsymbol{r}_0)$ implies that the limit of the value $\int_E R_\varepsilon^m P_\varepsilon^0 c_0(\boldsymbol{r}_0)d\boldsymbol{r}_0$ as $\varepsilon\to 0$ exists with probability 1 or 0. If the limit of the mean $\langle R_\varepsilon^m p_\varepsilon^0\rangle$ exists, then we have convergence with probability one. In this case it can be shown by standard methods that the limit

$$\lim_{\varepsilon\to 0}\int_E R_\varepsilon^m P_\varepsilon^0 c_o(\boldsymbol{r}_0)d\boldsymbol{r}_0 \tag{4.39}$$

is nonrandom. It would be quite obvious if the field under the integral was a homogeneous field independent of ε. In our case the fact that limit (4.39) is nonrandom can be deduced from the closeness of the field $\xi_{\varepsilon,m}(\varepsilon\boldsymbol{r}_0)$ to a homogeneous one when $\varepsilon\to 0$. If we hypothesize this, then (4.39) coincides with the limit of its mean. The same assertion holds for linear combinations of functions of the form indicated above and corresponding to various c and V from which the existence of limit (4.39) can be established for all the functions $c_0(\boldsymbol{r}_0)$ from $L_2(E)$.

The next step of the proof is computing the average of $\xi_{\varepsilon,m}$ defined by (4.35). First of all, we use representation (4.37) to obtain the following form of expression (4.35)

$$\xi_{m,\varepsilon}\equiv R^m P_\varepsilon^0=\frac{i^m}{(2\pi)^{dm}}\int\limits_{G_m}\int\limits_{E^m}\int\limits_{E^{m+1}}\Big(\prod_{j=1}^{m}(\boldsymbol{u}_\varepsilon(s_j,\boldsymbol{z}_j)$$

$$-\boldsymbol{u}_\varepsilon(s_j,\boldsymbol{z}_{j-1}))\cdot\boldsymbol{k}_j\Big)\exp\{\sum_{j=0}^{m}[i\boldsymbol{k}_{j+1}\cdot(\boldsymbol{z}_{j+1}-\boldsymbol{z}_j \tag{4.40}$$

$$-\int\limits_{s_j}^{s_{j+1}}\boldsymbol{u}_\varepsilon(s,\boldsymbol{z}_j)ds)-\boldsymbol{k}_{j+1}^2\kappa(s_{j+1}-s_j)]\}(d\boldsymbol{k})^{m+1}\times(d\boldsymbol{z})^m(ds)^m,$$

where

$$\begin{aligned}&s_0\equiv 0,\quad s_{m+1}\equiv t,\quad \boldsymbol{z}_0\equiv\boldsymbol{r}_0,\\&\boldsymbol{z}_{m+1}\equiv\boldsymbol{r},\quad \boldsymbol{k}_{m+1}\equiv\boldsymbol{k},\quad (d\boldsymbol{k})^{m+1}=d\boldsymbol{k}_1d\boldsymbol{k}_2\cdots d\boldsymbol{k}_m d\boldsymbol{k}.\end{aligned} \tag{4.41}$$

The expectation of $\xi_{m,\varepsilon}$ is written as follows

$$\langle\xi_{m,\varepsilon}\rangle=\langle R_\varepsilon^m P_\varepsilon^0\rangle=i^m\int\limits_{G_m}\int\limits_{E^m}\int\limits_{E^{m+1}}\langle\xi_1\cdots\xi_m e^{-i\eta}\rangle$$

$$\times(2\pi)^{-dm}\exp\{\sum_{j=0}^{m}[i(\boldsymbol{k}_{j+1},\boldsymbol{z}_{j+1}-\boldsymbol{z}_j)-\kappa\boldsymbol{k}_{j+1}^2(s_{j+1}-s_j)\} \tag{4.42}$$

$$\times(d\boldsymbol{k})^{m+1}(d\boldsymbol{z})^m(ds)^m,$$

where

$$\xi_j = \left(\boldsymbol{u}_\varepsilon(s_j, \boldsymbol{z}_j) - \boldsymbol{u}_\varepsilon(s_j, \boldsymbol{z}_{j-1})\right) \cdot \boldsymbol{k}_j, \quad j = 1, \cdots, m,$$

and

$$\eta = \sum_{j=0}^{m} \int_{s_j}^{s_{j+1}} \boldsymbol{u}_\varepsilon(s, \boldsymbol{z}_j) ds \cdot \boldsymbol{k}_{j+1} \tag{4.43}$$

are the Gaussian random values with zero mean. Note that

$$\langle \xi_1 \cdots \xi_m e^{i\eta} \rangle = e^{-\frac{1}{2}\langle \eta^2 \rangle} \times \sum_{\substack{i, \cdots, i_k \\ j_1, \cdots, j_{m-k}}} \langle \xi_{i_1} \cdots \xi_{i_k} \rangle \langle \eta \xi_{j_1} \rangle \cdots \langle \eta \xi_{j_{m-k}} \rangle, \tag{4.44}$$

where the sum is taken over all $k = 0, \cdots, m$ and all sets of indices $I = (i_1, \cdots, i_k)_1$ and $J = (j_1, \cdots, j_{m-k})$ such that $I \cap J = \emptyset$, $I \cup J = (1, \cdots, m)$ and if $k = m$ or $k = 0$, then the corresponding sets J or I are considered to be empty. It can easily be seen that $\langle \xi_{\varepsilon,m} \rangle = 0$ for odd m. For even m we first compute the limit ($\varepsilon \to 0$) of the term on the right-hand side of (4.42) corresponding to the term

$$\langle \xi_1 \xi_2 \rangle \cdots \langle \xi_{m-1} \xi_m \rangle e^{-\frac{1}{2}\langle \eta^2 \rangle} \tag{4.45}$$

and then show that the rest of terms go to zero. To compute $\langle \eta^2 \rangle$ and correlations $\langle \eta \xi_j \rangle$ we need the following statement

Lemma 4.2. *Under condition (4.33), the following relationships hold asymptotically :*

$$\frac{1}{\varepsilon^2} \int_s^t \int_s^t \boldsymbol{R_u} \left(\frac{s_1 - s_2}{\varepsilon^2}, \boldsymbol{r} \right) ds_1 ds_2 = 2\pi(t-s) \boldsymbol{S_r}(0) + 2\varepsilon^2 \int_{-\infty}^{\infty} \omega^{-2} (\boldsymbol{S_r}(\omega) - \boldsymbol{S_r}(0)) d\omega + O(\varepsilon^{2N+2}), \tag{4.46}$$

$$\frac{1}{\varepsilon^2} \int_0^s \int_s^t \boldsymbol{R_u} \left(\frac{s_1 - s_2}{\varepsilon^2}, \boldsymbol{r} \right) ds_1 ds_2 = O(\varepsilon^{2N+2}). \tag{4.47}$$

Proof. Let us denote the expression on the left-hand side of (4.46) by $\gamma(\varepsilon)$. Then from the definition of $\boldsymbol{S_r}(\omega)$ it follows that

$$\gamma(\varepsilon) = 2\varepsilon^2 \int_{-\infty}^{\infty} \boldsymbol{S_r}(\omega) \frac{1 - \cos[\varepsilon^{-2}\omega(t-s)]}{\omega^2} d\omega. \tag{4.48}$$

Let us make the change of variables $\omega = \lambda\varepsilon^2$ in the last integral. This yields,

$$\gamma(0) = 2\boldsymbol{S_r}(0)\int_{-\infty}^{\infty}\frac{1-\cos\lambda(t-s)}{\lambda^2}d\lambda = 2\pi(t-s)\boldsymbol{S_r}(0). \tag{4.49}$$

Then

$$\gamma(\varepsilon)-\gamma(0) = 2\varepsilon^2\int_{-\infty}^{\infty}\omega^{-2}(\boldsymbol{S_r}(\omega)-\boldsymbol{S_r}(0))(1-\cos[\varepsilon^2\omega(t-s)])d\omega. \tag{4.50}$$

Using well-known results on asymptotic behavior of oscillatory integrals (see e.g. Fedoruk 1986), we obtain (4.46) and relation (4.47) can be proved similarly. ∘

Let us continue the proof of Proposition 4.1. According to (4.46) and (4.47), we have

$$\langle\eta^2\rangle \sim 2\sum_{j=0}^{m}(s_{j+1}-s_j)\boldsymbol{S}\boldsymbol{k}_{j+1}\cdot\boldsymbol{k}_{j+1}, \tag{4.51}$$

up to the terms of order ε^2, where

$$\boldsymbol{S} = \pi\boldsymbol{S_0}(0) = \pi\int\boldsymbol{E}(0,\boldsymbol{k})d\boldsymbol{k}. \tag{4.52}$$

Henceforth the tilde means that the difference between both sides of the equation is small compared with each side of the equation. By substituting (4.51) into (4.45) we obtain that for even $m = 2p$ and $\varepsilon \to 0$

$$\begin{aligned}
a_{p,\varepsilon} &\equiv (-1)^p\int_{G_m}\int_{E^m}\int_{E^{m+1}}\langle\xi_1\xi_2\rangle\cdots\langle\xi_{m-1}\xi_m\rangle e^{-\frac{1}{2}\langle\eta^2\rangle}\\
&\times(2\pi)^{-dm}\exp\{\sum_{j=0}^{m}[i(\boldsymbol{k}_{j+1},\boldsymbol{z}_{j+1}-\boldsymbol{z}_j)-\kappa\boldsymbol{k}_{j+1}^2(s_{j+1}-s_j)\}\\
&\times(d\boldsymbol{k})^{m+1}(d\boldsymbol{z})^m(ds)^m \sim\\
&(-1)^p\int_{G_m}\int_{E^m}\int_{E^{m+1}}\prod_{l=1}^{p}(\boldsymbol{Q}_\varepsilon(s_{2l},s_{2l-1},\boldsymbol{z}_{2l},\boldsymbol{z}_{2l-1})\boldsymbol{k}_{2l}\cdot\boldsymbol{k}_{2l-1})\\
&\times\exp\left\{\sum_{j=0}^{m}\left[i\boldsymbol{k}_{j+1}\cdot(\boldsymbol{z}_{j+1}-\boldsymbol{z}_j)-\boldsymbol{D}_0(s_{j+1}-s_j)\boldsymbol{k}_{j+1}\cdot\boldsymbol{k}_{j+1}\right]\right\}\\
&(d\boldsymbol{k})^{m+1}(d\boldsymbol{z}^m(ds)^m,
\end{aligned} \tag{4.53}$$

where $\boldsymbol{D}_0 = \boldsymbol{S} + \kappa \boldsymbol{I}$ and

$$\boldsymbol{Q}_\varepsilon(s_l, s_j, \boldsymbol{z}_l, \boldsymbol{z}_j) \equiv \langle \xi_l \xi_j \rangle =$$

$$\frac{1}{\varepsilon^2}\left[\boldsymbol{Ru}\left(\frac{s_j - s_l}{\varepsilon^2}, \frac{\boldsymbol{z}_j - \boldsymbol{z}_l}{\varepsilon}\right) - \boldsymbol{Ru}\left(\frac{s_j - s_l}{\varepsilon^2}, \frac{\boldsymbol{z}_{j-1} - \boldsymbol{z}_l}{\varepsilon}\right) - \right. \tag{4.54}$$

$$\left. \boldsymbol{Ru}\left(\frac{s_j - s_l}{\varepsilon^2}, \frac{\boldsymbol{z}_j - \boldsymbol{z}_{l-1}}{\varepsilon}\right) + \boldsymbol{Ru}\left(\frac{s_j - s_l}{\varepsilon^2}, \frac{\boldsymbol{z}_{j-1} - \boldsymbol{z}_{l-1}}{\varepsilon}\right)\right].$$

From $\nabla \cdot \boldsymbol{u} = 0$ it follows that in (4.53) we can replace $\boldsymbol{k}_{2p}$ by $\boldsymbol{k}$, $\boldsymbol{k}_{2p-2}$ by $\boldsymbol{k}_{2p-1}$, $\ldots$, $\boldsymbol{k}_2$ by $\boldsymbol{k}_3$ in the product preceding the exponential since $i(\boldsymbol{k}_{2j} - \boldsymbol{k}_{2j-1})$ is the Fourier image of the operator ∇_{2j}.

After changing and renumbering the variables of integration according to

$$s'_j = \frac{s_{2j} - s_{2j-1}}{\varepsilon^2}, \quad \boldsymbol{z}'_j = \frac{\boldsymbol{z}_{2j} - \boldsymbol{z}_{2j-1}}{\varepsilon}, \quad \boldsymbol{k}'_j = \varepsilon \boldsymbol{k}_{2j}$$
$$s''_j = s_{2j-1}, \quad \boldsymbol{z}''_j = \boldsymbol{z}_{2j-1}, \quad \boldsymbol{k}''_j = \boldsymbol{k}_{2j-1}, \quad j = 1, \cdots, p \tag{4.55}$$

we obtain

$$a_{p,\varepsilon} \sim (-1)^p \varepsilon^{2p} \int_E \int_{G'_p} \int_{E^p} \int_{E^p} \prod_{j=1}^{p} \boldsymbol{Q}_\varepsilon(s'_j, \boldsymbol{z}'_j) \boldsymbol{k} \cdot \boldsymbol{k} \exp\{i(\boldsymbol{k}'_j \cdot \boldsymbol{z}'_j) - s'_j \boldsymbol{D}_0 \boldsymbol{k}'_j \cdot \boldsymbol{k}'_j\}$$

$$\times (d\boldsymbol{k}')^p (d\boldsymbol{z}')^p \exp\{-\sum_{j=1}^{p+1} (s''_j - s''_{j-1}) \boldsymbol{D}_0 \boldsymbol{k} \cdot \boldsymbol{k} + i\boldsymbol{k} \cdot (\boldsymbol{r} - \boldsymbol{r}_0)\}$$

$$\times (ds'_1)^p (ds''_2)^p d\boldsymbol{k}, \tag{4.56}$$

where the time integration region is given by

$$G'_p = \{0 \le s''_1 \le \ldots \le s''_p < t, \; 0 \le s'_1 \le \frac{s''_2 - s''_1}{\varepsilon^2}, \ldots,$$
$$0 \le s'_{p-1} \le \frac{s''_p - s''_{p-1}}{\varepsilon^2}, \; 0 \le s'_p \le \frac{t - s''_p}{\varepsilon^2}\}, \tag{4.57}$$

the auxiliary function $\boldsymbol{Q}_\varepsilon$ is defined by

$$\boldsymbol{Q}_\varepsilon(s, \boldsymbol{z}) = \frac{1}{\varepsilon^2}(\boldsymbol{Ru}(s, \boldsymbol{z}) - \boldsymbol{Ru}(s, 0)) \tag{4.58}$$

and the approximation $s_{2j-1} - s_{2j-2} = s''_j - s''_{j-1} - \varepsilon^2 s'_{j-1} \sim s''_j - s''_{j-1}$ was used at the second exponent of (4.57).

By setting

$$G(t, \boldsymbol{k}) = \exp\{-t\boldsymbol{D}_0\boldsymbol{k} \cdot \boldsymbol{k}\} \tag{4.59}$$

and

$$H_\varepsilon(t, \boldsymbol{k}) = \frac{\varepsilon^2}{(2\pi)^d} G(t, \boldsymbol{k}) \int_E \int_0^{t/\varepsilon^2} (\hat{\boldsymbol{Q}}_\varepsilon(s, \boldsymbol{k}')\boldsymbol{k} \cdot \boldsymbol{k}) G(s, \boldsymbol{k}') ds d\boldsymbol{k}', \tag{4.60}$$

where $\hat{\boldsymbol{Q}}_\varepsilon(s, \boldsymbol{k}') = \int_E e^{i\boldsymbol{k}' \cdot \boldsymbol{z}} \boldsymbol{Q}_\varepsilon(s, \boldsymbol{z}) d\boldsymbol{z}$, we can rewrite (4.57) as follows

$$a_{p,\varepsilon} \sim (-1)^p \int_E G_\varepsilon * (H_\varepsilon)^{p*} e^{i\boldsymbol{k} \cdot (\boldsymbol{r} - \boldsymbol{r}_0)} d\boldsymbol{k}, \tag{4.61}$$

where the asterisk stands for convolution with respect to the time variable and $f^{p*} = \underbrace{f * \ldots * f}_{p}$.

Let us introduce the Laplace transform

$$f^*(\lambda) = \int_0^\infty e^{-\lambda t} f(t) dt. \tag{4.62}$$

Proceeding to the limit in (4.61), we get

$$a_p^*(\lambda) \equiv \lim_{\varepsilon \to 0} a_{p,\varepsilon}^*(\lambda) = (-1)^p \int_E G^*(\lambda)(H^*(\lambda))^p e^{i\boldsymbol{k} \cdot (\boldsymbol{r} - \boldsymbol{r}_0)} d\boldsymbol{k}, \tag{4.63}$$

where

$$G^*(\lambda) = \frac{1}{\lambda + \boldsymbol{D}_0 \boldsymbol{k} \cdot \boldsymbol{k}} \tag{4.64}$$

Dominant diagram, $m = 10, p = 5, q = 5$

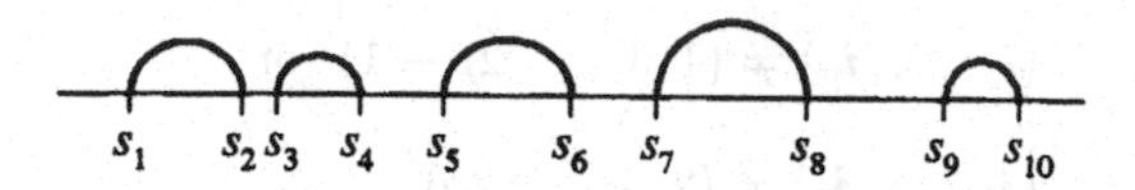

Neglected diagram, $m = 10, p = 5, q = 2$

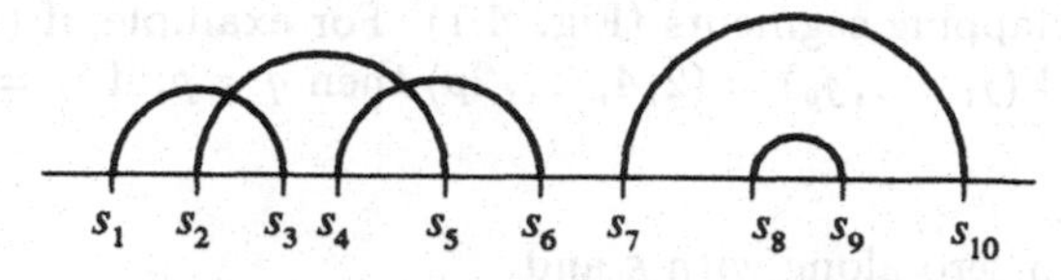

Figure 4.1 Examples of the leading term from (4.45) which is dominant in (4.42) (top) and the second order term from the same expansion (bottom).

and

$$H^*(\lambda) = \frac{(2\pi)^{-d}}{(\lambda + \boldsymbol{D}_0\boldsymbol{k}\cdot\boldsymbol{k})}\int\limits_E\int\limits_0^\infty (\hat{\boldsymbol{R}}_{\boldsymbol{u}}(s,\boldsymbol{k}') - (2\pi)^d\delta(\boldsymbol{k}')\boldsymbol{R}_{\boldsymbol{u}}(s,0))\boldsymbol{k}\cdot\boldsymbol{k})$$

$$\times\exp\{-s\boldsymbol{D}_0\boldsymbol{k}'\cdot\boldsymbol{k}'\}dsd\boldsymbol{k}' =$$

$$\frac{1}{(\lambda + \boldsymbol{D}_0\boldsymbol{k}\cdot\boldsymbol{k})}[(2\pi)^{-d}\int\limits_E\int\limits_o^\infty \hat{\boldsymbol{R}}_{\boldsymbol{u}}(s,\boldsymbol{k}')\exp\{-s\boldsymbol{D}_0\boldsymbol{k}'\cdot\boldsymbol{k}'\}dsd\boldsymbol{k}'$$

$$-\int\limits_o^\infty \boldsymbol{R}_{\boldsymbol{u}}(s,0)ds]\boldsymbol{k}\cdot\boldsymbol{k} = \frac{1}{\lambda + \boldsymbol{D}_0\boldsymbol{k}\cdot\boldsymbol{k}}(\boldsymbol{D} - \boldsymbol{S})\boldsymbol{k}\cdot\boldsymbol{k}. \tag{4.65}$$

Thus, by taking into account $\boldsymbol{D}_0 - \boldsymbol{S} = \kappa\boldsymbol{I}$ we get from (4.63)

$$\sum_{p=0}^\infty a_p^*(\lambda) = \int\limits_E \frac{G^*(\lambda)}{1 + H^*(\lambda)}e^{i\boldsymbol{k}\cdot(\boldsymbol{r}-\boldsymbol{r}_0)}d\boldsymbol{k} = \int\limits_E \frac{1}{\lambda + \boldsymbol{D}\boldsymbol{k}\cdot\boldsymbol{k}}e^{i\boldsymbol{k}\cdot(\boldsymbol{r}-\boldsymbol{r}_0)}d\boldsymbol{k}. \tag{4.66}$$

Now we return to the expansion (4.44) and show that the term (4.45) gives the main contribution in (4.42). First, let us estimate the contribution to $\langle\xi_{m,\varepsilon}\rangle$ in (4.42) due to the term

$$\langle\xi_{i_1}\xi_{j_1}\rangle\cdots\langle\xi_{i_p}\xi_{j_p}\rangle e^{-\frac{1}{2}\langle\eta^2\rangle}, \tag{4.67}$$

where $(i_1, j_1, \ldots, i_p, j_p)$, $i_k < j_k$, is a permutation of $(1, 2, \ldots, 2p-1, 2p)$ such that either

$$(i_1, \ldots, i_p) \neq (1, 3, \ldots, 2p-1) \text{ or} \\ (j_1, \ldots, j_p) \neq (2, 4, \ldots, 2p). \tag{4.68}$$

Let $\mathbf{s} = (s_1, \ldots, s_m)$ be a fixed point from G_m and $S = \cup_{k=1}^{p} [s_{i_k}, s_{j_k}]$ be a subset of the real line. This set, of course, can be represented as a union of q, $(q \le p)$ non-overlapping segments (Fig. 4.1). For example, if $(i_1, \ldots, i_p) = (1, 3, \ldots, 2p-1)$ and $(j_1, \ldots, j_p) = (2, 4, \ldots, 2p)$ then $q = p$, if $i_1 = 1$, $j_p = 2p$ then $q = 1$, etc.

Let $\delta = \delta(\varepsilon)$ tends to zero along with ε and

$$M_\delta = \{\mathbf{s} : \ |s_{i_k} - s_{j_k}| \le \delta, \ k = 1, \ldots, p\}. \tag{4.69}$$

Lemma 4.3. *For any bounded function* $f_\varepsilon(\mathbf{s})$

$$\int_{G_m \cap M_\delta} f_\varepsilon(\mathbf{s}) d\mathbf{s} \le C \delta^{m-q}. \tag{4.70}$$

Proof. Let

$$S = [s_1, s_{k_1}] \cup [s_{k_1+1}, s_{k_2}] \cup \ldots \cup [s_{k_{q-1}+1}, s_{2p}] \tag{4.71}$$

be a decomposition of S into non-overlapping segments. If $\mathbf{s} \in G_m \cap M_\delta$ then $|s_{k_l} - s_{k_{l-1}+1}| \le C_l \delta$, since all original segments included in $[s_{k_{l-1}+1}, s_{k_l}]$ are chained, where $l = 1, \ldots, q$ and for the sake of convenience we suppose $k_0 = 0$, $k_q = 2p$. Thus,

$$|G_m \cap M_\delta|_m \le |V_1|_{k_1} |V_2|_{k_2 - k_1} \cdots |V_q|_{m - k_{q-1}}, \tag{4.72}$$

where $|V|_k$ is the volume of V in k-dimensional Euclidean space and $V_l = \{|s_{k_l} - s_{k_{l-1}+1}| \le C_l \delta\}$. Obviously,

$$|V_l|_{k_l - k_{l-1}} \le c_l \delta^{k_l - k_{l-1} - 1}. \tag{4.73}$$

From (4.72) and (4.73) it follows that

$$|G_m \cap M_\delta|_m \le C_l \delta^{m-q}. \tag{4.74}$$

The statement of the lemma readily follows from (4.74). ∘

Let us set

$$b_{m,\varepsilon} = i^m \int\limits_{Gm} \int\limits_{E^m} \int\limits_{E^{m+1}} \langle \xi_{i_1} \xi_{j_1} \rangle \cdots \langle \xi_{i_p} \xi_{j_p} \rangle e^{-\frac{1}{2}\langle \eta^2 \rangle}$$

$$\times (2\pi)^{-dm} \exp\{\sum_{j=0}^{m} [i(\boldsymbol{k}_{j+1}, \boldsymbol{z}_{j+1} - \boldsymbol{z}_j) - \kappa \boldsymbol{k}_{j+1}^2 (s_{j+1} - s_j)\} \tag{4.75}$$

$$\times (d\boldsymbol{k})^{m+1} (d\boldsymbol{z})^m (ds)^m,$$

Using (4.55) we can represent $b_{m,\varepsilon}$ in the following form

$$b_{m,\varepsilon} = \varepsilon^{-2p} \int\limits_{G_m} F(\frac{s_{i_1} - s_{j_1}}{\varepsilon^2}, \ldots, \frac{s_{i_p} - s_{j_p}}{\varepsilon^2}; \ \boldsymbol{r}, \boldsymbol{r}_0)(ds)^m, \tag{4.76}$$

where F is a bounded function. Using Lemma 4.3 with $\delta = \varepsilon^2$ we get

$$b_{m,\varepsilon} \leq C\varepsilon^{-2p} \int\limits_{G_m \cap M_{\varepsilon^2}} F(\frac{s_{i_1} - s_{j_1}}{\varepsilon^2}, \ldots, \frac{s_{i_p} - s_{j_p}}{\varepsilon^2}; \ \boldsymbol{r}, \boldsymbol{r}_0)(ds)^m \leq$$
$$\leq C_1 \varepsilon^{2(p-q)}. \tag{4.77}$$

From (4.77) it can be seen that only if $q = p$, i.e. if $(i_1, \ldots, i_p) = (1, 3, \ldots, 2p-1)$ and $(j_1, \ldots, j_p) = (2, 4, \ldots, 2p)$, does $b_{m,\varepsilon}$ not tend to zero when $\varepsilon \to 0$. If (4.70) holds then necessarily $\lim_{\varepsilon \to 0} b_{m,\varepsilon} = 0$.

Now we proceed to the terms in the expansion (4.44) containing correlations such as $\langle \xi_j \eta \rangle$. For the sake of definiteness let us consider the following term

$$d_{m,\varepsilon} \equiv i^m \int\limits_{Gm} \int\limits_{E^m} \int\limits_{E^{m+1}} \langle \xi_1 \xi_2 \rangle \cdots \langle \xi_{2p-3} \xi_{2p-2} \rangle \langle \xi_{2p-1} \eta \rangle \langle \xi_{2p} \eta \rangle e^{-\frac{1}{2}\langle \eta^2 \rangle}$$

$$\times (2\pi)^{-dm} \exp\{\sum_{j=0}^{m} [i\boldsymbol{k}_{j+1} \cdot (\boldsymbol{z}_{j+1} - \boldsymbol{z}_j) - \kappa \boldsymbol{k}_{j+1}^2 (s_{j+1} - s_j)\} \tag{4.78}$$

$$\times (d\boldsymbol{k})^{m+1} (d\boldsymbol{z})^m (ds)^m.$$

The correlation $r_\varepsilon(\boldsymbol{z}_j - \boldsymbol{z}_{j-1}, \boldsymbol{k}_j, \boldsymbol{k}_{j+1}) \equiv \langle \xi_j \eta \rangle$ can be estimated from (4.43) as follows

$$r_\varepsilon(\boldsymbol{z}_j - \boldsymbol{z}_{j-1}, \boldsymbol{k}_j, \boldsymbol{k}_{j+1}) \sim \int_0^\infty \left[\boldsymbol{Ru}(s,0) - \boldsymbol{Ru}\left(\frac{\boldsymbol{z}_{j-1} - \boldsymbol{z}_j}{\varepsilon}\right)\right] \boldsymbol{k}_j \cdot \boldsymbol{k}_{j+1} ds - \int_{-\infty}^0 \left[\boldsymbol{Ru}(s,0) - \boldsymbol{Ru}\left(\frac{\boldsymbol{z}_j - \boldsymbol{z}_{j-1}}{\varepsilon}\right)\right] \boldsymbol{k}_j \cdot \boldsymbol{k}_j ds. \tag{4.79}$$

Let us change variables in (4.78) by setting $\boldsymbol{v}_j = \boldsymbol{z}_j - \boldsymbol{z}_{j-1}, \quad j = 1, \ldots, m$. As the result (4.78) can be written as follows

$$d_{m,\varepsilon} = \int_{E^m} \int_{E^{m+1}} \exp\{\sum_{j=0}^{m-1} i(\boldsymbol{k}_{j+1} - \boldsymbol{k}) \cdot \boldsymbol{v}_{j+1} r_\varepsilon(\boldsymbol{v}_{2p-1}, \boldsymbol{k}_{2p-1}, \boldsymbol{k}_{2p}) \times r_\varepsilon(\boldsymbol{v}_{2p}, \boldsymbol{k}_{2p}, \boldsymbol{k}) F(\boldsymbol{v}_1, \boldsymbol{v}_2, \ldots, \boldsymbol{v}_{2p-2}; \boldsymbol{k}_1, \ldots, \boldsymbol{k}_{2p}, \boldsymbol{k}; \boldsymbol{r}, \boldsymbol{r}_0) (d\boldsymbol{k})^{m+1} (d\boldsymbol{v})^m, \tag{4.80}$$

where F is a bounded rapidly decaying function as $|\boldsymbol{k}_j| \to \infty$. Integrating with respect to $\boldsymbol{v}_{2p}$ in (4.80) yields

$$d_{m,\varepsilon} \sim \varepsilon^d \int_{E^{m-1}} \int_{E^{m+1}} \exp\{\sum_{j=0}^{2p-1} i(\boldsymbol{k}_{j+1} - \boldsymbol{k}) \cdot \boldsymbol{v}_{j+1}\} r_\varepsilon(\boldsymbol{v}_{2p-1}, \boldsymbol{k}_{2p-1}, \boldsymbol{k}_{2p}) \times [\int_{-\infty}^0 \hat{\boldsymbol{Ru}}(s, \boldsymbol{k}_j) \boldsymbol{k}_j \cdot \boldsymbol{k} ds - \int_0^\infty \hat{\boldsymbol{Ru}}(s, -\boldsymbol{k}_j) \boldsymbol{k}_j \cdot \boldsymbol{k}_j ds] \times F(\boldsymbol{v}_1, \boldsymbol{v}_2, \ldots, \boldsymbol{v}_{2p-2}; \boldsymbol{k}_1, \ldots, \boldsymbol{k}_{2p}, \boldsymbol{k}; \boldsymbol{r}, \boldsymbol{r}_0)(d\boldsymbol{k})^{m+1} (d\boldsymbol{v})^m, \tag{4.81}$$

where we used the property $\boldsymbol{Ru}(s,0) = \boldsymbol{Ru}(-s,0)$. From (4.81) it follows that $d_{m,\varepsilon}$ disappears when $\varepsilon \to 0$. We believe that in the general case, the quantities $d_{m,\varepsilon}$ are also negligible. Thus

$$\langle \xi_{p,\varepsilon} \rangle \sim a_{m,\varepsilon}, \tag{4.82}$$

where $a_{m,\varepsilon}$ is defined in (4.53). Let us denote by $\hat{P}_\varepsilon^*(\lambda, \boldsymbol{k})$ the Laplace transform in t and Fourier transform in $\boldsymbol{r} - \boldsymbol{r}_0$ of the Green's function $P_\varepsilon(t, \boldsymbol{r}; s, \boldsymbol{r}_0)$ determined by (4.7) with the velocity renormalization given by (4.2). Then

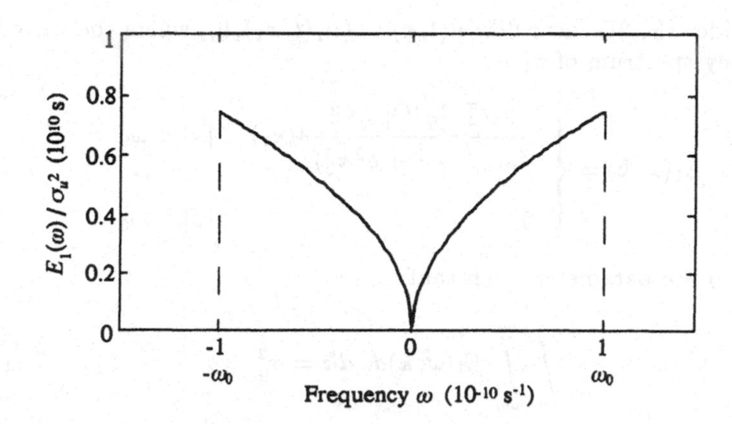

Figure 4.2 The frequency spectrum $E_1(\omega)$ of a random shear flow computed with the parameters $\sigma_u^2 = 0.05m^2s^{-1}, \omega_0 = 2 \cdot 10^{-3}s^{-1}$.

from (4.12), (4.82) and (4.66) it follows that

$$\lim_{\varepsilon \to 0} \hat{P}_\varepsilon^*(\lambda, \boldsymbol{k}) = \frac{1}{\lambda + \boldsymbol{D}\boldsymbol{k} \cdot \boldsymbol{k}}. \tag{4.83}$$

Obviously (4.83) is equivalent to (4.34). Proposition 4.1 is proven. ∘

4.2 ENHANCEMENT OF TURBULENT DIFFUSION BY MOLECULAR DIFFUSION

We point out an important consequence of the considered scale separation: turbulent diffusivity depends on molecular diffusivity. This fact sharply distinguishes the homogenization case from the δ-correlation approximation. From (4.5) it follows that the molecular diffusion is able to affect the turbulent diffusion. Let us give an example of where this effect is dramatic.

Let us consider the 2D-shear flow $\boldsymbol{u}(t,\boldsymbol{r}) = (u_1(t,x_2),0)$, where the wave number frequency spectrum of u_1 is

$$E_1(\omega,\boldsymbol{k}) = \begin{cases} \dfrac{3\sqrt{2}\sigma_u^2 q^{3/2}|\omega|k_2^2}{4\pi\omega_0^{3/2}(\omega^2+q^2k_2^4)}\delta(k_1), & |\omega|<\omega_0 \\ 0, & |\omega|>\omega_0 \end{cases} \tag{4.84}$$

and ω_0, σ_u, q are parameters such that

$$\int\int_{-\omega_0}^{\omega_0} E_1(\omega,\boldsymbol{k})d\omega d\boldsymbol{k} = \sigma_u^2. \tag{4.85}$$

Fig. 4.2 shows the corresponding frequency spectrum $E_1(\omega) = \int E_1(\omega,\boldsymbol{k})d\boldsymbol{k}$.

Assume that molecular diffusion acts in the direction x_2 only. In agreement with homogenization theory, the limiting equation of the passive scalar is as follows

$$\frac{\partial\bar{c}}{\partial t} = D_1(\kappa)\frac{\partial^2\bar{c}}{\partial x_1^2} + \kappa\frac{\partial^2\bar{c}}{\partial x_2^2} \tag{4.86}$$

where the effective diffusivity in x_1-direction is easily computed from the general formula (4.5)

$$D_1(\kappa) = \frac{3\sigma_u^2}{\omega_0}\,\frac{q\sqrt{\kappa}}{(\kappa+q)(\sqrt{\kappa}+\sqrt{q})} \tag{4.87}$$

The plot of D_1 is given in Fig. 4.3.

From (4.87) one can see that $D_1(0) = 0$, but if $\kappa \neq 0$ we can choose q in such a way that $D_1 = 3\sigma_u^2/4\omega_0$. To do this we should set $q = \kappa$. For example if $\kappa = 10^3\ cm^2s^{-1}$ and $\sigma_u^2/\omega_0 = 10^7\ cm^2s^{-1}$ we can find a spectrum in the form of (4.84) such that the turbulent diffusion in x_1-direction increases from 0 to $0.75\times 10^7\ cm^2s^{-1}$. Hence, at least theoretically molecular diffusion can enhance turbulent diffusion dramatically. The physical mechanism, clearly, is quite simple. If $\kappa = 0$ there is no relation between streamlines and since $E_1(\omega)\mid_{\omega=0} = 0$, the turbulent diffusivity in the x_1-direction is zero. Indeed, it can be shown that before turning on the molecular diffusion, the displacement of a Lagrangian particle in the x_1-direction satisfies $\langle X_1(t)^2\rangle \sim t^{1/2}$. This slow growth regime of the mean square displacement is called subdiffusion (Isichenko 1992).

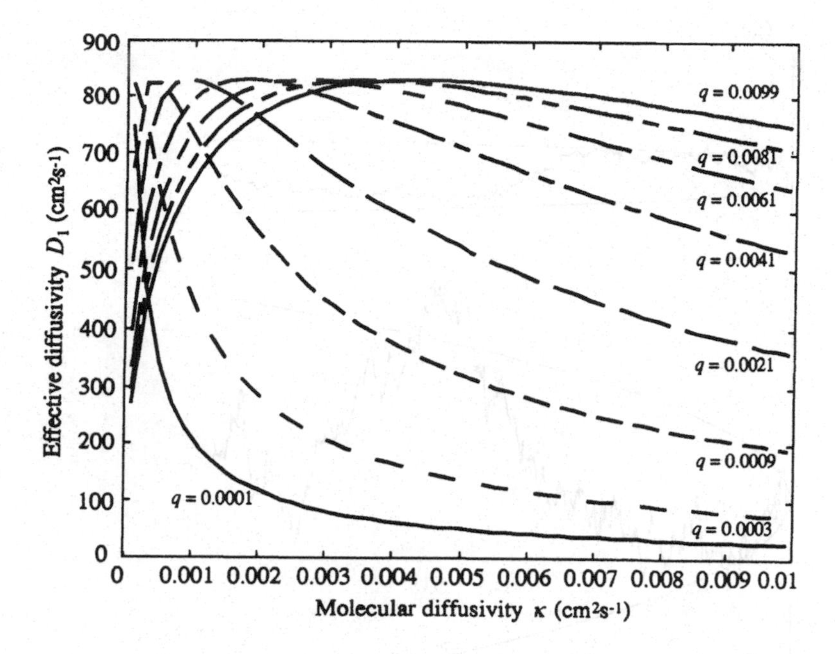

Figure 4.3 Dependence of the effective diffusion D_1 on the molecular diffusivity κ for various values of the parameter q.

When the molecular diffusion in the x_2-direction is turned on, Lagrangian particles can possibly jump from one stream line to another. Due to this interaction, the particles can travel quite far in the x_1-direction. More exactly, after switching on the molecular diffusion one will have $M\langle\xi_1^2(t)\rangle \sim D_1(\kappa)t$, where the Lagrangian trajectory $\xi_1(t)$ is found from the stochastic equation $\dot{\xi}_1 = u_1(t, \sqrt{2\kappa}w_2(t))$ and $w_2(t)$ is a Wiener process, (Fig. 4.4).

Let us note that for big κ (or small τ_D), the formula (4.87) gives the following asymptotic

$$D_1(\kappa) \sim \frac{3\sigma_u^2 q}{\omega_0 \kappa}, \tag{4.88}$$

i.e. the eddy diffusivity is inversely proportional to the molecular diffusivity. A similar result was obtained by Taylor (1953) for a deterministic flow through a tube. Let us note that the Taylor's reasoning fails for small molecular diffusion and does not concern the enhancement of turbulent diffusion by molecular diffusion.

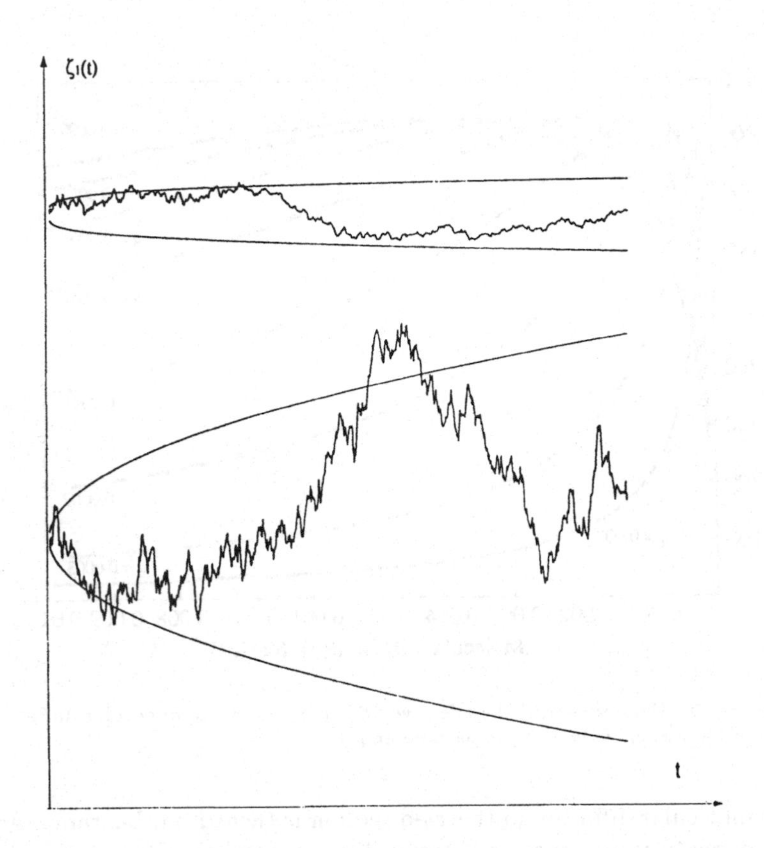

Figure 4.4 Typical Lagrangian trajectories for $\kappa = 0$ (top) and $\kappa > 0$ (bottom).

4.3 PARAMETRIX METHOD: INFINITE LOW-FREQUENCY ENERGY

The choice of the rescaling factor ε^{-2} for t in (4.2) is due to the fact that under the condition on the spectrum

$$0 < \int \|\boldsymbol{E}(0, \boldsymbol{k})\| d\boldsymbol{k} < \infty, \tag{4.89}$$

where $\|\cdot\|$ is a matrix norm, the Lagrangian particle displacement satisfies the scaling $\langle \boldsymbol{X}^2(t)\rangle \sim t$. In this section we focus on the velocity spectrum

$\boldsymbol{E}^{(\varepsilon)}(\omega, \boldsymbol{k})$ with a large scale cutoff

$$\boldsymbol{E}^{(\varepsilon)}(\omega, \boldsymbol{k}) = \begin{cases} \boldsymbol{E}(\omega, \boldsymbol{k}), & |\boldsymbol{k}| > \varepsilon k_0 \\ 0, & |\boldsymbol{k}| < \varepsilon k_0. \end{cases} \tag{4.90}$$

We assume that $\boldsymbol{E}^{(\varepsilon)}(0, \boldsymbol{k})$ satisfies (4.89) for any ε and a fixed cutoff k_0, but

$$\lim_{\varepsilon \to 0} \int ||\boldsymbol{E}^{(\varepsilon)}(0, \boldsymbol{k})|| d\boldsymbol{k} = \infty. \tag{4.91}$$

This behavior is typical of most geophysical and hydrodynamical spectra (Monin and Yaglom 1975). If (4.91) holds, then the rescaling (4.2) fails in the sense that the corresponding solution $c_\varepsilon(t, \boldsymbol{r})$ of (4.1) goes to zero for all $t, \boldsymbol{r}$. To overcome this trouble we first find an appropriate rescaling. Intuitively it is clear that the variability on time scales shorter than ε^{-2} should be considered. Thus, a reasonable problem statement is as follows.

Assuming (4.91), find a rescaling function $\rho(\varepsilon)$ such that under the normalization

$$\boldsymbol{u}_\varepsilon = \frac{\varepsilon}{\rho^2} \boldsymbol{u}\left(\frac{t}{\rho^2}, \frac{\boldsymbol{r}}{\varepsilon}\right), \qquad \kappa_\varepsilon = \frac{\varepsilon^2}{\rho^2} \kappa. \tag{4.92}$$

the limit

$$\bar{c}(t, \boldsymbol{r}) = \lim_{\varepsilon \to 0} \langle c_\varepsilon(t, \boldsymbol{r}) \rangle \tag{4.93}$$

is not degenerate, where $c_\varepsilon(t, \boldsymbol{r})$ is the solution of the equation

$$\frac{\partial}{\partial t} c_\varepsilon + \boldsymbol{u}_\varepsilon \cdot \nabla c_\varepsilon = \kappa_\varepsilon \nabla^2 c_\varepsilon. \tag{4.94}$$

To our knowledge, this problem statement first appeared in (Avellaneda and Majda 1990) regarding 2D shear flow. In this section we give a complete solution to this problem in the short-correlated case for the special velocity spectrum (2.18) with $C_0 = C_0 \varepsilon$. First let us introduce notation for three regions in the quarteplane $\{(\alpha, z),\ \alpha \geq 0,\ z \geq 0\}$

$$\begin{aligned} G_1 &= \{\alpha, z :\ \alpha + z > 2,\ \ \alpha + 2z < 4\} \\ G_2 &= \{\alpha, z :\ \alpha + 2z > 4,\ \ 2z(1 - \alpha) < (2 - \alpha)^2, \\ &\qquad (6 - 2\alpha - z)(\alpha + 2z - 2) > 2z\} \\ G_3 &= \{\alpha, z :\ 2z(1 - \alpha) > (2 - \alpha)^2,\ \ z < 4 - \alpha\} \end{aligned} \tag{4.95}$$

(see Fig. 4.5). The reason for this division will become clear later.

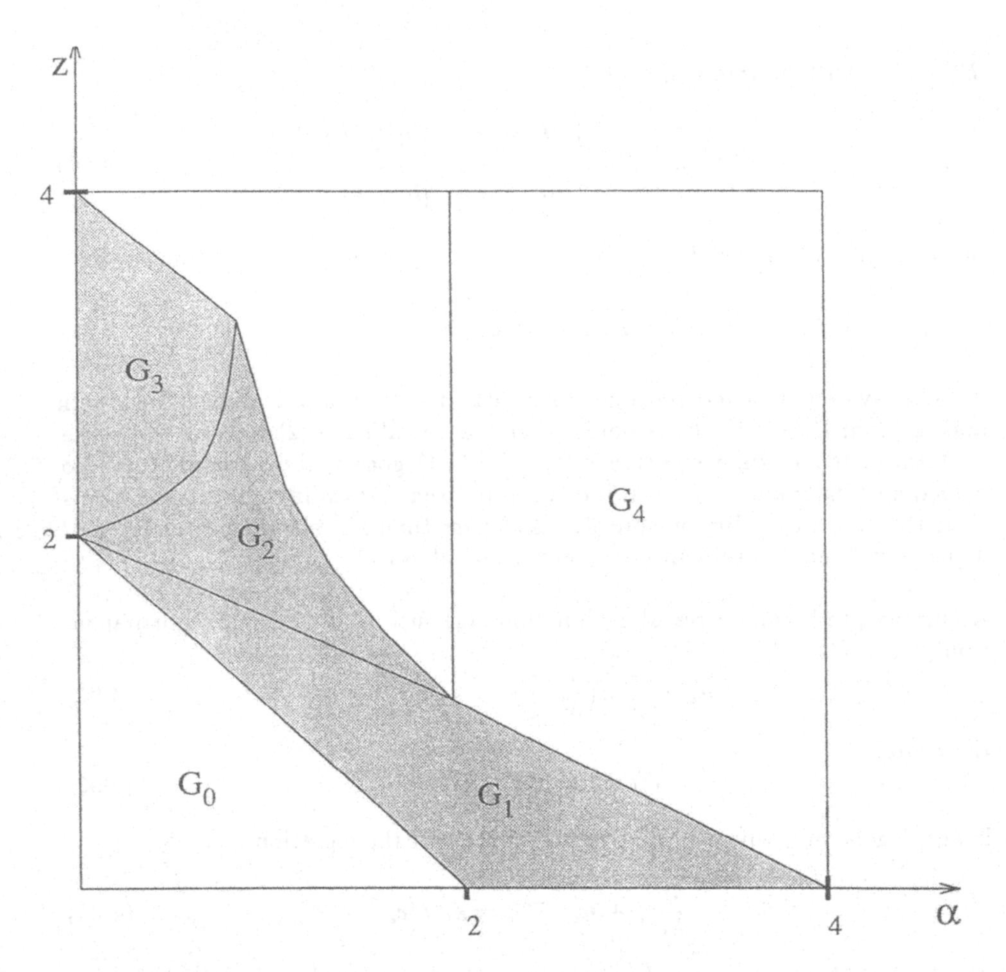

Figure 4.5 The phase diagram in (α, z) plane for isotropic velocity field. In the dashed area $\tau_T \ll \tau_E \ll 1$, in $G_0 : \tau_T \sim \tau_E \ll 1$ and $G_u : \tau_T \ll 1 \ll \tau_E$.

Proposition 4.2. *Let $\rho = \rho_j$ be the rescaling function in region G_j, $j = 1, 2, 3$ determined by*

$$\rho_1 = \varepsilon^{\frac{4-\alpha-z}{2}}, \quad \rho_2 = \varepsilon^{2-\frac{\alpha}{2}-\beta}, \quad \rho_3 = \varepsilon, \tag{4.96}$$

where $\beta = z/(\alpha+2z-2)$, then the limit (4.93) exists for the solution of equation (4.94) which satisfies the equation

$$\frac{\partial \bar{c}}{\partial t} = D_j \nabla^2 \bar{c} \tag{4.97}$$

for $(\alpha, z) \in G_j$ *with the effective diffusivity given by*

$$D_1 = \frac{C}{a(\alpha+z-2)}(C_0\bar{k})^{2-\alpha-z},$$
$$D_2 = q_{\alpha,z}(C_0\bar{k})^{2-\alpha-2\beta}, \tag{4.98}$$
$$D_3 = q_{\alpha,z}(C_1\bar{k})^{2-\alpha-2\beta} + \kappa,$$

where

$$q_{\alpha,z} = \frac{A^{-\beta}\Gamma(\beta)\beta}{|2-\alpha-2\beta|}, \tag{4.99}$$

and

$$A = C\pi\vartheta_d(d-1)a^{\frac{\alpha+z-2}{2}}\frac{1}{z}\int_0^\infty x^{-1/\beta}\left(1-\frac{1}{x}+\frac{1}{x}e^{-x}\right)dx. \tag{4.100}$$

A crucial point of the proof is that the short-correlation condition $\tau_E \ll 1$ must hold. Thus, let us first make sure that the short-correlation approximation holds for regions $G_j, j = 1,2,3$. For region G_1, the result was shown in section 2.3. From (2.21) and (2.51) we have that the Eulerian correlation time in G_2 and G_3 satisfies

$$\tau_E \sim \rho^2\varepsilon^{2-\alpha-z} \tag{4.101}$$

and, hence, to guarantee $\tau_E \ll t$ one must require that in G_2

$$4-\alpha-\frac{2z}{\alpha+2z-2}+2-\alpha-z>0, \tag{4.102}$$

and in G_3

$$4-\alpha-z>0.$$

Obviously (4.102) follows from definition (4.95) and therefore the formulated proposition gives a description of turbulent diffusion in the case when the turnover time is much smaller then the Eulerian correlation time, which in turn is much smaller than the observation time

$$\tau_T \ll \tau_E \ll t. \tag{4.103}$$

Notice that for G_1 and G_2 the observation time is much smaller than the molecular diffusion time, $t \ll \tau_D$, whereas for G_3 the molecular diffusion time is of order one.

The next remark is that the region $G_0: \ \alpha+z<2$ in Fig.4.5 gives an example of the separation

$$\tau_T \sim \tau_E \ll t \sim \tau_D, \tag{4.104}$$

considered in the general case of section 4.1. Also let us mention that the turbulent diffusion in region G_4 (Fig.4.5) can be described for $z < 1$ (Avellaneda and Majda 1992). In this region the relation

$$\tau_T \ll t \ll \tau_D \ll \tau_E \tag{4.105}$$

holds. One can check by using the fact that $\rho = \varepsilon^{1-\alpha/4}$ in G_4. This separation is not covered by the short-correlation approximation.

Proof of Proposition 4.2. As in the homogenization case we use the parametrix method. The essential difference is that the relation (4.46) of Lemma 4.2 does not hold. As before, we obtain from the short-correlation condition that the main contribution to the quantity $\langle \xi_{m,\varepsilon} \rangle$ given by (4.42) is due to the term corresponding (4.45). In doing so we apply Lemma 4.3 with $\delta = \rho^2 \varepsilon^{2-\alpha-z}$ in regions G_2 and G_3 and with the corresponding Eulerian scale correlation for G_1. In the infinite energy case the mentioned term cannot be reduced to (4.53). Instead we will use the following expression, which is the starting point for the proof

$$a_{p,\varepsilon} \equiv (-1)^p \int\limits_{G_m} \int\limits_{E^m} \int\limits_{E^{m+1}} \langle \xi_1 \xi_2 \rangle \cdots \langle \xi_{m-1} \xi_m \rangle e^{-\frac{1}{2}\langle \eta^2 \rangle}$$

$$\times (2\pi)^{-dm} \exp\{\sum_{j=0}^{m} [i(\boldsymbol{k}_{j+1}, \boldsymbol{z}_{j+1} - \boldsymbol{z}_j)$$

$$-\kappa_\varepsilon \boldsymbol{k}_{j+1}^2 (s_{j+1} - s_j)]\} (d\boldsymbol{k})^{m+1} (d\boldsymbol{z})^m (ds)^m \sim$$

$$\sim (-1)^p \int\limits_{G_m} \int\limits_{E^m} \int\limits_{E^{m+1}} \prod_{l=1}^{p} (\boldsymbol{Q}_\varepsilon(s_{2l}, s_{2l-1}, \boldsymbol{z}_{2l}, \boldsymbol{z}_{2l-1}) \boldsymbol{k}_{2l} \cdot \boldsymbol{k}_{2l-1})$$

$$\times \exp\{\sum_{j=0}^{m} [i\boldsymbol{k}_{j+1} \cdot (\boldsymbol{z}_{j+1} - \boldsymbol{z}_j) - (\boldsymbol{V}_\varepsilon(s_{j+1} - s_j) \tag{4.106}$$

$$+\kappa_\varepsilon (s_{j+1} - s_j) \boldsymbol{I}) \boldsymbol{k}_{j+1} \cdot \boldsymbol{k}_{j+1}]\} (d\boldsymbol{k})^{m+1} d\boldsymbol{z}^m (ds)^m,$$

where G_m and the other notation are explained in (4.36), κ_ε in (4.92),

$$Q_\varepsilon(s_i, s_j, z_i, z_j) = \langle \xi_i \xi_j \rangle =$$

$$\frac{\varepsilon^2}{\rho^4}\left[Ru\left(\frac{s_j - s_l}{\varepsilon^2}, \frac{z_j - z_l}{\varepsilon}\right) - Ru\left(\frac{s_j - s_l}{\varepsilon^2}, \frac{z_{j-1} - z_l}{\varepsilon}\right)\right. \tag{4.107}$$

$$\left. - Ru\left(\frac{s_j - s_l}{\varepsilon^2}, \frac{z_j - z_{l-1}}{\varepsilon}\right) + Ru\left(\frac{s_j - s_l}{\varepsilon^2}, \frac{z_{j-1} - z_{l-1}}{\varepsilon}\right)\right],$$

and

$$V_\varepsilon(t) = \frac{1}{2}\int_0^t\int_0^t \langle u_\varepsilon(s_1, z) u_\varepsilon^T(s_2, z)\rangle ds_1 ds_2. \tag{4.108}$$

After changing and renumbering the variables as in (4.55)

$$s'_j = \frac{s_{2j} - s_{2j-1}}{\rho^2}, \quad z'_j = \frac{z_{2j} - z_{2j-1}}{\varepsilon}, \quad k'_j = \varepsilon k_{2j},$$

$$s''_j = s_{2j-1}, \quad z''_j = z_{2j-1}, \quad k''_j = k_{2j-1}, \tag{4.109}$$

$$j = 1, \cdots, p,$$

we obtain

$$\langle \xi_{\varepsilon,2p}\rangle \cong (-1)^p \rho^{2p} \int_E \int_{G'_p} \int_{E^p} \int_{E^p} \prod_{j=1}^p (Q_\varepsilon(s'_j, z'_j) k \cdot k)$$

$$\times \exp\{i(k'_j \cdot z'_j) - \varepsilon^{-2}(V_\varepsilon(\rho^2 s'_j) k'_j \cdot k'_j) + \kappa_\varepsilon \rho^2 s'_j k'^2_j]\}$$

$$\times (dk')^p (dz')^p) \exp\{-\sum_{j=1}^{p+1}[(V_\varepsilon(s''_j - s''_{j-1}) k \cdot k) \tag{4.110}$$

$$+ \kappa_\varepsilon (s''_j - s''_{j-1}) k^2] + ik \cdot (r - r_0)\} \times (ds'_1)^p (ds''_2)^p dk,$$

where the time integration region is given by

$$G'_p = \{0 \le s''_1 \le \ldots \le s''_p < t, \ 0 \le s'_1 \le \frac{s''_2 - s''_1}{\rho^2}, \ldots,$$

$$0 \le s'_{p-1} \le \frac{s''_p - s''_{p-1}}{\rho^2}, \ 0 \le s'_p \le \frac{t - s''_p}{\rho^2}\}, \tag{4.111}$$

the auxiliary function $\boldsymbol{Q}_\varepsilon$ is defined by

$$\boldsymbol{Q}_\varepsilon(s, \boldsymbol{z}) = \frac{\varepsilon^2}{\rho^4}(\boldsymbol{Ru}(s, \boldsymbol{z}) - \boldsymbol{Ru}(s, \boldsymbol{0})) \tag{4.112}$$

and the approximation $s_{2j-1} - s_{2j-2} = s''_j - s''_{j-1} - \rho^2 s'_{j-1} \approx s''_j - s''_{j-1}$ was used on the second exponent of (4.110).

By setting

$$G_\varepsilon(t, \boldsymbol{k}) = \exp\{-\boldsymbol{V}_\varepsilon(t)\boldsymbol{k} \cdot \boldsymbol{k} - \kappa_\varepsilon t \boldsymbol{k}^2\} \tag{4.113}$$

and

$$H_\varepsilon(t, \boldsymbol{k}) = \frac{\rho^2}{(2\pi)^d} G_\varepsilon(t, \boldsymbol{k}) \int_E \int_0^{t/\rho^2} (\hat{\boldsymbol{Q}}_\varepsilon(s, \boldsymbol{k}')\boldsymbol{k} \cdot \boldsymbol{k}) G_\varepsilon(\rho^2 s, \varepsilon^{-1}\boldsymbol{k}') ds d\boldsymbol{k}', \tag{4.114}$$

where $\hat{\boldsymbol{Q}}_\varepsilon(s, \boldsymbol{k}') = \int_E e^{i\boldsymbol{k}' \cdot \boldsymbol{z}} \boldsymbol{Q}_\varepsilon(s, \boldsymbol{z}) d\boldsymbol{z}$, we rewrite (4.110) as follows

$$\langle \xi_{\varepsilon,2p} \rangle \cong (-1)^p \int_E G_\varepsilon * (H_\varepsilon)^{p*} e^{i\boldsymbol{k} \cdot (\boldsymbol{r} - \boldsymbol{r}_0)} d\boldsymbol{k}, \tag{4.116}$$

Thus, we have for the Green's function of equation (4.94)

$$\hat{P}^*_\varepsilon(\lambda, \boldsymbol{k}) \cong \frac{G^*_\varepsilon(\lambda, \boldsymbol{k})}{1 + H^*_\varepsilon(\lambda, \boldsymbol{k})} \tag{4.117}$$

Noting that

$$\frac{\partial G_\varepsilon(t, \boldsymbol{k})}{\partial t} = -G_\varepsilon(t, \boldsymbol{k}) \left(\kappa_\varepsilon \boldsymbol{k}^2 + \frac{\varepsilon^2}{\rho^2} \int_0^t \boldsymbol{Ru}\left(\frac{s}{\rho^2}, \boldsymbol{0}\right) \boldsymbol{k} \cdot \boldsymbol{k} \, ds \right), \tag{4.118}$$

we have

$$H^*_\varepsilon(\lambda, \boldsymbol{k}) = L^*_\varepsilon(\lambda, \boldsymbol{k}) + \lambda G^*_\varepsilon(\lambda, \boldsymbol{k}) - 1, \tag{4.119}$$

where

$$L_\varepsilon(t, \boldsymbol{k}) = G_\varepsilon(t, \boldsymbol{k}) S_\varepsilon(t, \boldsymbol{k}), \tag{4.120}$$

and

$$S_\varepsilon(t, \boldsymbol{k}) = \frac{\varepsilon^2}{(2\pi)^d \rho^2} \int_E \int_E \int_0^{t/\rho^2} \left(e^{i\boldsymbol{k}' \cdot \boldsymbol{z}} \boldsymbol{Ru}(s, \boldsymbol{z}) \boldsymbol{k} \cdot \boldsymbol{k} \right) \tag{4.121}$$

$$\times G_\varepsilon(\rho s, \varepsilon^{-1}\boldsymbol{k}') ds d\boldsymbol{z} d\boldsymbol{k}' + \kappa_\varepsilon \boldsymbol{k}^2.$$

By substituting (4.119) into (4.117) we obtain

$$\hat{P}^*_\varepsilon(\lambda, \boldsymbol{k}) = \frac{1}{\lambda + L^*_\varepsilon(\lambda, \boldsymbol{k})/G^*_\varepsilon(\lambda, \boldsymbol{k})}. \tag{4.122}$$

Comparing (4.120) and (4.122), one can see that if $\lim_{\varepsilon \to 0} S_\varepsilon(t, \boldsymbol{k}) = \boldsymbol{Dk} \cdot \boldsymbol{k}$, where $\boldsymbol{D}$ is independent of t then

$$\lim_{\varepsilon \to 0} \hat{P}^*_\varepsilon(\lambda, k) = \frac{1}{\lambda + \boldsymbol{Dk} \cdot \boldsymbol{k}} \tag{4.123}$$

and hence the limiting equation is

$$\frac{\partial \bar{c}}{\partial t} = \nabla \cdot \boldsymbol{D} \nabla \bar{c} \tag{4.124}$$

In the isotropic case with spectrum (2.18) straightforward computations yields

$$G_\varepsilon(t, \boldsymbol{k}) = \exp\{-\boldsymbol{k}^2(V_\varepsilon(t) + \kappa_\varepsilon t)\}, \quad S_\varepsilon(t, \boldsymbol{k}) = S_\varepsilon(t)\boldsymbol{k}^2, \tag{4.125}$$

where

$$V_\varepsilon(t) = \frac{B\varepsilon^2}{\rho^2} \int_{k_0\varepsilon}^{k_1} k^{1-z-\alpha} \left(t - \frac{\rho^2}{ak^z} + \frac{\rho^2}{ak^z} e^{-\frac{ak^z t}{\rho^2}}\right) dk,$$

$$S_\varepsilon(t) = \frac{B\varepsilon^2}{\rho^4} \int_0^t \int_{\varepsilon k_0}^{k_1} k^{1-\alpha} \exp\{-\frac{ask^z}{\rho^2} - \frac{k^2}{\varepsilon^2}(D_\varepsilon(s) + s\kappa_\varepsilon)\} dk ds + \kappa_\varepsilon, \tag{4.126}$$

and $k_0 = C_0 \bar{k}$, $k_1 = C_1 \bar{k}$, $B = C\pi\vartheta_d(d-1)/d$.

When an infrared catastrophe occurs, i.e. $\int_{\varepsilon k_0}^{k_1} E_L(0, k) k^{d-1} dk$ tends to infinity as $(\varepsilon \to 0)$, then the choice of ρ and the limiting equation for $\bar{c}$ are determined by the asymptotic behavior of $D_\varepsilon(t)$ and $S_\varepsilon(t)$ which is described in the following statement.

Lemma 4.4. (1) *If* $(\alpha, z) \in G_1$ *and* $\rho = \rho_1$ *then*

$$\lim_{\varepsilon \to 0} V_\varepsilon(t) = D_1 t, \quad \lim_{\varepsilon \to 0} S_\varepsilon(t) = D_1, \tag{4.127}$$

where D_1 *is given by (4.98)*

(2) *If* $(\alpha, z) \in G_2 \cup G_3$ *and* $\rho = \varepsilon^m$ *such that* $\varepsilon^z/\rho^2 \to 0$ *then*

$$\lim_{\varepsilon \to 0} \frac{V_\varepsilon(t)}{\varepsilon^\gamma} = At^{1/\beta}, \tag{4.128}$$

where A *is given by (4.100) and*

$$\gamma = 2 - \frac{2m}{\beta} < 0, \tag{4.129}$$

and if $\rho = \rho_2$ *or* ρ_3 *then*

$$\lim_{\varepsilon \to 0} S_\varepsilon(t) = \lim_{\varepsilon \to 0} \left(\frac{B\varepsilon^2}{\rho^4} \int_0^t \int_{k_0\varepsilon}^{k_1} k^{1-\alpha} exp\{-\frac{Ak^2 s^{1/\beta}}{\varepsilon^{2-\gamma}}\} dkds + \kappa_\varepsilon \right) =$$

$$= \begin{cases} D_2 & (\alpha, z) \in G_2 \\ D_3 + \kappa & (\alpha, z) \in G_3, \end{cases} \tag{4.130}$$

where D_2, D_3 *are given by (4.98).*

From this Lemma and (4.120, 4.123) Proposition 4.2 follows immediately.∘

Proof of Lemma 4.4. After substituting $\rho = \varepsilon^m$, $0 < m \le 1$ into $V_\varepsilon(t)$ given in (4.126) we obtain

$$V_\varepsilon(t) = Bz^{-1} a^{(z+\alpha-2)/2} t^{1/\beta} \varepsilon^{2-2m+2m(2-\alpha-z)/z}$$

$$\times \int_{b_1(\varepsilon)}^{b_2(\varepsilon)} x^{-1/\beta} \left(1 - \frac{1}{x} + \frac{1}{x} e^{-x}\right) dx, \tag{4.131}$$

where

$$b_1(\varepsilon) = ak_0^z t\varepsilon^{z-2m},$$
$$b_2(\varepsilon) = ak_0^z t\varepsilon^{-2m} \to \infty. \tag{4.132}$$

Note that the function under the integral looks like $x^{(2-\alpha-z)/z}$ at zero and like $x^{-1+(2-\alpha-z)/z}$ at infinity. Thus, the integral is always convergent at infinity.

If

$$2m < z, \tag{4.133}$$

then $b_1(\varepsilon) \to 0$ and we have two cases: first, $\alpha < 2$, in which the integral converges and hence

$$V_\varepsilon(t) \sim Bz^{-1}a^{(z+\alpha-2)/2}t^{1/\beta}\varepsilon^{2-2m+2m(2-\alpha-z)/z}$$
$$\times \int_0^\infty x^{-1/\beta}\left(1-\frac{1}{x}+\frac{1}{x}e^{-x}\right)dx, \tag{4.134}$$

and second, $\alpha > 2$, in which the integral diverges and hence

$$V_\varepsilon(t) \sim B(\alpha-2)^{-1}k_0^{2-\alpha}t^2\varepsilon^{4-4m-\alpha} \tag{4.135}$$

Finally, if

$$2m > z, \tag{4.136}$$

then $b_1(\varepsilon) \to \infty$ and we arrive at

$$V_\varepsilon(t) \sim B(\alpha+z-2)^{-1}k_0^{2-\alpha-z}t\varepsilon^{4-2m-\alpha-z}. \tag{4.137}$$

In summary, we have

$$V_\varepsilon(t) \sim Ct^p\varepsilon^q, \tag{4.138}$$

where p, q are defined in (4.134), (4.135) , (4.137) for the different cases. After substituting (4.138) into formula (4.126) for $S_\varepsilon(t)$ and changing the variable of integration we get

$$S_\varepsilon(t) = B\varepsilon^{4-4m-\alpha}\int_0^t\int_{k_0}^{k_1/\varepsilon} k^{1-\alpha}\exp\{-ask^z\varepsilon^{z-2m}-Ck^2\varepsilon^p s^q-s\kappa\varepsilon^{2-2m}\}dkds+\kappa_\varepsilon, \tag{4.139}$$

If $(\alpha, z) \in G_1$, i.e. $\alpha + z > 2$ and $\alpha + 2z < 4$, and $m = (4 - \alpha - z)/2$ then (4.136) holds and hence (4.137) is true. By substituting the corresponding p, q into (4.139) and integrating with respect to s we have

$$S_\varepsilon(t) \sim B\varepsilon^{4-4m-\alpha}\int_{k_0}^{k_1/\varepsilon}\frac{k^{1-\alpha}}{ak^z\varepsilon^{z-2m}+Ck^2\varepsilon^p}$$
$$\times(1-\exp\{-t(ak^z\varepsilon^{z-2m}+Ck^2\varepsilon^p\})dk, \tag{4.140}$$

where the terms with κ_ε vanish because of $m < 1$. Since $\varepsilon^{z-2m} \to \infty$ we have

$$\lim_{\varepsilon\to 0} S_\varepsilon(t) = B \lim_{\varepsilon\to 0} \varepsilon^{4-4m-\alpha} \int_{k_0}^{k_1/\varepsilon} \frac{k^{1-\alpha}}{ak^z\varepsilon^{z-2m}} dk = \frac{B}{a(\alpha+z-2)} k_0^{2-\alpha-z}, \quad (4.141)$$

whereby (1) is proven.

Note, that if $(\alpha, z) \in G_2 \cup G_3$ and

$$m = \begin{cases} 2-\alpha/2-\beta & (\alpha,z) \in G_2 \\ 1 & (\alpha,z) \in G_3, \end{cases} \quad (4.142)$$

then (4.133) holds. Indeed, in the first case $z - 2m = (\alpha+2z-4)(\alpha+z-2)/(\alpha+2z-2)$ and in the second case $z - 2m = z - 2$. Since in both cases $\alpha < 2$ one should use the asymptotic (4.134). It can easily be seen that for $p = 2-2m+m(2-\alpha-z)$ prescribed by (4.134), the quantity $Ck^2\varepsilon^p s^q$, where C, p, and q come from (4.134), surpresses the other terms in the exponent in (4.139) since its value at $k = k_1/\varepsilon$ is much bigger than that of the other terms. Taking this into account, we obtain that for both regions

$$S_\varepsilon(t) \sim B\varepsilon^{4-4m-\alpha} \int_0^t \int_{k_0}^{k_1/\varepsilon} k^{1-\alpha} \exp\{-Ck^2\varepsilon^p s^q\} dk ds + \kappa_\varepsilon \sim$$
$$\sim q_{\alpha,z} |2-\alpha-2\beta| \varepsilon^{4-4m-\alpha-p\beta} \int_{k_0}^{k_1/\varepsilon} k^{1-\alpha-2\beta} dk + \kappa_\varepsilon. \quad (4.143)$$

The behavior of the latter integral depends on the sign of $v \equiv 2-\alpha-2\beta = (2z(1-\alpha)-(\alpha-2)^2)/(\alpha+2z-2)$. If $(\alpha, z) \in G_2$ and m is given by (4.142) then $v < 0$ and this integral converges when $\varepsilon \to 0$. Thus

$$S_\varepsilon(t) \sim q_{\alpha,z} \varepsilon^{4-4m-\alpha-p\beta} k_0^{2-\alpha-2\beta}. \quad (4.144)$$

By substituting $m = 2-\alpha/2-\beta$ and $p = 2-2m/\beta$ from (4.134) into the last equality we obtain that the exponent of ε is zero.

Thus, if m is given by (4.142), $S_\varepsilon(t)$ has the finite limit given in Proposition 4.2.

If $(\alpha, z) \in G_3$ and $m = 1$ then $v > 0$ and the integral on the right-hand side of (4.143) diverges when $\varepsilon \to 0$. Thus

$$S_\varepsilon(t) \sim q_{\alpha,z} \varepsilon^{2-4m+(2-p)\beta} k_1^{2-\alpha-2\beta} + \kappa. \quad (4.145)$$

Hence if $m = 1$, $S_\varepsilon(t)$ has the finite limit given in Proposition 4.2. and the lemma is proven. ∘

Let us make two remarks. First, inequalities $(6-2\alpha-z)(\alpha+2z-2) > 2z$ and $z < 4-\alpha$ appearing in (4.95) come from the short-correlation restriction. Second in region G_4, defined by $\alpha > 2$, $\alpha+2z > 4$ one should take $m = (4-\alpha)/4$ in order to have a finite limit for $S_\varepsilon(t)$, even though the short-correlation condition does not hold for this region.

4.4 SUMMARY

First, we considered the velocity rescaling (4.2) for which the Eulerian correlation time and the turnover time have the same order ε^2. Under this assumption the tracer field converges as $\varepsilon \to 0$ to a deterministic field satisfying the Fickian equation (4.4). In this approximation the effective diffusivity is a nonlinear function of the molecular diffusivity (4.5, 4.6). Formula (4.5) for diffusivity is obtained by rigorous summation of all diagrams in the expansion of the solution of the advection-diffusion equation through its parametrix, (4.12), (4.42).

A mathematical example is given illustrating that molecular diffusion can drastically enhance turbulent diffusion.

In the case of superdiffusion, $\tau_T \ll \tau_E$, (turnover time much less then the eddy lifetime) an exhausting analysis of all possible diffusion regimes is given (4.96-4.98) for an isotropic velocity field with a Kolmogorov type spectrum (2.18). As in the homogenization case, the derivation is based on a complete diagram summation which can be done only if the short-correlation condition $\tau_E \ll t$ holds.

5

FINITE CORRELATION TIME

All the approximations thus far considered address situations with infinitesimally small Eulerian correlation times τ_E whereas the observation time t is of order 1. Here we will discuss the situation when τ_E is fixed and t goes to infinity. Naively the two approaches should be the same. However their asymptotics are similar but not exactly identical. A definite advantage of the second approach is that it can be checked by numerical simulations.

In this chapter, first we consider a time innovating homogeneous velocity field with $\tau \equiv \tau_E$ and study the asymptotic behavior of $\langle c(n\tau, \boldsymbol{r})\rangle$ when $n \to \infty$. Then we address the same issue for an innovating flow with linear shear, i.e. velocity field $\boldsymbol{u}_n(\boldsymbol{r})$ at the n-th time step is given by $\boldsymbol{u}_n(\boldsymbol{r}) = \boldsymbol{B}_n \boldsymbol{r}$,where $\{\boldsymbol{B}_n\}$ is a sequence of independent identically distributed matrices. Finally, we discuss some non-rigorous models proposed by other authors for describing turbulent diffusion when both the correlation time and t are finite.

5.1 TIME INNOVATING HOMOGENEOUS VELOCITY

Here we again consider the advection-diffusion equation

$$\frac{\partial c}{\partial t} + \boldsymbol{u} \cdot \nabla c = \kappa \nabla^2 c \tag{5.1}$$

with the innovating velocity field discussed earlier in Chapter 3

$$\boldsymbol{u}(t, \boldsymbol{r}) = \boldsymbol{u}_{[t/\tau]}(\boldsymbol{r}), \tag{5.2}$$

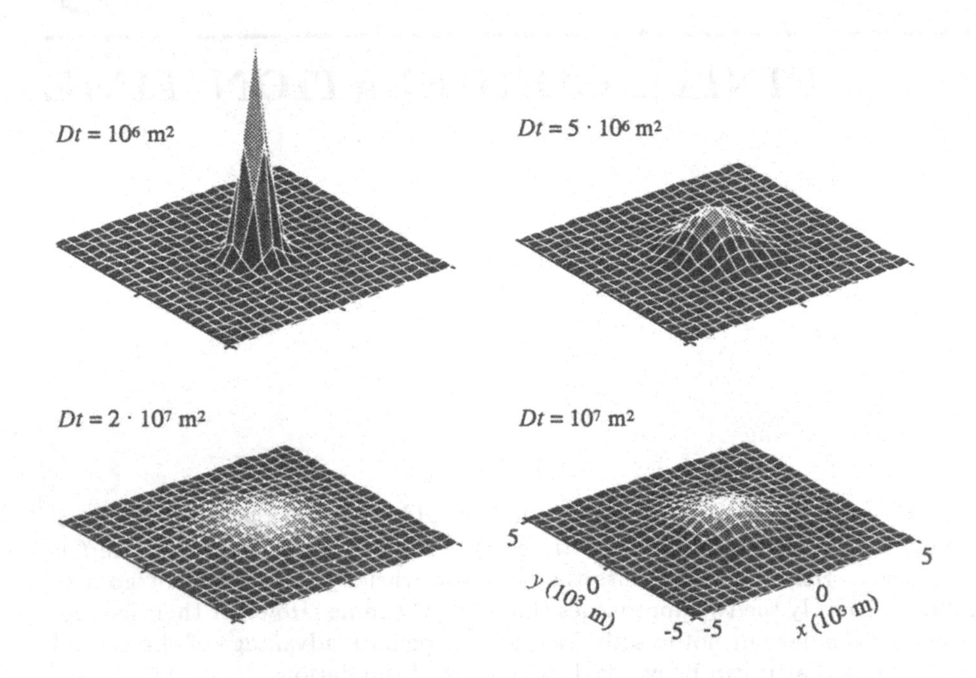

Figure 5.1 The time evolution of tracer distribution from an initial point source. Clockwise from upper left $Dt = 10^6 m^2$, $Dt = 5 \cdot 10^6 m^2$, $Dt = 10^7 m^2$, $Dt = 2 \cdot 10^6 m^2$.

where $\{\boldsymbol{u}_n(\boldsymbol{r})\}$, $n = 1, 2, \ldots$ is a sequence of independent identically distributed (i.i.d.) fields with zero mean, $[x]$ is the integer part of x and τ is the innovation time. If t is fixed and $\tau \ll t$ then equation (3.42) for the mean concentration can be written as

$$\frac{\partial}{\partial t}\langle c \rangle = \nabla \cdot \boldsymbol{D}_\tau \nabla \langle c \rangle, \tag{5.3}$$

where

$$\boldsymbol{D}_\tau = \kappa \boldsymbol{I} + \tau \langle \boldsymbol{u}_n(\boldsymbol{r}) \boldsymbol{u}_n(\boldsymbol{r})^T \rangle. \tag{5.4}$$

In the homogeneous case, $\boldsymbol{D}_\tau$ does not depend on $\boldsymbol{r}$ and equation (5.3) can be easily solved. As a result we obtain a Gaussian form for the mean tracer concentration due to an initial point source (Fig. 5.1)

$$\langle c(t, \boldsymbol{r}) \rangle = [4\pi t]^{-d/2} |\boldsymbol{D}_\tau|^{-1/2} \exp\{-\boldsymbol{D}_\tau^{-1} \boldsymbol{r} \cdot \boldsymbol{r}/4t\} \tag{5.5}$$

or, equivalently, the corresponding Fourier transform

$$\langle \hat{c}(t,\boldsymbol{k})\rangle = \exp\{-t\boldsymbol{D}_\tau \boldsymbol{k}\cdot\boldsymbol{k}\}. \tag{5.6}$$

Later we will obtain an asymptotic similar to (5.6) when τ is fixed and t goes to infinity. In order to see the difference between these asymptotics let us consider first the simplest situation, when $\boldsymbol{u}_n$ is independent of $\boldsymbol{r}$. In this case, from (1.78), we have the exact formula for the Fourier transform of the mean tracer at the moment $n\tau$

$$\langle \hat{c}_n(\boldsymbol{k})\rangle = \exp\{-n\tau\boldsymbol{D}_\tau \boldsymbol{k}\cdot\boldsymbol{k}\}, \tag{5.7}$$

where as before $\boldsymbol{D}_\tau = \tau\boldsymbol{R}_u + \kappa\boldsymbol{I}$ and $\boldsymbol{R}_u$ is the covariance matrix of the vector $\boldsymbol{u}_n$. Thus,

$$\lim_{n\to\infty}\langle \hat{c}_n(\boldsymbol{k}/\sqrt{n})\rangle = \exp\{-\tau\boldsymbol{D}_\tau \boldsymbol{k}\cdot\boldsymbol{k}\}. \tag{5.8}$$

After the corresponding renormalization we obtain from (5.6) the same right-hand side when τ goes to 0, with t is fixed

$$\langle \hat{c}(t,\boldsymbol{k}\sqrt{\tau/t})\rangle \sim \exp\{-\tau\boldsymbol{D}_\tau \boldsymbol{k}\cdot\boldsymbol{k}\}. \tag{5.9}$$

The essential difference is that the right-hand side in (5.9) tends to zero while the right-hand side of (5.8) is a non-zero constant.

Now let us establish a relation of type (5.8) in the general case, i.e. without the assumption that $\boldsymbol{u}_n$ is independent of $\boldsymbol{r}$. For this purpose, the representation (1.50) of the solution of (5.4) is not sufficient and one must use the representation (1.56), which we rewrite here for the sake of convenience

$$c(t,\boldsymbol{r}) = \boldsymbol{E}\{\exp[\frac{1}{\sqrt{2\kappa}}\int_s^t \boldsymbol{u}(v,\boldsymbol{y}_{tv})\cdot d\boldsymbol{w}(v) - \frac{1}{4\kappa}\int_s^t \boldsymbol{u}^2(v,\boldsymbol{y}_{tv})dv]c(s,\boldsymbol{y}_{ts})\}, \tag{5.10}$$

where $t > s$ are arbitrary,

$$\boldsymbol{y}_{ts} = \boldsymbol{r} + \sqrt{2\kappa}(\boldsymbol{w}(t) - \boldsymbol{w}(s)), \tag{5.11}$$

and $\boldsymbol{E}$ represents averaging with respect to the Wiener process $\boldsymbol{w}(\cdot)$. By passing to the conditional expectation and averaging over the velocity ensemble one can obtain

$$\begin{aligned}\langle c(t,\boldsymbol{r})\rangle = \langle\int \boldsymbol{E}\{\exp[\frac{1}{\sqrt{2\kappa}}\int_s^t \boldsymbol{u}(v,\boldsymbol{y}_{tv})\cdot d\boldsymbol{w}(v) \\ -\frac{1}{4\kappa}\int_s^t \boldsymbol{u}^2(v,\boldsymbol{y}_{tv})dv]|\boldsymbol{y}_{ts} = \boldsymbol{r}'\}G(t-s,\boldsymbol{r}-\boldsymbol{r}')c(s,\boldsymbol{r}')d\boldsymbol{r}'\rangle,\end{aligned} \tag{5.12}$$

where

$$G(t,\boldsymbol{r}) = [4\pi\kappa t]^{-d/2}\exp\{-\boldsymbol{r}^2/4\kappa t\}. \tag{5.13}$$

For the innovating velocity field (5.1) we set in (5.12) $s = n\tau$, $t = (n+1)\tau$, $c_n(\boldsymbol{r}) = c(n\tau, \boldsymbol{r})$. As a result we obtain

$$\langle c_{n+1}(\boldsymbol{r})\rangle = \int K_\tau(\boldsymbol{r},\boldsymbol{r}')\langle c_n(\boldsymbol{r}')\rangle d\boldsymbol{r}', \tag{5.14}$$

where

$$K_\tau(\boldsymbol{r},\boldsymbol{r}') = \langle \boldsymbol{E}\{\exp[\frac{1}{\sqrt{2\kappa}}\int_{n\tau}^{(n+1)\tau} \boldsymbol{u}_n(\boldsymbol{y}_{(n+1)\tau,v})\cdot d\boldsymbol{w}(v) \\ -\frac{1}{4\kappa}\int_{n\tau}^{(n+1)\tau} \boldsymbol{u}_n^2(\boldsymbol{y}_{(n+1)\tau,v})dv]|\boldsymbol{y}_{(n+1)\tau,n\tau} = \boldsymbol{r}'\}\rangle G(\tau,\boldsymbol{r}-\boldsymbol{r}'). \tag{5.15}$$

Obviously, $K_\tau(\boldsymbol{r},\boldsymbol{r}')$ is independent of n and so

$$K_\tau(\boldsymbol{r},\boldsymbol{r}') = \langle \boldsymbol{E}\{\exp[\frac{1}{\sqrt{2\kappa}}\int_0^\tau \boldsymbol{u}(\boldsymbol{r}'-\sqrt{2\kappa}\boldsymbol{w}(v))\cdot d\boldsymbol{w}(v) \\ -\frac{1}{4\kappa}\int_0^\tau \boldsymbol{u}^2(\boldsymbol{r}'-\sqrt{2\kappa}\boldsymbol{w}(v))dv]|\sqrt{2\kappa}\boldsymbol{w}(\tau) = \boldsymbol{r}'-\boldsymbol{r}\}\rangle G(\tau,\boldsymbol{r}-\boldsymbol{r}'). \tag{5.16}$$

From homogeneity it follows that $K_\tau(\boldsymbol{r},\boldsymbol{r}') = K_\tau(\boldsymbol{r}-\boldsymbol{r}')$ and finally we obtain by setting $\boldsymbol{r}' = 0$ in (5.16) and changing $\boldsymbol{w}$ to $-\boldsymbol{w}$

$$K_\tau(\boldsymbol{r}) = \langle \boldsymbol{E}\{\exp[-\frac{1}{\sqrt{2\kappa}}\int_0^\tau \boldsymbol{u}(\sqrt{2\kappa}\boldsymbol{w}(v))\cdot d\boldsymbol{w}(v) - \\ \frac{1}{4\kappa}\int_0^\tau \boldsymbol{u}^2(\sqrt{2\kappa}\boldsymbol{w}(v))dv]|\sqrt{2\kappa}\boldsymbol{w}(\tau) = \boldsymbol{r}\}\rangle G(\tau,\boldsymbol{r}). \tag{5.17}$$

Note, that for any random vectors ξ, η and any function $f(\cdot)$

$$\int \langle \xi | \eta = \boldsymbol{r}\rangle f(\boldsymbol{r}) p_\eta(\boldsymbol{r}) d\boldsymbol{r} = \langle \xi f(\eta)\rangle, \tag{5.18}$$

where $p_\eta(\boldsymbol{r})$ is the probability density for η. By using (5.18) with $\eta = \boldsymbol{w}(\tau)$, $f(\boldsymbol{r}) = \exp[i\boldsymbol{k}\cdot\boldsymbol{r}]$ one can obtain for the Fourier transform of $K_\tau(\boldsymbol{r})$

$$\hat{K}_\tau(\boldsymbol{k}) = \boldsymbol{E}\{\exp[i\boldsymbol{k}\cdot\sqrt{2\kappa}\boldsymbol{w}(\tau)]\langle\exp[\frac{1}{\sqrt{2\kappa}}\int_0^\tau \boldsymbol{u}(\sqrt{2\kappa}\boldsymbol{w}(v))\cdot d\boldsymbol{w}(v) \\ -\frac{1}{4\kappa}\int_0^\tau \boldsymbol{u}^2(\sqrt{2\kappa}\boldsymbol{w}(v))dv]\rangle\}. \tag{5.19}$$

Thus, from (5.14) it follows that

$$\langle \hat{c}_{n+1}(\boldsymbol{k})\rangle = \hat{K}_\tau(\boldsymbol{k})\langle \hat{c}_n(\boldsymbol{k})\rangle, \tag{5.20}$$

where $\hat{K}_\tau(\boldsymbol{k})$ is given by (5.19).

Now, by conventional arguments when deriving the Central Limit Theorem, from (5.20) and (5.19) we obtain the following Proposition 5.1.

Proposition 5.1.

$$\lim_{n\to\infty} \langle \hat{c}_n(\boldsymbol{k}/\sqrt{n})\rangle = \exp\{-\tau \boldsymbol{D}_\tau \boldsymbol{k}\cdot\boldsymbol{k}\}, \tag{5.21}$$

where

$$\begin{aligned}\boldsymbol{D}_\tau = \frac{1}{\tau}\boldsymbol{E}\{\kappa \boldsymbol{w}^2(\tau)\langle \exp[\frac{1}{\sqrt{2\kappa}}\int_0^\tau \boldsymbol{u}(\sqrt{2\kappa}\boldsymbol{w}(v))\cdot d\boldsymbol{w}(v)\\ -\frac{1}{4\kappa}\int_0^\tau \boldsymbol{u}^2(\sqrt{2\kappa}\boldsymbol{w}(v))dv]\rangle\}.\end{aligned} \tag{5.22}$$

In the constant velocity case considered before we get

$$\begin{aligned}D_\tau &= \frac{1}{\tau}\boldsymbol{E}\{\kappa \boldsymbol{w}^2(\tau)\langle \exp[\frac{1}{\sqrt{2\kappa}}\boldsymbol{u}\cdot\boldsymbol{w}(\tau) - \frac{1}{4\kappa}\boldsymbol{u}^2\tau]\rangle\} =\\ &= \frac{\kappa}{\tau}\langle \tau(\boldsymbol{I} + \frac{\tau}{2\kappa}\boldsymbol{u}\boldsymbol{u}^T)\rangle,\end{aligned} \tag{5.23}$$

which coincides with the previous expression for D_τ given after (5.7).

Finally let us note that one can obtain (5.3) from (5.20) by writing the latter as

$$\frac{\langle \hat{c}_{n+1}(\boldsymbol{k})\rangle - \langle \hat{c}_n(\boldsymbol{k})\rangle}{\tau} = \frac{\hat{K}_\tau(\boldsymbol{k}) - 1}{\tau}\langle \hat{c}_n(\boldsymbol{k})\rangle, \tag{5.24}$$

and by letting τ approach zero.

5.2 RANDOM SHEAR FLOW

Here we continue to consider the advection-diffusion equation with an innovating velocity field (5.2), but now the field $\boldsymbol{u}_n(\boldsymbol{r})$ is assumed to be strongly inhomogeneous. We use the term 'strong' in order to distinguish the case discussed below from an inhomegenous velocity field with a constant variance matrix, i.e. when $\langle \boldsymbol{u}_n(\boldsymbol{r})\boldsymbol{u}_n(\boldsymbol{r})^T\rangle$ is independent of $\boldsymbol{r}$. In the last case the limiting distribution of the mean concentration ($\tau \to 0$) is Gaussian as in the homogeneous case (chapter 3, equations (3.24), (3.41)). Thus, concerning the weakly inhomogeneous case we deal with the Central Limit Theorem (CLT) for vector valued random processes, while in the strongly inhomogeneous case considered below we use the CLT for a product of random matrices (Tutubalin 1976). It leads to the limiting distribution of a quite different type.

Let us restrict the consideration to the 2D linear shear flow which is of a physical importance in oceanographic applications and allows an exact solution of the discussed problem. Thus, it is assumed that

$$\boldsymbol{u}_n(\boldsymbol{r}) = \boldsymbol{B}_n\boldsymbol{r}, \tag{5.25}$$

where $\boldsymbol{B}_n = (b_{ij}^{(n)})$ are i.i.d. random matrices with zero mean and zero trace, i.e.

$$b_{11} + b_{22} = 0. \tag{5.26}$$

Notice, here and below the index n is omitted unless it causes an ambiguity. Relation (5.26) implies that the considered velocity field is non-divergent and determined by the following stream function

$$\psi(\boldsymbol{r}) = \frac{1}{2}(-b_{21}x^2 + (b_{11} - b_{22})xy + b_{12}y^2). \tag{5.27}$$

The form of stream-lines varies depending on the relationship between the b_{ij} (Fig. 5.2). For the Lagrangian displacement $\boldsymbol{X}_n = \boldsymbol{X}(n\tau)$ corresponding to the initial point $\boldsymbol{X}_0$ we have

$$\boldsymbol{X}_n = \boldsymbol{A}_n\boldsymbol{X}_{n-1}, \tag{5.28}$$

where $\boldsymbol{A}_n = e^{\tau\boldsymbol{B}_n}$ is the new sequence of i.i.d. random matrices with unit determinants. Note that the entries of $\boldsymbol{A}$ can be expressed in terms of the entries of $\boldsymbol{B}$ in the following way

$$a_{11} = \cos\lambda\tau + \frac{b_{11}}{\lambda}\sin\lambda\tau, \quad a_{12} = \frac{b_{12}}{\lambda}\sin\lambda\tau,$$

$$a_{21} = \frac{b_{21}}{\lambda}\sin\lambda\tau, \quad a_{22} = \cos\lambda\tau + \frac{b_{22}}{\lambda}\sin\lambda\tau, \tag{5.29}$$

where $\lambda = \sqrt{b_{11}b_{22} - b_{12}b_{21}}$.

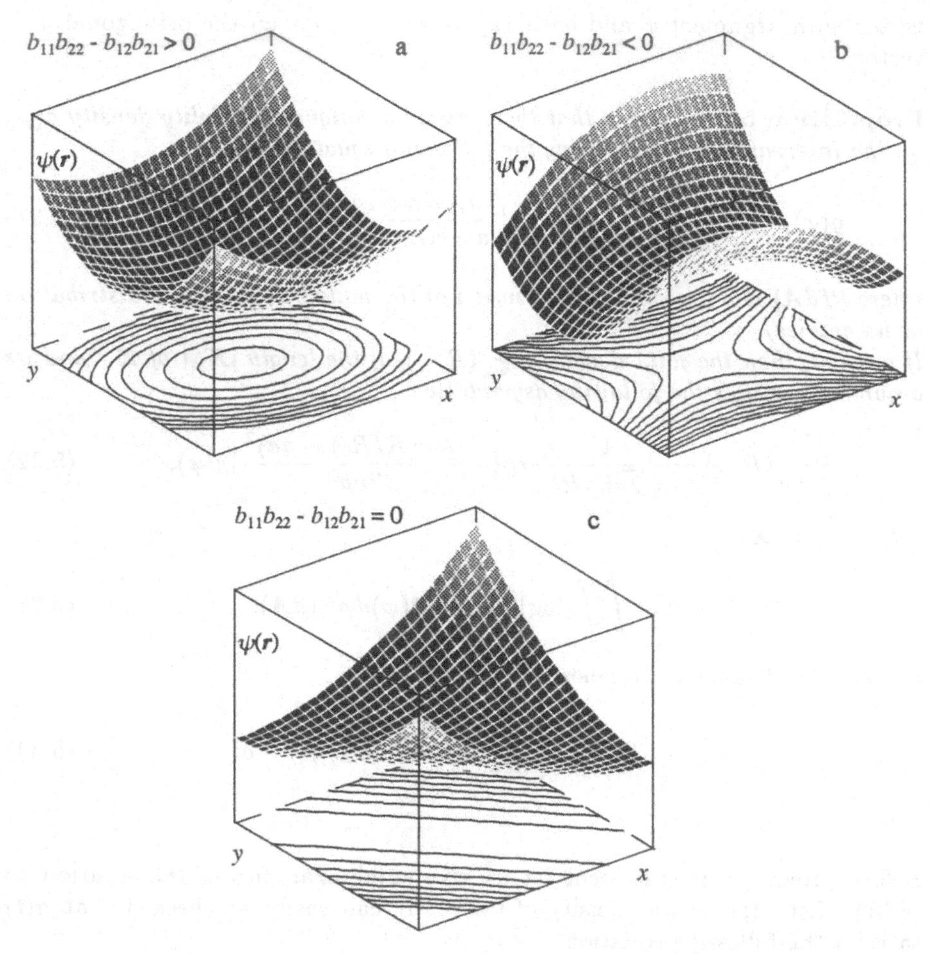

Figure 5.2 The three types of the stream functions: *a*) elliptical, *b*) hyperbolic, *c*) parabolic.

Thus, the problem is to find the asymptotic distribution of the random vector

$$\boldsymbol{X}_n = \boldsymbol{A}_n \boldsymbol{A}_{n-1} \dots \boldsymbol{A}_1 \boldsymbol{X}_0, \tag{5.30}$$

when n goes to infinity while τ remains fixed, for any initial position $\boldsymbol{X}_0$. The answer to this question is contained in the mentioned paper by Tutubalin (1976), (see also references given there) in the general situation. For 2×2 matrices the corresponding computations are simple and the result can be formulated in an explicit form. Let us denote by $\boldsymbol{e}(\varphi) = (\cos\varphi,\ \sin\varphi)$ the unit vector with argument φ and by $e^*(\varphi) = (\sin\varphi, \ -\cos\varphi)$ the orthogonal unit vector.

Proposition 5.2. *Assume that there exists a unique probability density $g(\varphi)$ on the interval $[-\pi, \pi]$ satisfying the following equation*

$$g(\varphi) = \int \frac{1}{|A^T e^*(\varphi)|^2} g(\tan^{-1} \frac{a_{11}\sin\varphi - a_{21}\cos\varphi}{a_{22}\cos\varphi - a_{12}\sin\varphi}) P(dA), \tag{5.31}$$

where $P(dA)$ is a probability distribution of the matrix A (the joint distribution of its entries).
If $n \to \infty$, then the joint density $P_{X_n}(R, \varphi)$ of the length $|X_n|$ of X_n and its argument satisfies the following asymptotic

$$P_{X_n}(R, \varphi) \sim \frac{1}{\sqrt{2\pi n}\sigma R^2} \exp\{-\frac{(\log(R/R_0) - na)^2}{2n\sigma^2}\} g(\varphi), \tag{5.32}$$

where $R_0 = |X_0|$,

$$a = \int\int (\log|Ae(\varphi)|) g(\varphi) d\varphi P(dA), \tag{5.33}$$

is the upper Lyapunov exponent and

$$\sigma^2 = \lim_{n\to\infty} n^{-1}\langle (\sum_1^n \log|A_k e(\varphi_k)|)^2\rangle - a^2. \tag{5.34}$$

Before proving this statement let us give some examples of the solution to (5.31). Let $h(t)$ be the density of $\tan\varphi$. It can easily be checked that $h(t)$ satisfies the following equation

$$h(t) = \int \frac{1}{(a_{22} - a_{12}t)^2} h\left(\frac{a_{11}t - a_{21}}{a_{22} - a_{12}t}\right) P(dA), \tag{5.35}$$

First, let us find conditions under which the limiting distribution of the angle is uniform, i.e.,

$$h(t) = \frac{1}{\pi}\frac{1}{1+t^2}. \tag{5.36}$$

After substituting this expression into (5.35) we obtain

$$\frac{1}{1+t^2} = \int \frac{1}{t^2(a_{11}^2 + a_{12}^2) - 2t(a_{11}a_{21} + a_{22}a_{12}) + a_{21}^2 + a_{22}^2} P(dA). \tag{5.37}$$

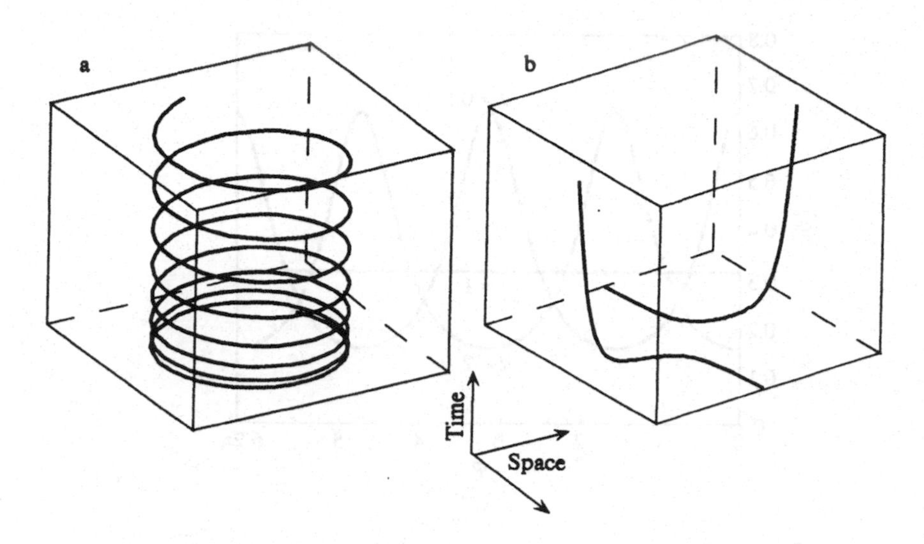

Figure 5.3 Lagrangian trajectories in a velocity field where the limiting distribution of tracer is uniform in the polar angle: *a*) elliptic stream function (5.38), *b*) hyperbolic stream function (5.39).

Hence, the matrix $\boldsymbol{A}$ must be orthogonal. This is possible if either

$$b_{11}b_{22} - b_{12}b_{21} = \pi^2 m^2/\tau \tag{5.38}$$

for an integer m, or

$$b_{12} = b_{21} = 0. \tag{5.39}$$

Both cases are not very interesting because in the first case the stream lines are circles centered at the origin and only the speed of particles along those circles varies (Fig. 5.3). Thus, the distribution of the $|\boldsymbol{X}_n|$ is degenerate and concentrated at the single point $R_0 = |\boldsymbol{X}_n|$. In the second case the stream lines are hyperbolas $xy = C$. Let us note that this case can be realized if and only if

$$\int \frac{1}{\cosh^2 b_{11}\tau} P\{d\boldsymbol{B}\} = 1. \tag{5.40}$$

If $b_{12} = 0$ and $b_{11} = -b_{22} = 1/\tau_0$ are nonrandom constants then equation (5.31) is reduced to

$$h(t) = \int q h(qt - b\tau_0(q-1)/2) p(b) db, \tag{5.41}$$

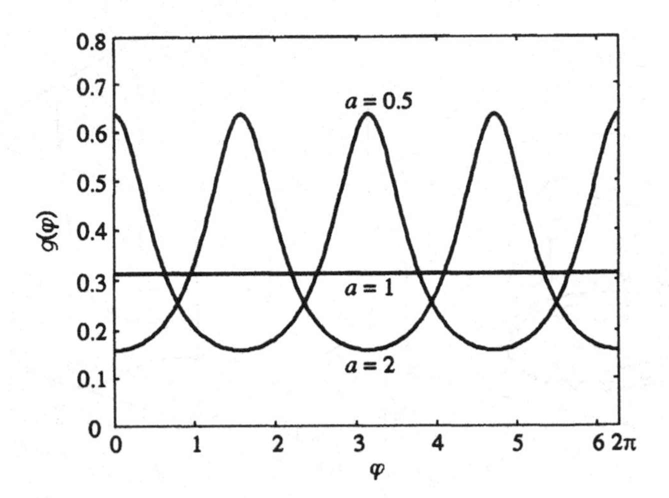

Figure 5.4 The limiting angular tracer distribution, a, for the Cauchy distribution of $b = b_{21}$.

where $p(b)$ is the density of $b = b_{21}$. For the corresponding characteristic functions we obtain

$$\hat{h}(\omega) = \hat{h}(\omega/q)\hat{p}(\tau_0\omega(q-1)/2q). \tag{5.42}$$

For the Cauchy distribution

$$\hat{p}(\omega) = e^{-q_0|\omega|} \tag{5.43}$$

one can determine that

$$\hat{h}(\omega) = e^{-q_0\tau_0|\omega|}, \quad \text{i.e.} \quad h(t) = \frac{1}{\pi}\frac{a}{a^2+t^2}, \tag{5.44}$$

where $a = q_0\tau_0$ and hence

$$g(\varphi) = \frac{a}{\pi(a^2\cos^2\varphi + \sin^2\varphi)}. \tag{5.45}$$

The graph of $g(\varphi)$ is shown in Fig. 5.4.

As was mentioned above, Proposition 5.2 follows from general results concerning CLT for the product of independent matrices. Here we focus only on the derivation of (5.31) and computing a, σ^2.

Proof. If (5.28) is written in coordinate form, we have

$$x_n = a_{11}^{(n)} x_{n-1} + a_{12}^{(n)} y_{n-1}, \quad y_n = a_{21}^{(n)} x_{n-1} + a_{22}^{(n)} y_{n-1}, \tag{5.46}$$

where $X_n = (x_n, y_n)$, $A_n = (a_{ij}^{(n)})$. By dividing the second equality by the first we get

$$t_n = \frac{a_{21} + a_{22} t_{n-1}}{a_{11} + a_{12} t_{n-1}}, \tag{5.47}$$

where $t_n = y_n / x_n = \tan(\arg X_n)$ and the upper indices are omitted. Then squaring and adding both equalities in (5.46) yields

$$R_n = R_{n-1} |A_n e(\varphi_n)|, \tag{5.48}$$

where $R_n = |X_n|$. From (5.47) we obtain

$$h_n(t) = \langle \delta(t_n - t) \rangle = \langle \delta(\frac{a_{21} + a_{22} t_{n-1}}{a_{11} + a_{12} t_{n-1}} - t) \rangle =$$

$$\int\int h_{n-1}(s) \delta\left(\frac{a_{21} + s a_{22}}{a_{11} + s a_{12}} - t\right) ds P(dA) = \tag{5.49}$$

$$\int \frac{1}{(a_{22} - a_{12} t)^2} h_{n-1}\left(\frac{a_{11} t - a_{21}}{a_{22} - a_{12} t}\right) P(dA).$$

This equation implies equation (5.31) for a stationary angle density. The asymptotic distribution for R_n can be computed by applying the Central Limit Theorem to $\log R_n$ because it can be expanded into a sum of weakly independent identically distributed random values as follows. From (5.48),

$$\log R_n = \sum_1^n \log |A_k e(\varphi_k)| + \log R_0. \tag{5.50}$$

To compute the parameters of the limiting distribution note that

$$a = \langle \log |A_k e(\varphi_k)| \rangle = \int \langle \log |A_k e(\varphi_k)| | \varphi_k = \varphi \rangle g(\varphi) d\varphi =$$

$$\int\int (\log |Ae(\varphi)|) g(\varphi) d\varphi P(dA). \tag{5.51}$$

Formula (5.34) for the variance readily follows from (5.50). ∘

By taking into account the fact that the mean tracer field $\langle c_n(\boldsymbol{r})\rangle \equiv \langle c_n(\boldsymbol{r};\boldsymbol{r}_0)\rangle$ corresponding to the initial condition $\delta(\boldsymbol{r}-\boldsymbol{r}_0)$ is equal to the Lagrangian particle density $P_{\boldsymbol{X}_n}(r,\varphi)$, where $r=|\boldsymbol{r}|$, $\varphi=\arg\boldsymbol{r}$, we obtain from (5.32)

$$\langle c_n(\boldsymbol{r})\rangle \sim \frac{1}{\sqrt{2\pi n\sigma r^2}}\exp\{-\frac{(\log(r/r_0)-na)^2}{2n\sigma^2}\}g(\varphi), \tag{5.52}$$

where $r_0=|\boldsymbol{r}_0|$. Examples of the radial tracer distribution are given in Fig. 5.5.

Suppose that

$$\sigma^2 > a. \tag{5.53}$$

Then it can easily be checked that if

$$r_1(n) = r_0 e^{-\alpha_1 n-\beta\log n-s}$$

or

$$r_2(n) = r_0 e^{-\alpha_2 n+\beta\log n+s} \tag{5.54}$$

then

$$\lim_{n\to\infty}\langle c_n(r_j,\varphi)\rangle = \frac{1}{\sqrt{2\pi r_0^2}}e^{-4\beta s},\quad j=1,2, \tag{5.55}$$

where

$$\begin{aligned}\alpha_1 &= 2\sigma^2-a+2\sigma\sqrt{\sigma^2-a}<0,\\ \alpha_2 &= 2\sigma^2-a-2\sigma\sqrt{\sigma^2-a}<0,\\ \beta &= 2\sigma/\sqrt{\sigma^2-a}>0,\end{aligned} \tag{5.56}$$

and s is arbitrary. Thus, under condition (5.53) there exist two rings centered at $\boldsymbol{r}_0$ with exponentially small radius where the tracer concentration remains finite when time increases indefinitely. Notice that, in fact, the discussed effect has a mathematical meaning only because the integral of $c(\cdot)$ over any finite neighborhood of the rings definitely goes to zero. Nevertheless this fact distinguishes diffusion in linear shear flow from the diffusion in the homogeneous velocity field, where $\langle c_n(\boldsymbol{r})\rangle$ decays for all $\boldsymbol{r}$ due to the fact that

$$|\langle c_n(\boldsymbol{r})\rangle| < \frac{c}{n^{d/2}}, \tag{5.57}$$

from (5.5).

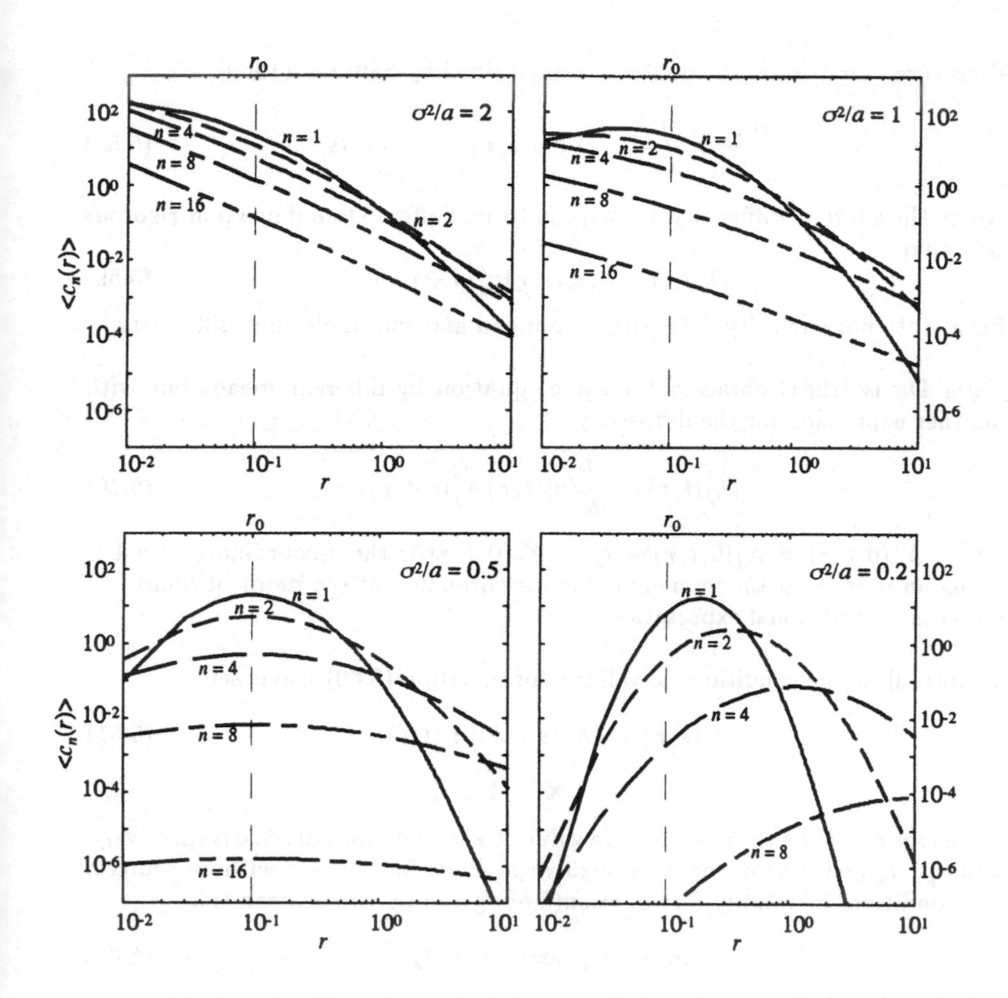

Figure 5.5 The evolution of the radial tracer distribution. See text for explanation.

5.3 THE FINITE CORRELATION AND OBSERVATION TIMES

From a mathematical viewpoint it is hopeless to find out an exact closed equation for the mean tracer concentration when both τ_E and t are finite. Some equations for the mean tracer regarding this case were discussed in the physical literature. First, such an equation was obtained by Saffman (1969)

$$\frac{\partial \langle c \rangle}{\partial t} = \int_0^t \nabla \cdot \boldsymbol{D}(t-s, \boldsymbol{r}) \nabla \langle c(s, \boldsymbol{r}) \rangle ds, \qquad (5.58)$$

where the effective diffusivity depends on time. Saffman found in a non-rigorous way that

$$D_{ij}(t, \boldsymbol{r}) = \langle u'_i(s, \boldsymbol{r}) u'_j(s+t, \boldsymbol{r}) \rangle \qquad (5.59)$$

for a stationary velocity field with zero mean and zero molecular diffusivity.

Later Davis (1987) obtained the same equation by different means but with another expression for the diffusivity

$$D_{ij}(t, \boldsymbol{r}) = \frac{\partial}{\partial t} \langle u'_i(t, \boldsymbol{r}) X'_j(0; t, \boldsymbol{r}) \rangle, \qquad (5.60)$$

where $X'_j(0;t,\boldsymbol{r}) = X_j(0;t,\boldsymbol{r}) - x_j$, $X_j(0;t,\boldsymbol{r})$ is the j-coordinate of a Lagrangian particle at the moment 0 passing through $\boldsymbol{r}$ at the moment t and $\langle \cdot | \cdot \rangle$ represents conditional expectation.

To formulate the conditions of validity for equation (5.60) Davis set

$$\boldsymbol{Y}(s; \boldsymbol{r}) = \langle \boldsymbol{X}'(s; t, \boldsymbol{r}) | \boldsymbol{u}'_i(t, \boldsymbol{r}) \rangle, \qquad (5.61)$$

$$\boldsymbol{Z} = \boldsymbol{X}' - \boldsymbol{Y}.$$

Let us set σ_z, l_z be the mean square value of $\boldsymbol{Z}$ and its spatial scale respectively, and σ_y, l_y be the same for $\boldsymbol{Y}$. Roughly speaking, for the case without sources, the conditions of validity for equation(5.58) given by Davis mean that

$$\sigma_y \ll \sigma_z, \quad \text{and} \quad \sigma_z \ll l_z, \qquad (5.62)$$

or

$$\boldsymbol{u}' \text{ is Gaussian and } \sigma_y \ll l_z. \qquad (5.63)$$

Lipscombe, Frenkel and ter Haar (1991) included molecular diffusion in equation (5.58). Their version of the turbulent diffusivity is

$$D_{ij}(t, \boldsymbol{r}) = \kappa \delta_{ij} + G(t, \boldsymbol{r}) \langle u'_i(s, \boldsymbol{r}) u'_j(s+t, \boldsymbol{r}) \rangle, \qquad (5.64)$$

where $G(t, \boldsymbol{r})$ is the Green's function for the molecular diffusion equation given by (5.13). The validity conditions for (5.64) were formulated as follows

$$\tau_E \ll \tau_T, \quad \tau_E \ll \tau_D, \qquad (5.65)$$

i.e. the velocity correlation time is much less than both, the turnover time and the molecular diffusion time, while nothing is said about the observation time t.

One unresolved issue in these approaches is that they do not hold for the simplest case of the space coordinate free Gaussian velocity fields considered in Chapter 2, even though all of the conditions (5.62), (5.63), (5.65) are obviously satisfied. It seems that an approximation is not very good if the resulting equation does not hold exactly, when all the conditions of the approximation are exactly satisfied. We conjecture that the non-locality of the mean tracer equation in time is not directly related to the finiteness of the Eulerian velocity correlation time.

5.4 SUMMARY

We consider asymptotics of the mean tracer concentration when the Eulerian correlation time τ_E is fixed and t goes to infinity. Asymptotics for the mean tracer field are obtained in two important cases: homogeneous innovating flow (5.1) and innovating flow with linear shear (5.25). In the first case the limiting mean tracer is described by a Gaussian distribution due to the Central Limit Theorem (5.21), (5.22). In the second case the problem is reduced to the asymptotic behavior of the product of random matrices. As a result, the mean tracer field corresponding to an initial point source is logarithmically normal in the radial direction (5.52). Its angular distribution can be found from equation (5.31). Examples of the exact solution to this equation are given, (5.36), (5.45).

There have been efforts to give an equation for the mean tracer concentration in the physical literature when both τ_E and t are fixed. Unfortunately, such attempts are not satisfactory in describing the simplest case: a Gaussian velocity field which is independent of spatial coordinates.

6

THE INVERSE PROBLEM: MAXIMUM LIKELIHOOD METHOD

This and the next two chapters address the problem of estimating the deterministic parameters $\boldsymbol{u} = (u_1, \ldots, u_d)$, $\boldsymbol{D} = (D_{ij})$, $i, j = 1, \ldots, d$, and λ in the following stochastic equation describing transport and dissipation of a passive scalar (tracer)

$$\frac{\partial c}{\partial t} + \boldsymbol{u} \cdot \nabla c + \lambda c = \nabla \cdot \boldsymbol{D} \nabla c + S(t, \boldsymbol{r}) \tag{6.1}$$

given observations of $c = c(t, \boldsymbol{r})$, where $S(t, \boldsymbol{r})$ is an unobserved source assumed to be a statistically stationary in time Gaussian random field. One can see that equation (6.1) is a particular case of the following general model

$$\frac{\partial c}{\partial t} + (A_0 + \theta_1 A_1 + \ldots + \theta_K A_K) c = S(t, \boldsymbol{r}), \tag{6.2}$$

where $A_1, \ldots, A_K$ are linear differential operators and $\theta_1, \ldots, \theta_K$ are unknown parameters.

For example, the 2-dimensional problem of estimating λ and D given $\boldsymbol{u} = (u_1, u_2)$ with a diagonal diffusion tensor

$$\frac{\partial c}{\partial t} + \boldsymbol{u} \cdot \nabla c + \lambda c = D \nabla^2 c + S(t, \boldsymbol{r}) \tag{6.3}$$

can be embedded into the general scheme by setting

$$A_0 = \boldsymbol{u} \cdot \nabla, \quad A_1 = I, \quad A_2 = -\nabla^2, \quad \theta_1 = \lambda, \quad \theta_2 = D, \tag{6.4}$$

where I is the identity operator. Definitely, other estimation problems can be posed in this way also. The point is that it is not much more difficult to deal with the general stochastic equation (6.2) than the specific model (6.1). In addition, studying the general model provides a technique applicable to more complex models of tracer propagation. Moreover, other phenomena in the ocean can be described by stochastic equations of type (6.2), where the operators $A_1, \dots, A_K$ are not necessarily differential (Piterbarg and Rozovskii 1996).

Let us concentrate on the problem with only one unknown parameter, i.e. $K = 1$ in (6.2). This parameter can represent the feedback parameter, diffusivity or a component of velocity. In this case, asymptotic properties of the maximum likelihood (ML) estimators can be studied rigorously.

In what follows we will change the notation for the observed field since a number of constants below will be denoted by c with a subscript. Another reason is to stress that the following model (6.5) actually covers many other phenomena besides a passive scalar transport. Thus, we consider the problem of estimating the unknown scalar parameter θ in the equation

$$\frac{\partial u}{\partial t} + (A_0 + \theta A_1)u = S(t, \boldsymbol{r}). \tag{6.5}$$

Here $u = u(t, \boldsymbol{r})$ is the observed random field, $\boldsymbol{r} \in G$, G is a bounded region in the d-dimensional Euclidean space E; A_k, $k = 0, 1$ are linear operators defined on a convenient functional space included into $L_2(G)$, $S(t, \boldsymbol{r})$ is the Gaussian white noise in t.

Let us assume that there exists a complete orthonormal system of eigenfunctions $\{\varphi_m(\boldsymbol{r})\}_1^\infty$ common to the operators A_0, A_1 and the spatial covariance operator of $S(t, \boldsymbol{r})$. Further assume that the following quantities are available

$$u_m(t_n), \quad n = 1, \dots, N \quad m = 1, \dots, M, \tag{6.6}$$

where

$$u_m(t) = \int_G u(t, \boldsymbol{r}) \varphi_m(\boldsymbol{r}) d\boldsymbol{r}, \tag{6.7}$$

and $t_1, \dots, t_N$ are fixed time moments. Thus, we observe the amplitudes of the first M harmonics in a finite number of time moments. In practice a trigonometric basis is often used, thus (6.7) are the Fourier coefficients which can easily be estimated from grid observations (see chapter 7). This statement of the problem is in part motivated by applications to satellite data. Satellite measurements enable us to obtain information concerning the temperature field, sea level and other hydrophysical characteristics of the sea surface with high accuracy and spatial resolution. Many satellite measurements result in a short time series of observations. Thus, the observation model (6.6)-(6.7) with small N and large M may reasonably describe satellite measurements.

In this chapter we study the asymptotic behavior of the maximum likelihood estimator (MLE) $\hat{\theta}_M$ of θ when M goes to infinity and N remains to be fixed. First of all, we are interested in checking the consistency, asymptotic normality and efficiency of $\hat{\theta}_M$.

An estimator $\hat{\theta}_M$ is *consistent (asymptotically unbiased)* if it converges in a suitable sense to the true value of the parameter as $M \to \infty$.

If $\hat{\theta}_M$ is a consistent estimator of parameter θ it is often possible to find a normalizing function $\sigma_M(\theta)$ such that the probability distribution $(\hat{\theta}_M - \theta)/\sigma_M(\theta)$ converges to one of the classical distributions (e.g. Gaussian, in this case the estimator $\hat{\theta}_M$ is referred to as asymptotically normal). This property is very important for the derivation of approximate confidence intervals. Finally, one of the most important concepts of the asymptotic approach to statistical analysis is *asymptotic efficiency*. A sequence of estimators $\hat{\theta}_M$ of the parameter θ is said to be asymptotically efficient with respect to the loss (risk) function $w(\cdot,\cdot)$ if these estimators minimize the quantity

$$\lim_{M\to\infty} \langle w(\tilde{\theta}_M, \theta)\rangle. \tag{6.8}$$

The distance between $\hat{\theta}_M$ and θ is a good example of a loss function $w(\hat{\theta}_M, \theta)$.

Now, to study asymptotic properties of the ML estimator, let

$$A_k \varphi_m = \lambda_{km} \varphi_m, \quad k = 0, 1, \tag{6.9}$$

where

$$\lambda_{0m} = \alpha_{0m} + i\beta_{0m}, \ \lambda_{1m} = \alpha_{1m} + i\beta_{1m} \tag{6.10}$$

are the eigenvalues of the operators A_0 and A_1 respectively, and set

$$\alpha_m(\theta) = \alpha_{0m} + \alpha_{1m}\theta. \tag{6.11}$$

As we will soon see, roughly speaking, the condition

$$\sum_{m=1}^{\infty} \left[\frac{\alpha_{1m}^2}{\alpha_m(\theta)^2} + h^2 \beta_{1m}^2 e^{-2h\alpha_m(\theta)} \right] = \infty \tag{6.12}$$

is necessary and sufficient for the consistency of $\hat{\theta}_M$, where $h = \min_n(t_{n+1} - t_n)$. It also implies asymptotic normality and efficiency.

In particular for self-adjoint operators, (6.12) becomes

$$\sum_{m=1}^{\infty} \frac{\alpha_{1m}^2}{\alpha_m(\theta)^2} = \infty. \tag{6.13}$$

If additionally A_0, A_1 are elliptic differential operators then (6.13) is equivalent to

$$order(A_\theta) - order(A_1) \leq d/2. \tag{6.14}$$

Particularly for equation (6.3), the MLE of D is consistent because $order(A_\theta) = order(A_1) = d = 2$, while the MLE of λ is not consistent since in this case $order(A_1) = 0$.

Returning to the general condition (6.12), let us suppose that for some non-negative r, s

$$\alpha_m(\theta) \sim m^r, \quad \lim_{m\to\infty} \inf |\beta_{1m}| > 0, \quad \lim_{m\to\infty} \sup (|\beta_{1m}| m^{-s}) < \infty. \tag{6.15}$$

These conditions are typical for non-self-adjoint differential operators used in applications. In this case if $r > 0$, (6.12) is equivalent to (6.13) and we again arrive at (6.14). In turn, if $r = 0$, (6.12) automatically holds. Consequently, any velocity component in equation (6.3) is estimated consistently if and only if $D = 0$.

To our knowledge, the first rigorous results concerning ML estimation for SPDE's were obtained in (Hubner, Khas'minskii and Rozovskii 1993) and (Hubner and Rozovskii 1995) where properties of the MLE's were studied for self-adjoint operators and continuous observations in time for the same model (6.2). In this case the condition (6.14) for elliptic differential operators should be replaced by

$$order(A_\theta)/2 - order(A_1) \leq d/2 \tag{6.16}$$

which is less restrictive. In (Hubner 1993) the mentioned results were extended to the case of multi-dimensional parameter.

This chapter is organized as follows. In section 6.1, we specify the problem statement and explain the main results. In section 6.2, an elementary derivation of criterion (6.13) is given in the simplest case of a single observation in time. Finally, in section 6.3, our results are proven in the general form.

6.1 ASYMPTOTICAL PROPERTIES OF ML ESTIMATORS IN SPDE'S: MAIN RESULTS

Here we formulate and explain the main results. First, let us specify constraints imposed on $S(t, \boldsymbol{r})$ and the eigenvalues of operator

$$A_\theta = A_0 + \theta A_1 \tag{6.17}$$

for model (6.2).

Assume that θ belongs to a bounded interval Θ of the real line and set

$$\underline{\alpha}_m = \inf_{\Theta} \alpha_m(\theta) \qquad \overline{\alpha}_m = \sup_{\Theta} \alpha_m(\theta). \tag{6.18}$$

Recall that $\alpha_m(\theta)$ given by (6.11) is the real part of the eigenvalue $\lambda_m(\theta)$ of the operator A_θ. We assume that

$$\overline{\alpha}_m / \underline{\alpha}_m \leq c_1, \tag{6.19}$$

where c_1 is independent of m, and that the eigenfunctions $\varphi_m = 1, 2, \ldots$ are arrange such that

$$\overline{\alpha}_1 \leq \overline{\alpha}_2 \leq \ldots \leq \overline{\alpha}_m \leq \ldots. \tag{6.20}$$

Further, we suppose that

$$S(t, \boldsymbol{r}) = \sum_1^\infty \sigma_m \dot{w}_m(t) \varphi_m(\boldsymbol{r}), \tag{6.21}$$

where $w_m = w_m^1 + i w_m^2$, w_m^1 and w_m^2 are independent Wiener processes, and w_m are independent for different $m = 1, 2, \ldots$. Suppose that an infinite number of the amplitudes σ_m^2 are non-zero and

$$\sum_1^\infty \sigma_m^2 < \infty, \quad \sum_1^\infty \sigma_m^2 / 2\alpha_m(\theta) < \infty, \tag{6.22}$$

providing finite variances for $u(t, \boldsymbol{r})$ and $S(t, \boldsymbol{r})$. Note that equation (6.2) and the discussed problem make a sense under much wider conditions (Hubner and

Rozovskii 1995). From the orthonormality of $\{\varphi_m\}$ it readily follows that the amplitudes

$$u_m(t) = \int_G u(t, \boldsymbol{r})\varphi_m(\boldsymbol{r})d\boldsymbol{r} \tag{6.23}$$

are independent complex-valued processes obeying the equation

$$\dot{u}_m + \lambda_m(\theta)u_m = \sigma_m \dot{w}_m, \quad m = 1, 2, \ldots. \tag{6.24}$$

which is equivalent to the following pair of real equations

$$\begin{aligned} \dot{u}_m^1 + \alpha_m u_m^1 - \beta_m u_m^2 &= \sigma_m w_m^1(t), \\ \dot{u}_m^2 + \beta_m u_m^1 + \alpha_m u_m^2 &= \sigma_m w_m^2(t), \end{aligned} \tag{6.25}$$

where $\alpha_m = \alpha_m(\theta)$ is given by (6.11), $\beta_m = \beta_m(\theta) = \beta_{0m} + \theta\beta_{1m}$, and

$$u_m^1(t) = \text{Re } u_m(t), \quad u_m^2(t) = \text{Im } u_m(t). \tag{6.26}$$

Let $P_M^\theta(\boldsymbol{u})$ be the probability density of the sample

$$\left\{u_m^1(t_j), u_m^2(t_j)\right\}, \quad m = 1, \ldots, M; \ j = 1, \ldots, N, \tag{6.27}$$

and

$$\hat{\theta}_M = \arg\max_\theta P_M^\theta(\boldsymbol{u}_M^{\theta_0}) \tag{6.28}$$

be the MLE of θ, where $\boldsymbol{u}_M^{\theta_0}$ is the sample (6.27) that corresponds to the true parameter θ_0. Below we will give conditions for the consistency, asymptotic normality and efficiency of $\hat{\theta}_M$

Further we will consider only the stationary in time solution $u(t, \boldsymbol{r})$ of (6.2). This solution can be obtained by solving a Cauchy problem with an initial condition corresponding to an invariant distribution. In terms of the amplitudes (6.23), this invariant distribution is given by

$$u_m|_{t=0} \sim N\left(0, \frac{\sigma_m^2}{2\alpha_m(\theta)}\right) \tag{6.29}$$

In doing so, we assume that the operator A_θ is uniformly strictly positive, i.e.

$$\inf_m \underline{\alpha_m} \geq c_0 \tag{6.30}$$

for some positive constant c_0. The assumption of stationarity is made in order to simplify further computations and the final formulas. We believe that the main results remain true under reasonably smooth arbitrary initial conditions and without any positivity assumption.

Let us define

$$q_m(\theta) = \frac{\alpha_{1m}^2}{\alpha_m(\theta)^2} + h^2 \beta_{1m}^2 e^{-2h\alpha_m(\theta)}, \tag{6.31}$$

where

$$h = \min_{1 \le i \le N-1} (t_{i+1} - t_i). \tag{6.32}$$

Theorem 6.1. *Let (6.19) and (6.31) hold.*

1^0 *If*

(i)

$$\lim_{m \to \infty} \frac{\alpha_m \beta_{1m} e^{-2h\underline{\alpha}_m}}{\alpha_{1m}} = 0 \tag{6.33}$$

or

$$\limsup_{m \to \infty} \overline{\alpha}_m < \infty \tag{6.34}$$

(ii) for some $0 < \gamma < 1$

$$\lim_{M \to \infty} \frac{\max_{1 \le m \le M} q_m}{(\sum_1^M q_m)^{1-\gamma}} = 0 \tag{6.35}$$

where q_m *is given by (6.32),*

then the estimator $\hat{\theta}_M$ *is consistent with probability 1, asymptotically normal and efficient when* $M \to \infty$.

2^0 *If*

$$\lim_{M \to \infty} \langle |\hat{\theta}_M - \theta_0| \rangle = 0 \tag{6.36}$$

then

$$\sum_1^\infty q_m(\theta_0) = \infty. \tag{6.37}$$

Let us explain conditions (i) and (ii). First, since (6.19) and (6.31) are assumed to hold, condition (i) holds for all θ if it holds for some $\theta \in \Theta$. Condition (i) implies that either the contribution of the second term on the right-hand side of (6.32) is small or its asymptotic is determined by β_{1m}. In both cases we have

$$\sup_\theta q_m / \inf_\theta q_m \leq c_2. \tag{6.38}$$

Hence under (6.34), the relation (6.35) is true for all θ if it is true for one θ. For this reason we omit the variable θ in $\alpha_m(\theta)$ and $q_m(\theta)$ when the context allows no room for ambiguity.

Obviously, (6.35) implies (6.37). Moreover, we conjecture that (6.37) is necessary and sufficient for consistency and that the following condition (together with (6.19) and (6.31))

$$\lim_{M\to\infty} \frac{\max_{1\leq m\leq M} q_m}{\sum_1^M q_m} = 0 \tag{6.39}$$

is sufficient for asymptotic normality. As will be shown later, (6.39) is valid in the case of a single observation in time. More restrictive conditions (6.34)-(6.35) are needed for the derivation of consistency, asymptotic normality and efficiency in the general case.

For typical elliptic differential operators found in applications, (6.34) holds automatically and condition (6.37) is equivalent to (6.35). In particular, if A is an elliptic linear differential operator of order p_A with a self-adjoint main part then under wide conditions (Keldysh 1951, Markus and Matsaev 1979) the eigenfunctions of A form a complete orthogonal system in $L_2(G)$ and the asymptotics of its eigenvalues are given by

$$\text{Re } \lambda_m = O(m^{p_A/d}) \quad |\text{Im } \lambda_m| = o(|\text{Re } \lambda_m|). \tag{6.40}$$

If A_1 and A_θ are operators of this type then the sufficiency condition for consistency can be expressed in terms of their orders.

Theorem 6.2. *Let A_1, A_θ be elliptic differential operators of order p_1, p respectively, such that (6.19) and (6.40) hold. The following statements are equivalent*

(i)

$$\sum_m \frac{\alpha_{1m}^2}{\alpha_m^2} = \infty, \tag{6.41}$$

(ii)

$$p - p_1 \leq d/2, \tag{6.42}$$

(iii)

$$\lim_{M\to\infty} \langle |\hat{\theta}_M - \theta_0| \rangle = 0. \tag{6.43}$$

Each of these statements implies consistency, asymptotical normality and efficiency of the MLE of θ.

Let us apply the formulated results to the transport equation (6.3) in d-dimensional space. In this case

$$A_\theta = -D\nabla^2 + \boldsymbol{u}\cdot\nabla + \lambda \tag{6.44}$$

and for a wide class of boundary conditions

$$\alpha_m \sim \begin{cases} m^{2/d} & D > 0 \\ \text{const} & D = 0. \end{cases} \tag{6.45}$$

Let us estimate λ. In this case $\alpha_{1m} = \lambda = \text{const}$, $\beta_{1m} = 0$ and from Theorem 6.2 it follows that the MLE of λ is consistent in the case $D > 0$ if and only if

$$d \geq 4. \tag{6.46}$$

Of course, (6.46) also provides asymptotic normality and efficiency. If $D = 0$ then this estimator is consistent for any d in accordance with Theorem 6.1.

If D is the unknown parameter then $\alpha_{1m} \sim m^{2/d}$ and Theorems 6.1, 6.2 yield that the MLE of D is consistent, asymptotically normal and efficient for any d.

Finally, consider the problem of estimating one of the velocity components. Let us agree that in addition to (6.20), φ_m are numbered in such a way that for any large m_0 there is non-zero β_{1m} for $m > m_0$. In this case obviously $\alpha_{1m} = 0$ and the series

$$\sum_m \beta_{1m}^2 e^{-2h\alpha_m(\theta)} \tag{6.47}$$

converges if and only if $D > 0$. Thus, the condition

$$D = 0 \tag{6.48}$$

is necessary and sufficient for the consistency of the MLE of the velocity component if the other velocity components together with λ and D are given.

6.2 PROOF OF THE MAIN RESULTS FOR A SINGLE OBSERVATION IN TIME

Here we consider the case of single observation in time i.e. the sample

$$\{u_m(t_1), \quad m = 1, \ldots, M\} \tag{6.49}$$

is available for some $t_1 > t_0$. From condition (6.29) it follows that the distribution $u_m(t_1)$ depends only on the real part of the eigenvalues ofA_θ. In other words, the distribution of sample (6.49) corresponding to the operator $(A_\theta + A_\theta^*)/2$ is the same. For this reason this consideration can be restricted to a self-adjoint operator without any loss of generality. Therefore eigenvalues $\lambda_m(\theta) = \lambda_{0m} + \lambda_{1m}\theta$ are real and $u_m(t_1)$ is Gaussian with zero mean and variance

$$Var\{u_m(t_1)\} = \frac{1}{2\lambda_m(\theta)}\sigma_m^2. \tag{6.50}$$

Let P_M^θ be the measure generated by the sample (2.1) corresponding to the model(1) with parameter θ, and θ_0 be its true value. Using the sample independence, one can easily obtain that the likelihood ratio (the density of measure P_M^θ with respect to $P_M^{\theta_0}$) is given by

$$\frac{dP_M^\theta}{dP_M^{\theta_0}} = \exp\left\{-\frac{1}{2}\left[(\theta-\theta_0)\sum_{m=1}^{M} 2\frac{\lambda_{1m}u_m^2}{\sigma_m^2} - \sum_{m=1}^{M} \ln\frac{\lambda_m(\theta)}{\lambda_m(\theta_0)}\right]\right\}, \tag{6.51}$$

where $u_m = u_m(t_1)$.

Note that P_M^θ goes to measure P^θ generated by $u(t_1, \boldsymbol{r})$ when M tends to infinity. Let $\theta \neq \theta_0$. From the general results concerning absolute continuity of the Gaussian measures (Gihman and Skorokhod 1979) it follows that P^θ and P^{θ_0} are absolutely continuous if and only if

$$\sum_{m=1}^{\infty} \frac{\lambda_{1m}^2}{\lambda_m(\theta)\lambda_m(\theta_0)} < \infty. \tag{6.52}$$

In this case

$$\frac{dP^\theta}{dP^{\theta_0}} = \exp\left\{-\sum_{m=1}^{\infty}\left[(\theta-\theta_0)\frac{\lambda_{1m}u_m^2}{\sigma_m^2} - \frac{1}{2}\ln\frac{\lambda_m(\theta)}{\lambda_m(\theta_0)}\right]\right\}. \tag{6.53}$$

If (6.52) is fulfilled then

$$G_{\theta_0}(\theta) = (\theta - \theta_0) \sum_{m=1}^{\infty} \frac{\lambda_{1m}^2}{\lambda_m(\theta)\lambda_m(\theta_0)} \tag{6.54}$$

is a continuously differentiable function of $\theta \in \Theta$ for any fixed θ_0 and

$$\frac{dG_{\theta_0}(\theta)}{d\theta} \geq \delta > 0, \tag{6.55}$$

where δ is not dependent on θ or θ_0.

Indeed, from (6.52) and (6.19), which we rewrite for the sake of convenience

$$0 < C_1 < \lambda_m(\theta)/\lambda_m(\theta_0) < C_2, \quad \text{for all} \quad \theta, \theta_0, m, \tag{6.56}$$

it follows that the series on the right side of (6.54) converges uniformly with respect to $\theta \in \Theta$ and can be differentiated because the series

$$\frac{dG}{d\theta} = \sum_{m=1}^{\infty} \frac{\lambda_{1m}^2}{\lambda_m(\theta)^2} \tag{6.57}$$

converges uniformly as well. Thus, to prove (6.55) one can set

$$\delta = \inf_{\theta} \sum_{m=1}^{\infty} \frac{\lambda_{1m}^2}{\lambda_m(\theta)^2}. \tag{6.58}$$

In the following statement we set $\lambda_m = \lambda_m(\theta)$.

Theorem 6.3. *Suppose (6.56) holds. If*

$$\sum_{m=1} \frac{\lambda_{1m}^2}{\lambda_m^2} = \infty \tag{6.59}$$

then

(i)

$$\lim_{M \to \infty} \hat{\theta}_M = \theta_0 \quad \textit{with probability } 1. \tag{6.60}$$

If additionally

$$\lim_{M \to \infty} \frac{\max_{1 \leq m \leq M} (\lambda_{1m}^2/\lambda_m^2)}{\sum_1^M (\lambda_{1m}^2/\lambda_m^2)} = 0 \tag{6.61}$$

then

(ii)

$$(\hat{\theta}_M - \theta_0)\sqrt{2\sum_{m=1}^{M} \frac{\lambda_{1m}^2}{\lambda_m(\theta_0)^2}} \xrightarrow{\mathcal{D}} N(0,1)\,. \tag{6.62}$$

On the other hand, if the series $\sum_m \lambda_{1m}^2/\lambda_m^2$ *converges then* $\hat{\theta}_M$ *converges to the* ML *estimate* θ^* *with respect to* $\frac{dP^\theta}{dP^{\theta_0}}$, *given by*

$$\theta^* = G_{\theta_0}^{-1}\left(\sum_{m=1}^{\infty} \lambda_{1m}(u_m^2 - Eu_m^2)\right). \tag{6.63}$$

Proof. From (6.51) we get

$$\ln \frac{dP_M^\theta}{dP_M^{\theta_0}} = -(\theta - \theta_0)\sum_{m=1}^{M} \frac{\lambda_{1m} u_m^2}{\sigma_m^2} + \frac{1}{2}\sum_{m=1}^{M} \ln \frac{\lambda_m(\theta)}{\lambda_m(\theta_0)}. \tag{6.64}$$

Differentiating with respect to θ we obtain that $\hat{\theta}_M$ is the solution of equation

$$2\sum_{m=1}^{M} \frac{\lambda_{1m} u_m^2}{\sigma_m^2} = \sum_{m=1}^{M} \frac{\lambda_{1m}}{\lambda_m(\theta)}. \tag{6.65}$$

Let us put

$$\xi_m = \frac{u_m^2 - Eu_m^2}{\sqrt{Var(u_m^2 - Eu_m^2)}} = \frac{\sqrt{2}u_m^2 \lambda_m(\theta_0)}{\sigma_m^2} - \frac{1}{\sqrt{2}} \tag{6.66}$$

and note that

$$E\xi_m = 0, \quad E\xi_m^2 = 1, \quad E\xi_m^4 < \infty, \tag{6.67}$$

$$\frac{2u_m^2}{\sigma_m^2} = \frac{1 + \xi_m\sqrt{2}}{\lambda_m(\theta_0)}. \tag{6.68}$$

Substituting (6.68) into (6.65), we obtain the equivalent equation

$$\sum_{m=1}^{M} \frac{\lambda_{1m}\sqrt{2}}{\lambda_m(\theta_0)}\xi_m = (\theta - \theta_0)\sum_{m=1}^{M} \frac{\lambda_{1m}^2}{\lambda_m(\theta)\lambda_m(\theta_0)}, \tag{6.69}$$

where ξ_m are i.i.d. random variables satisfying properties (6.67). Let us divide both sides of (6.68) by $\sum_{m=1}^{M} \frac{2\lambda_{1m}^2}{\lambda(\theta_0)^2}$. As a result, we have

$$\frac{\sum_{m=1}^{M} \zeta_m}{\sum_{m=1}^{M} E\zeta_m^2} = (\theta - \theta_0) Q_M(\theta, \theta_0) , \tag{6.70}$$

where $\zeta_m = \frac{\lambda_{1m}\sqrt{2}}{\lambda_m(\theta_0)} \xi_m$ and

$$Q_M(\theta, \theta_0) = \frac{\sum_{m=1}^{M} \lambda_{1m}^2 / \lambda_m(\theta)\lambda_m(\theta_0)}{2\sum_{m=1}^{M} \lambda_m^2 / \lambda_m(\theta_0)^2} > \frac{1}{2C_2}. \tag{6.71}$$

Applying the law of large numbers to the left-hand side of (6.70) we obtain that $\hat{\theta}_M \to \theta_0$ with probability 1. Further from(6.70) we have

$$Q_M(\hat{\theta}_M, \theta_0)(\hat{\theta}_M - \theta_0)\sqrt{2\sum_{m=1}^{M} \frac{\lambda_{1m}^2}{\lambda(\theta_0)^2}} = \frac{\sum_{m=1}^{M} \zeta_m}{\sqrt{\sum_{m=1}^{M} E\zeta_m^2}}. \tag{6.72}$$

The Central Limit Theorem holds for the right-hand side of (6.72) since (6.61) is assumed. To prove (6.62) we will show that $Q_M(\hat{\theta}_M, \theta_0)$ goes to 1. Since $Q_M(\theta_0, \theta_0) = 1$, it is enough to prove that the derivative $\partial Q_M / \partial \theta$ is uniformly bounded. Indeed,

$$\frac{\partial Q_M}{\partial \theta} = -\frac{\sum_{m=1}^{M} \lambda_{1m}^3 / \lambda_m(\theta)^2 \lambda_m(\theta_0)}{2\sum_{m=1}^{M} \lambda_m^2 / \lambda_m(\theta_0)^2} \tag{6.73}$$

and hence

$$|\partial Q_M / \partial \theta| \le (2C_1)^{-1} \max_m (|\lambda_{1m}| / \lambda_m(\theta)). \tag{6.74}$$

Let $\theta_1 \in \Theta$ be arbitrary. Then

$$|\lambda_{1m}| / \lambda_m(\theta) = (\theta_1 - \theta)^{-1} |\lambda_m(\theta_1) / \lambda_m(\theta) - 1|. \tag{6.75}$$

Due to (6.56) it follows from (6.75) that

$$|\lambda_{1m}| / \lambda_m(\theta) < |\theta_1 - \theta|^{-1} (C_2 + 1). \tag{6.76}$$

Arbitrariness of θ_1 and (6.74) imply the uniform boundness of $|\partial Q_M / \partial \theta|$.

Let us prove the second part. Denote the right hand side of (6.69) by $G_M(\theta)$. Similarly to (6.55) one can get

$$\left|\frac{dG_M(\theta)}{d\theta}\right| \geq \delta_1 > 0. \tag{6.78}$$

Assume that the series $\sum_m \lambda_{1m}^2/\lambda_m^2$ converges and let M tend to infinity in (6.69). As a result we get

$$2\sum_1^\infty \frac{\lambda_{1m}\sqrt{2}}{\lambda_m(\theta_0)}\xi_m = G_M(\theta^*) + (\theta^* - \theta_0)\sum_M^\infty \frac{\lambda_{1m}^2}{\lambda_m(\theta^*)\lambda_m(\theta_0)}. \tag{6.79}$$

Subtracting (6.69) from (6.79) yields

$$2\sum_M^\infty \frac{\lambda_{1m}\sqrt{2}}{\lambda_m(\theta_0)}\xi_m - (\theta^* - \theta_0)\sum_M^\infty \frac{\lambda_{1m}^2}{\lambda_m(\theta^*)\lambda_m(\theta_0)} = G_M(\theta^*) - G_M(\hat{\theta}_M). \tag{6.80}$$

Since the left side of (6.80) goes to zero and (6.78) holds we conclude that

$$\lim_{M\to\infty} \hat{\theta}_M = \theta^* \quad \text{with probability 1} \tag{6.81}$$

and the theorem is proved. ∘

6.3 PROOF OF THE MAIN RESULTS IN THE GENERAL CASE

The above elementary computations fail in the general case. For this reason we use a powerful technique developed in (Ibragimov and Khasminskii 1981) which provides sufficient conditions for the consistency, asymptotic normality and efficiency of ML-estimates for arbitrarily distributed observations. These conditions are then rewritten for the particular case of the Gaussian observations.

Let $f_m(u,\theta)$, $m = 1, 2, \ldots$ be probability densities with respect to some measure $\nu(du)$ in a measurable space. We are interested in the ML estimate $\hat{\theta}_M$ of θ with respect to the density

$$f^{(M)}(u,\theta) = f_1(u,\theta)\ldots f_M(u,\theta)\,. \tag{6.82}$$

The Fisher information corresponding to the density $f^{(M)}(u,\theta)$ is

$$\psi^2(M,\theta) = \sum_{k=1}^{M} I_k(\theta) , \tag{6.83}$$

where $I_k(\theta)$ is the information corresponding to the density $f_k(u,\theta)$. Let

$$g_m(u,\theta) = f_m^{1/2}(u,\theta), \tag{6.84}$$

and

$$\eta_m(\theta) = \partial \ln f_m(u_m,\theta)/\partial\theta, \tag{6.85}$$

where u_m is the m-th observation. Further, let us denote $U_M(\theta) = \psi(M,\theta)(\Theta - \theta)$, where Θ is the parametric interval.

Theorem 6.4 *(Ibragimov and Khasminskii 1981). If*

1)

$$\limsup_{M\to\infty} \frac{\sup_\theta \psi^2(M,\theta)}{\inf_\theta \psi^2(M,\theta)} < \infty, \tag{6.86}$$

2) for some $\delta > 0$

$$\lim_{M\to\infty} \sup_\theta \psi(M,\theta)^{-2-\delta} \sum_1^M \langle \eta_m^{2+\delta} \rangle = 0, \tag{6.87}$$

3) exists $z(M) \to \infty$ *such that*

$$\sup_\theta \sup_{\substack{|v|<z(M)\\ v\in U_M(\theta)}} \psi^{-4}(M,\theta) \sum_{m=1}^{M} \int \left(\frac{\partial^2 g_m(x,\theta+\psi^{-1}(M,\theta)v)}{\partial\theta} \right)^2 \nu(dr) =$$
$$o(z^{-2}(M)), \tag{6.88}$$

4) for some constant $\beta > 0$

$$\inf_{\theta\in\Theta} \inf_{\substack{|v|>z(M)\\ v\in U_M(\theta)}} |\psi(M,\theta)|^{-\beta} \sum_{m=1}^{M} \int \left[g_m(x,\theta+\psi^{-1}(M,\theta)v) - g_m(x,\theta)\right]^2 \nu(dr) > 0\,; \tag{6.89}$$

then

(i) $\lim_{M\to\infty} \hat{\theta}_M = \theta_0$ *with probability 1.*

(ii) The random value $\psi(M,\theta_0)(\hat{\theta}_M - \theta_0)$ *is asymptotically normal with mean 0 and unit variance uniformly in* Θ.

(iii) The estimate $\hat{\theta}_M$ *is asymptotically efficient for any loss function of the form* $w(\psi(M,\theta)v)$ *having a polynomial majorant.*

Under Gaussian observations the conditions of Theorem 6.4 can be reduced to constraints on the Fisher information.

Lemma 6.1. *If*

$$f_m(\boldsymbol{u},\theta) = \frac{1}{(2\pi)^{N/2}\sqrt{|\boldsymbol{R}_m(\theta)|}} e^{-\frac{1}{2}(\boldsymbol{u}^T \boldsymbol{R}_m^{-1}(\theta)\boldsymbol{u})} \tag{6.90}$$

where $\boldsymbol{u} = (u_1,\ldots,u_N)$, *and* $\{\boldsymbol{R}_m(\theta)\}$, $m = 1,2,\ldots$ *is a sequence of positive, continuously differentiable in* θ *matrices such that the corresponding Fisher informations satisfy the following*

(i)

$$\frac{\sup_\theta \ I_m(\theta)}{\inf_\theta \ I_m(\theta)} < C. \tag{6.91}$$

(ii) For some $0 < \gamma < 1$

$$\frac{\sup_\theta \ \max_{1\le m\le M} \ I_m(\theta)}{\inf_\theta \left(\sum_{m=1}^M \ I_m(\theta)\right)^{1-\gamma}} \to 0 \text{ as } M \to \infty \tag{6.92}$$

(iii)

$$J_m(\theta) \le C_0 \ I_m(\theta), \tag{6.93}$$

where

$$J_m(\theta) = 2Sp\left(\frac{d^2\boldsymbol{R}_m^{-1}}{d\theta^2}\boldsymbol{R}_m\right)^2 + \left[Sp\left(\frac{d\boldsymbol{R}_m^{-1}}{d\theta}\frac{d\boldsymbol{R}_m}{d\theta}\right)\right]^2 \tag{6.94}$$

then the conditions (1) through (4) of Theorem 2.1 are fulfilled.

Here $Sp(A) = \sum a_{ii}$ for matrix $A = (a_{ij})$. Before the proof let us notice that (6.92) implies

$$\psi^2(M,\theta) = \sum_{k=1}^{M} I_k(\theta) \to \infty. \tag{6.95}$$

Then one can see that, in the Gaussian case, η_m in (6.85) is given by

$$\eta_m = -\frac{1}{2}\frac{\dot{D}_m(\theta)}{D_m(\theta)} - \frac{1}{2}\, \boldsymbol{u}_m^T\, \dot{\boldsymbol{R}}_m^{-1}(\theta)\boldsymbol{u}_m, \tag{6.96}$$

where the dot indicates differentiation with respect to θ and $D(\theta) = |\boldsymbol{R}(\theta)|$ is the determinant.

Finally, it can be easily checked that

$$\langle\eta_m\rangle = -\frac{1}{2}\frac{\dot{D}_m}{D_m} - \frac{1}{2}\, Sp(\dot{\boldsymbol{R}}_m^{-1}\boldsymbol{R}_m) = 0. \tag{6.97}$$

Therefore, the equation for the maximum likelihood estimator of θ is expressed in the form

$$\sum_{m=1}^{M} \boldsymbol{u}_m^T \dot{\boldsymbol{R}}_m^{-1}(\theta)\boldsymbol{u}_m = \sum_{m=1}^{M} Sp(\dot{\boldsymbol{R}}_m^{-1}(\theta)\boldsymbol{R}_m(\theta)). \tag{6.98}$$

The computations below are nothing more than an investigation of the solution to (6.98). It should be stressed that in the case of single observation discussed above the left-had side of (6.98) does not include θ at all. This point crucially simplifies the investigation. In the general situation this direct approach is impossible.

Proof of Lemma 6.1. Obviously (6.86) follows from (6.90). Thus, we proceed to check (6.87). First, let us show that there exists a constant C_1 such that

$$\langle\eta^4\rangle \le C_1\langle\eta^2\rangle^2, \tag{6.99}$$

where the subscript m is omitted for the sake of brevity.

Let $\nu_1, \ldots, \nu_N$ be the eigenvalues of $\dot{\boldsymbol{R}}^{-1}\boldsymbol{R}$ and $S_j = \nu_1^j + \cdots + \nu_N^j$, $j = 1, 2, \ldots$. Then from (6.56) it follows that $\dot{D}(\theta)/D(\theta) = -S_1$ and hence $\tilde{\eta} \equiv -2\eta + S_1 = \boldsymbol{u}^T\, \dot{\boldsymbol{R}}^{-1}\boldsymbol{u}$ which is a quadratic form of the Gaussian vector u. It is well known that its characteristic function $\Phi(t) = e^{it\tilde{\eta}}$ is given by

$$\Phi(t) = \prod_{n=1}^{N} \frac{1}{\sqrt{1 - 2i\nu_n t}}. \tag{6.100}$$

Straightforward computations yield

$$\langle\tilde{\eta}^2\rangle = S_1^2 + 2S_2,\ \langle\tilde{\eta}^3\rangle = S_1^3 + 14S_3,$$
$$\langle\tilde{\eta}^4\rangle = S_1^4 - 3S_1^2S_2 + 17S_1S_3 + 9S_2^2 + 81S_4. \tag{6.101}$$

From the latter one can find

$$\langle\eta^4\rangle = \frac{1}{16}\langle(\tilde{\eta} - S_1)^4\rangle = 81S_4 + 9S_1^2S_2 - 39S_1S_3 + 9S_2^2. \tag{6.102}$$

Repeatedly using the Cauchy inequality, we arrive at (6.89), where C_1 is dependent on N only.

Setting $\delta = 2$ in (6.87) and noting that $\langle\eta_m^2\rangle = I_m$, we obtain

$$\frac{\sum_{m=1}^M \langle\eta_m^4\rangle}{\psi^4(M,\theta)} \le \frac{C_1 \sum_1^M I_m^2}{\psi^4(M,\theta)} \le \frac{C_1 \max_{1\le m\le M} I_m}{\psi^2(M,\theta)}. \tag{6.103}$$

Thus (6.87) follows readily from condition (6.92) and remark (6.95). Now we will check (6.88), where $\nu(d\boldsymbol{r}) \equiv d\boldsymbol{r}$ is now Lebesgue measure in R^{2N}. Let us set

$$z(M) = \inf_\theta \psi(M,\theta)^\gamma \tag{6.104}$$

and note that

$$\int \left(\frac{\partial^2 g}{\partial\theta^2}(x,\theta)\right)^2 d\boldsymbol{r} = \frac{1}{16}\langle(2\eta_\theta + \eta^2)\rangle^2 \le$$
$$\le \frac{1}{2}\langle\eta_\theta^2\rangle + \frac{1}{8}\langle\eta^4\rangle, \tag{6.105}$$

where η_θ is the derivative of η with respect to θ.

One can check that $\langle\eta_\theta^2\rangle = \frac{1}{4}J_m$ where J is given by (6.94). Then it follows from (6.93) and (6.105) that

$$\int \left(\frac{\partial^2 g_m}{\partial\theta^2}(x,\theta)\right)^2 d\boldsymbol{r} \le C_2 I_m^2(\theta) \tag{6.106}$$

for some constant C_2. Hence the left hand side of (6.88) is less than

$$C_2 \sup_\theta \sup_{\theta_1} \psi^{-4}(M,\theta) \sum_{m=1}^M I_m^2(\theta_1) \le$$
$$\le C_2 \frac{\sup_\theta \max_{1\le m\le M} I_m(\theta)}{\inf_\theta \psi^2(M,\theta)} = o(z^{-2}(M)) \tag{6.107}$$

due to (6.104) and condition (6.92). Finally let us check (6.89). Taking into account the relation

$$\int \left(\frac{\partial g_m}{\partial \theta}\right)^2 dr = \frac{1}{4} I_m(\theta), \tag{6.108}$$

we have

$$\begin{aligned} &\sum_{m=1}^{M} \int \left[g_m(x, \theta + \psi^{-1}(M,\theta)v) - g_m(x,\theta)\right]^2 dr \geq \\ &\tfrac{1}{16}\psi^{-2}(M,\theta)v^2 \sum_{m=1}^{M} I_m(\theta_m) \geq C_3 v^2 \end{aligned} \tag{6.109}$$

The latter inequality follows from (6.91). Consequently for $|v| > z(M)$ the left-hand side of (6.109) is bigger than $C_3 z^2(M)$. By choosing $\beta = 2\gamma$ we find that condition (6.89) is fulfilled.

Lemma 6.1 is proved.∘

Lemma 6.2. *Let (6.91) and (6.93) hold. If*

$$\lim_{M\to\infty} \langle |\hat{\theta}_M - \theta_0| \rangle = 0, \tag{6.110}$$

where $\hat{\theta}_M$ is the MLE corresponding to the Gaussian distribution (6.90) then

$$\psi^2(M,\theta) \to \infty. \tag{6.111}$$

Proof. Due to (6.85) the equation for the MLE can be written in the form

$$F_M(\theta) \equiv \eta_1(\theta) + \ldots + \eta_M(\theta) = 0 \tag{6.112}$$

By applying the Cauchy inequality to the following trivial relation

$$F_M(\theta_0) = F_M(\theta_0) - F_M(\hat{\theta}_M) = \int_{\hat{\theta}_M}^{\theta_0} F_M'(\theta) d\theta, \tag{6.113}$$

we have

$$|F_M(\theta_0)| \leq |\hat{\theta}_M - \theta_0|^{1/2} \left(\int_\Theta (F_M'(\theta))^2 d\theta\right)^{1/2}, \tag{6.114}$$

from which it follows that

$$\langle |F_M(\theta_0)| \rangle \leq (\langle |\hat{\theta}_M - \theta_0| \rangle)^{1/2} \left(\int_\Theta \langle F_M'(\theta_0)^2 \rangle d\theta\right)^{1/2}. \tag{6.115}$$

Then we obtain from (6.93)

$$\langle F_M'(\theta)^2\rangle = \langle(\sum_1^M \partial\eta_m/\partial\theta)^2\rangle = \frac{1}{2}\sum_1^M J_m(\theta) \leq C_0\psi^2(M,\theta). \tag{6.116}$$

Further

$$\langle F_M(\theta_0)^2\rangle = \sum_1^M \langle\eta_m^2\rangle = \psi^2(M,\theta) \tag{6.117}$$

and from (6.99)

$$\langle F_M(\theta_0)^4\rangle \leq C\sum_1^M I_m^2. \tag{6.118}$$

Assume that (6.110) holds, but

$$\lim_{M\to\infty} \psi^2(M,\theta) < \infty. \tag{6.119}$$

From (6.115) it follows that

$$\lim_{M\to\infty} \langle|F_M(\theta_0)|\rangle = 0, \tag{6.120}$$

but this contradicts

$$\lim_{M\to\infty} \langle F_M(\theta_0)^2\rangle > 0, \quad \lim_{M\to\infty} \langle F_M(\theta_0)^4\rangle < \infty, \tag{6.121}$$

from (6.117) and (6.118) respectively. This contradiction proves Lemma 6.2. ∘

The proof of Theorem 6.1 is based on an explicit expression for the Fisher information and the quantity $J_m(\theta)$ defined in Lemma 6.1. To derive these expressions, let $u(t)$ be a complex-valued Ornstein-Uhlenbeck process, i.e. $u(t)$ is stationary and satisfies

$$\dot{u}(t) + ((\alpha_1 + i\beta_1)\theta + \alpha_0 + i\beta_0)\, u(t) = \sigma\left(\dot{w}^1(t) + i\dot{w}^2(t)\right) \tag{6.122}$$

where $w^1(t), w^2(t)$ are independent Wiener processes. Let $\boldsymbol{u}(t) = (u^1(t), u^2(t))$, where u^1, u^2 are the real and imaginary part of u respectively.

Lemma 6.3. *The Fisher information due to the following sample from (6.122)*

$$\{\boldsymbol{u}(t_1), \ldots, \boldsymbol{u}(t_N)\} \tag{6.123}$$

is given by

$$I(\theta) = \frac{\alpha_1^2}{\alpha(\theta)^2}\varphi(t_1, t_2, \ldots, t_N; \alpha(\theta)) + \beta_1^2\psi(t_1, t_2, \ldots, t_N; \alpha(\theta)), \tag{6.124}$$

where

$$\alpha(\theta) = \alpha_0 + \alpha_1\theta \tag{6.125}$$

and

$$\varphi(t_1, \ldots t_N; \boldsymbol{r}) = \sum_{i=1}^{N} (1 - g(x_{i-1}) - g(x_i))^2$$

$$+2\sum_{j>i} (f(x_i) - g(x_{i-1}))\,(f(x_{j-1}) - g(x_j))\,e^{-2x(t_j - t_i)},$$

$$\psi(t_1, \ldots, t_N; \boldsymbol{r}) = 2\sum_{j>i} e^{-2x(t_j-t_i)}\,(f(x_i)$$

$$-g(x_{i-1}))\,(f(x_{j-1}) - g(x_j)) - \sum_{i=1}^{N} (g(x_i) - g(x_{i-1}))^2 . \tag{6.126}$$

Here $f(u) = u/(1 - e^{-2u})$, $g(u) = f(u)e^{-2u}$, $x_i = x\Delta t_i$, $\Delta t_i = t_{i+1} - t_i$, $i = 1, \ldots, N-1$, and for convenience we set $\Delta t_0 = \Delta t_{N+1} = \infty$.

In particular when $\Delta t_i = h$ are the same, we obtain

$$I(\theta) = \frac{\alpha_1^2}{\alpha(\theta)^2}\Big\{1 + (N-1)\Big[\left(\frac{1 - e^{-2h\alpha(\theta)} - 2h\alpha(\theta)e^{-2h\alpha(\theta)}}{1 - e^{-2h\alpha(\theta)}}\right)^2$$

$$+\frac{2(h\alpha(\theta))^2 e^{-2h\alpha(\theta)}}{1 - e^{-2h\alpha(\theta)}}\Big]\Big\} + \beta_1^2 h^2 (N-1)\frac{e^{-2\alpha(\theta)}}{1 - e^{-2\alpha(\theta)}}. \tag{6.127}$$

Let us denote by $\boldsymbol{R}$ the $2N \times 2N$-correlation matrix of the vector (6.123).

Lemma 6.4. *The following equalities hold*

$$\frac{1}{2}Sp\left(\frac{d^2\boldsymbol{R}^{-1}}{d\theta^2}\boldsymbol{R}\right)^2 =$$

$$\frac{1}{\alpha(\theta)^4}\left[\sum_{i=1}^{N}(A_i^2 - B_i^2) + 2\sum_{j>i} e^{-2\alpha(\theta)(t_j - t_i)}(C_iE_j + D_iF_j)\right], \tag{6.128}$$

where

$$A_i = \alpha_1^2 \left(f_A(x_i) e^{-2x_i} + f_A(x_{i-1}) e^{-2x_{i-1}} \right)$$
$$+ \beta_1^2 \left(g_A(x_i) e^{-2x_i} + g_A(x_{i-1}) e^{-2x_{i-1}} \right),$$

$$f_A(x) = \frac{x(3x - 2 + 2e^{-2x} + xe^{-2x})}{(1 - e^{-2x})^2}, \quad g_A(x) = \frac{x^2}{1 - e^{-2x}},$$

$$B_i = 2\alpha_1 \beta_1 \left(f_B(x_i) e^{-2x_i} - f_B(x_{i-1}) e^{-2x_{i-1}} \right),$$

$$f_B(x) = \frac{x(1 - x - e^{-2x} - xe^{-2x})}{(1 - e^{-2x})^2},$$

$$C_i = \alpha_1^2 \left(f_C(x_i) + e^{-2x_{i-1}} f_A(x_{i-1}) \right) + \beta_1^2 \left(g_A(x_i) + e^{-2x_{i-1}} g_A(x_{i-1}) \right),$$

$$f_C(x) = \frac{x(2 - 2e^{-2x} - x - 3xe^{-2x})}{(1 - e^{-2x})^2},$$

$$D_i = 2\alpha_1 \beta_1 \left(f_B(x_i) - f_B(x_{i-1}) e^{-2x_{i-1}} \right),$$

$$E_i = \alpha_1^2 \left(f_C(x_{i-1}) + e^{-2x_i} f_A(x_i) \right) + \beta_1^2 \left(g_A(x_{i-1}) + e^{-2x_i} g_A(x_i) \right),$$

$$F_i = 2\alpha_1 \beta_1 \left(f_B(x_{i-1}) - f_B(x_i) e^{-2x_i} \right), \tag{6.129}$$

and

$$Sp\left(\frac{d\boldsymbol{R}^{-1}}{d\theta} \frac{d\boldsymbol{R}}{d\theta} \right) = -\frac{4}{\alpha(\theta)^2} \left[\alpha_1^2 \sum_{i=1}^{N-1} g_A(x_i) e^{-2x_i} + \right.$$
$$\left. + \beta_1^2 \sum_{i=1}^{N-1} f_B(x_i) e^{-2x_i} \left(1 + \frac{1}{x_i} \right) \right]. \tag{6.130}$$

Notice that the following upper bound holds

$$|f_A(x)|, \; |f_B(x)|, \; |f_C(x)|, \; |g_A(x)| \leq c_1 x^2 + c_2 \tag{6.131}$$

for some positive constants c_1, c_2.

Lemmas 6.3 and 6.4 are proven by straightforward computations. We show only some intermediate results.

Proof. First, the information $I(\theta)$ for the Gaussian vector $\boldsymbol{u} \sim N(0, \boldsymbol{R}(\theta))$

with non-degenerate correlation matrix $\boldsymbol{R}(\theta)$ can be expressed in the form

$$I(\theta) = \frac{1}{2} Sp \left(\left(\frac{d\boldsymbol{R}}{d\theta} \right)^{-1} \boldsymbol{R} \right)^2 \tag{6.132}$$

The correlation matrix of the vector process $\boldsymbol{u}(t)$ defined above in (6.122) is given by

$$\langle (\boldsymbol{u}(s)\boldsymbol{u}(s+t)^T) \rangle = \frac{\sigma^2}{2\alpha(\theta)} e^{-\alpha(\theta)|t|} \boldsymbol{V}(\beta(\theta)t), \tag{6.133}$$

where

$$\boldsymbol{V}(\beta) = \begin{pmatrix} \cos\beta & -\sin\beta \\ \sin\beta & \cos\beta \end{pmatrix} \tag{6.134}$$

is an orthogonal matrix and $\beta(\theta) = \beta_1\theta + \beta_0$. In particular

$$\boldsymbol{R}_{ij} = \langle (\boldsymbol{u}(t_i)\boldsymbol{u}(t_j)^T) \rangle = \frac{\sigma^2}{2\alpha(\theta)} e^{-\alpha(\theta)|t_i - t_j|} \boldsymbol{V}(\beta(\theta)(t_j - t_i)). \tag{6.135}$$

Therefore $\boldsymbol{R} = (\boldsymbol{R}_{ij})$, and it can easily be shown that the inverse matrix $\boldsymbol{R}^{-1}$ consists of the 2×2 blocks given by

$$(\boldsymbol{R}^{-1})_{ij} = \frac{2\alpha}{\sigma^2} \Bigg\{ \delta_{ij} \left(\frac{1}{1 - e^{-2\alpha\Delta t_{i-1}}} + \frac{1}{1 - e^{-2\alpha\Delta t_i}} - 1 \right) \boldsymbol{I} - \\ -\delta_{i,j-1} \frac{e^{-\alpha\Delta t_i}}{1 - e^{-2\alpha\Delta t_i}} \boldsymbol{V}(\beta\Delta t_i) - \delta_{i,j+1} \frac{e^{-\alpha\Delta t_{i-1}}}{1 - e^{-2\alpha\Delta t_{i-1}}} \boldsymbol{V}^T(\beta\Delta t_{i-1}) \Bigg\}, \tag{6.136}$$

where the dependence of α and β on θ is implied.

Differentiating of (6.136) with respect to θ yields

$$\left(\frac{d^\nu \boldsymbol{R}^{-1}}{d\theta^\nu} \right)_{ij} = \\ \frac{2}{\sigma^2} \left[\delta_{ij} \boldsymbol{I} a_i^{(\nu)} - \delta_{i,j-1} \left(b_i^{(\nu)} \boldsymbol{V}(\beta\Delta t_i) + c_i^{(\nu)} \dot{\boldsymbol{V}}(\beta\Delta t_i) \right) - \\ -\delta_{i,j+1} \left(b_j^{(\nu)} \boldsymbol{V}^T(\beta\Delta t_j) + c_j^{(\nu)} \dot{\boldsymbol{V}}^T(\beta\Delta t_j) \right) \right], \tag{6.137}$$

where

$$\dot{\boldsymbol{V}}(\beta) = \begin{pmatrix} -\sin\beta & -\cos\beta \\ \cos\beta & -\sin\beta \end{pmatrix},$$

$$a_i^{(1)} = \alpha_1 \left(f'(x_{i-1}) + f'(x_i) - 1\right),$$
$$b_i^{(1)} = \alpha_1 h'(x_i), \quad c_i^{(1)} = \beta_1 h(x_i), \tag{6.138}$$
$$f(x) = x/(1-e^{-2x}), \quad h(x) = f(x)e^{-x}$$

and

$$a_i^{(2)} = \frac{\alpha_1^2}{\alpha(\theta)}\left(x_{i-1}f''(x_{i-1}) + x_i f''(x_i)\right),$$
$$b_i^{(2)} = \frac{1}{\alpha(\theta)}\left(\alpha_1^2 x_i h''(x_i) - \beta_1^2 x_i h(x_i)\right), \tag{6.139}$$
$$c_i^{(2)} = \frac{2\alpha_1\beta_1}{\alpha(\theta)} h'(x_i).$$

From (6.137) it follows that

$$\left(\frac{d^\nu \boldsymbol{R}^{-1}}{d\theta^\nu}\boldsymbol{R}\right)_{ii} = \frac{1}{\alpha(\theta)}\begin{pmatrix} a_i - b_i e^{-x_i} - b_{i-1}e^{-x_{i-1}} & c_i e^{-x_i} - c_{i-1}e^{-x_{i-1}} \\ -c_i e^{-x_i} + c_{i-1}e^{-x_{i-1}} & a_i - b_i e^{-x_i} - b_{i-1}e^{-x_{i-1}} \end{pmatrix},$$

$$\left(\frac{d^\nu \boldsymbol{R}^{-1}}{d\theta^\nu}\boldsymbol{R}\right)_{ij} = \frac{1}{\alpha(\theta)}e^{-\alpha(\theta)(t_j-t_i)}\left[\left(a_i - b_i e^{x_i} - b_{i-1}e^{-x_{i-1}}\right)\right.$$
$$\left.\times \boldsymbol{V}\left(\beta(t_j-t_i)\right) + \left(c_i e^{x_i} - c_{i-1}e^{-x_{i-1}}\right)\dot{\boldsymbol{V}}\left(\beta(t_j-t_i)\right)\right], \quad i<j$$

$$\left(\frac{d^\nu \boldsymbol{R}^{-1}}{d\theta^\nu}\boldsymbol{R}\right)_{ij} = \frac{1}{\alpha(\theta)}e^{-\alpha(\theta)(t_i-t_j)}\left[\left(a_i - b_i e^{-x_i} - b_{i-1}e^{x_{i-1}}\right)\right. \tag{6.140}$$
$$\left.\times \boldsymbol{V}\left(\beta(t_j-t_i)\right)\left(c_{i-1}e^{x_{i-1}} - c_i e^{-x_i}\right)\dot{\boldsymbol{V}}\left(\beta(t_j-t_i)\right)\right], \quad j<i$$

where for the sake of convenience the superscripts on a, b, c have been dropped. From the latter formulas it is not difficult to derive (6.124) and (6.128).

To obtain (6.130) one can use (6.137) and the following

$$\begin{aligned}
\left(\frac{d\boldsymbol{R}}{d\theta}\right)_{ii} &= -\frac{\sigma^2}{2\alpha^2(\theta)}\alpha_1 \boldsymbol{I} \\
\left(\frac{d\boldsymbol{R}}{d\theta}\right)_{i-1,\, i} &= \frac{\sigma^2 x_{i-1}}{2\alpha(\theta)}\left[\alpha_1 s'(x_{i-1})\boldsymbol{V}(\beta\Delta t_{i-1}) + \beta_1 s(x_{i-1})\dot{\boldsymbol{V}}(\beta\Delta t_{i-1})\right], \\
\left(\frac{d\boldsymbol{R}}{d\theta}\right)_{i,\, i+1} &= \frac{\sigma^2 x_i}{2\alpha(\theta)}\left[\alpha_1 s'(x_i)\boldsymbol{V}^T(\beta\Delta t_i) + \beta_1 s(x_i)\dot{\boldsymbol{V}}^T(\beta\Delta t_i)\right],
\end{aligned} \tag{6.141}$$

where $s(x) = e^{-x}/x$. ∘

Proof of Theorem 6.1. 1^0 The first part will be deduced from Lemma 6.1. For each mode $u_m(t)$ given by (6.23) we have from Lemma 6.3

$$I_m(\theta) = \frac{\alpha_{1m}^2}{\alpha_m(\theta)^2}\varphi(t_1, t_2, \ldots t_N, \alpha_m(\theta)) + \beta_{1m}^2\psi(t_1, \ldots, t_N; \alpha_m(\theta)) \,. \tag{6.142}$$

Under conditions (6.19), (6.31) and from explicit expressions (6.126), one can conclude that

$$\begin{aligned}
0 < c_1 \leq |\varphi(t_1, t_2, \ldots, t_N; \alpha_m(\theta)| \leq c_2, \\
0 < c_3 \leq |\psi(t_1, t_2, \ldots, t_N; \alpha_m(\theta)| e^{2h\alpha_m(\theta)} \leq c_4
\end{aligned} \tag{6.143}$$

uniformly in $\theta \in \Theta$ for fixed $t_1, \ldots, t_N$.

Therefore

$$I_m \sim q_m, \tag{6.144}$$

when m goes to infinity, where q_m is given by (6.32), and hence (6.91) is the same as (6.38). In turn (6.38) follows from the conditions of Theorem 1.1. Also from (6.38) along with (6.35) it follows that (6.92) holds.

Let us check the last condition (6.93) of Lemma 6.1. From the explicit expressions (6.128) and (6.130) and estimates (6.131) it follows that

$$J_m(\theta) \leq c_5(\alpha_{1m}^2 + \beta_{1m}^2)^2 e^{-2h\alpha_m(\theta)}. \tag{6.145}$$

Due to (6.35) this inequality implies (6.93). Thus, the assertion of Theorem 1.1 follows readily from Lemma 3.1.

2^0. The second part of Theorem 6.1 follows from Lemma 6.2 since (6.144) holds.$\circ$

Proof of Theorem 6.2. For elliptic differential operators satisfying (6.41), the equivalence of (i) and (ii) is obvious. From the second part of Theorem 6.1 it follows that (iii)$\Rightarrow$ (i). Finally, since conditions (6.34), (6.35) are satisfied under assumption (6.41), the first part of Theorem 1.1 yields (i)$\Rightarrow$(iii). Theorem 6.2 is proved.

6.4 SUMMARY

The problem of estimating the unknown parameter in the evolution equation (6.5) based on discrete observations of the first M Fourier coefficients of $u(t, \boldsymbol{r})$ is considered. Necessary and sufficient conditions are given for the consistency, asymptotic normality and efficiency of the maximum likelihood estimator when $M \to \infty$ in terms of the eigenvalues of the linear operators A_0 and A_1, which are not assumed to be self-adjoint. For a wide class of elliptic differential operators these conditions are reformulated in terms of their orders.

It is shown that the diffusivity estimator in the advection-diffusion equation is always consistent, while the consistency of the feedback factor estimator holds iff $d \geq 4$, where d is the spatial dimension. In particular, in the most important case for applications, $d = 2$, this estimator is not consistent.

The MLE of the velocity is consistent if and only if the diffusivity is zero regardless to the spatial dimension.

7

MAXIMUM LIKELIHOOD ESTIMATORS: NUMERICAL SIMULATIONS

The theory presented in the previous chapter describes the behavior of the ML estimator only for a large number of observed modes. These asymptotic limits are too idealistic to be attained in reality. For this reason it is important to investigate numerically ML estimators under conditions where tracer data are limited in both time and space. In our simulations we concentrate on the 2-dimensional transport equation

$$\frac{\partial c}{\partial t} + \boldsymbol{u} \cdot \nabla c + \lambda c = \nabla \cdot \boldsymbol{D} \nabla c + S(t, \boldsymbol{r}), \tag{7.1}$$

where $\boldsymbol{u} = (u, v), \boldsymbol{r} = (x, y), \nabla = (\partial/\partial x,\ \partial/\partial y)$, and

$$\boldsymbol{D} = \begin{bmatrix} D_{xx} & D_{xy} \\ D_{yx} & D_{yy} \end{bmatrix}. \tag{7.2}$$

Below, we give an explicit form of the equations for the ML estimators of $\lambda, u, v, D_{xx}, D_{xy}, D_{yx}$, and D_{yy} when the tracer is observed over a rectangle with periodic boundary conditions (section 7.1). A description of the numerical procedure is given in section 7.2. Results are discussed in section 7.3.

7.1 CALCULUS

Suppose that the region where one observes the tracer distribution $c(t, \boldsymbol{r})$ is a rectangle with the center at the origin and with sides $2a$ and $2b$, along the x- and y-axes respectively

$$G = \{x, y : |x| < a, |y| < b\}. \tag{7.3}$$

Let $S(t, \boldsymbol{r})$ be white noise in time and periodically homogeneous random field in spatial coordinate. The latter means that $S(t, -a, y) = S(t, a, y)$, $S(t, x, -b) = S(t, x, b)$ for all x, y and $\langle S(t, x_1, y_1) S(t, x_2, y_2) \rangle$ is a function of the differences $x_1 - x_2$, $y_1 - y_2$ only. No loss of generality results by taking the coordinate axes to be directed along the axes of the correlation ellipse of the noise $S(t, x, y)$. From the above assumptions it follows that $S(t, \boldsymbol{r})$ can be written in the form

$$S(t, x, y) = \sum_{m,n=0}^{\infty} (S_{m,n}^{(1)}(t) \varphi_{m,n}^{(1)}(x, y) + \ldots + S_{m,n}^{(4)}(t) \varphi_{m,n}^{(4)}(x, y)), \tag{7.4}$$

where the orthogonal basis is given by

$$\begin{aligned} \varphi_{m,n}^{(1)}(x, y) &= \cos \frac{m\pi x}{a} \cos \frac{n\pi y}{b}, \quad \varphi_{m,n}^{(2)}(x, y) = \cos \frac{m\pi x}{a} \sin \frac{n\pi y}{b}, \\ \varphi_{m,n}^{(3)}(x, y) &= \sin \frac{m\pi x}{a} \cos \frac{n\pi y}{b}, \quad \varphi_{m,n}^{(4)}(x, y) = \sin \frac{m\pi x}{a} \sin \frac{n\pi y}{b}, \end{aligned} \tag{7.5}$$

$m, n = 0, 1, 2, \ldots$, and all $S_{m,n}^{(j)}(t)$ are independent white noise processes, i.e.

$$\langle S_{m,n}^{(j)}(t) S_{m+p,n+q}^{(l)}(t + s) \rangle = \sigma_{m,n}^2 \delta(s) \delta_{jl} \delta_{0p} \delta_{0q}, \tag{7.6}$$

where δ_{jl} is the Kronecker delta, $\delta(s)$ is the Dirac delta function, and s is the time lag. As follows from (7.4) - (7.6), the correlation function of S has the following representation

$$\langle S(t, x_1, y_1) S(t + s, x_2, y_2) \rangle = \delta(s) \sum_{m,n=0}^{\infty} \sigma_{m,n}^2 \cos \frac{m\pi(x_2 - x_1)}{a} \cos \frac{n\pi(y_2 \mp y_1)}{b}. \tag{7.7}$$

Next, consider the solution of (7.1) which is a statistically stationary process in time and expand it in terms of the basis (7.4)

$$c(t, x, y) = \sum_{m,n=0}^{\infty} \sum_{j=1}^{4} C_{m,n}^{(j)}(t) \varphi_{m,n}^{(j)}(x, y). \tag{7.8}$$

The stationarity of the amplitudes $C_{m,n}^{(j)}(t)$ as well as the stationarity of $c(t, x, y)$ itself are provided by the following initial conditions

$$C_{m,n}^{(j)}(0) \sim N(0, \sigma_{m,n}^2 / 2r_{m,n}), \tag{7.9}$$

where

$$r_{m,n} = \pi^2 (\frac{m^2}{a^2} D_{xx} + \frac{mn}{ab} D_{xy} + \frac{mn}{ab} D_{yx} + \frac{n^2}{b^2} D_{yy}) + \lambda. \tag{7.10}$$

In other words, the initial values of the amplitudes $C^{(j)}_{m,n}(0)$ are Gaussian with zero mean and variance given in (7.9).

Let us denote the vector of amplitudes as

$$\boldsymbol{C}_{m,n}(t) = (C^{(1)}_{m,n}(t), \ldots, C^{(4)}_{m,n}(t)). \tag{7.11}$$

If the time interval Δt is fixed then the discrete records of this vector

$$\boldsymbol{C}_{m,n}(i) = \boldsymbol{C}_{m,n}(i\Delta t), \quad i = 0, 1, 2, \ldots \tag{7.12}$$

satisfy a first order autoregressive equation of the following form

$$\boldsymbol{C}_{m,n}(i) = \exp\{-F_{m,n}\}\boldsymbol{\Phi}_{m,n}\boldsymbol{C}_{m,n}(i-1) + \tilde{\sigma}_{m,n}\boldsymbol{\varepsilon}_{m,n}(i), \tag{7.13}$$

where $F_{m,n}$ is given by

$$F_{m,n} = r_{m,n}\Delta t, \tag{7.14}$$

and the matrix $\boldsymbol{\Phi}_{m,n}$ is defined by

$$\boldsymbol{\Phi}_{m,n} = \begin{bmatrix} \varphi^{(1)}_{m,n} & \varphi^{(2)}_{m,n} & \varphi^{(3)}_{m,n} & \varphi^{(4)}_{m,n} \\ -\varphi^{(2)}_{m,n} & \varphi^{(1)}_{m,n} & -\varphi^{(4)}_{m,n} & \varphi^{(3)}_{m,n} \\ -\varphi^{(3)}_{m,n} & -\varphi^{(4)}_{m,n} & \varphi^{(1)}_{m,n} & \varphi^{(2)}_{m,n} \\ \varphi^{(4)}_{m,n} & -\varphi^{(3)}_{m,n} & -\varphi^{(2)}_{m,n} & \varphi^{(1)}_{m,n} \end{bmatrix} \tag{7.15}$$

with $\varphi^{(j)}_{m,n} = \varphi^{(j)}_{m,n}(u\Delta t, v\Delta t)$.

The quantity $\tilde{\sigma}_{m,n}$ is a normalized version of the standard deviation $\sigma_{m,n}$

$$\tilde{\sigma}_{m,n} = \frac{\sigma_{m,n}\sqrt{\Delta t(1 - \exp(-2F_{m,n}))}}{\sqrt{2F_{m,n}}}. \tag{7.16}$$

Finally,

$$\boldsymbol{\varepsilon}_{m,n}(i) = (\varepsilon^{(1)}_{m,n}(i), \ldots, \varepsilon^{(4)}_{m,n}(i)) \tag{7.17}$$

is the set of independent Gaussian vectors with zero mean and unit covariance matrix.

Let us show that the vector of the tracer amplitudes satisfies the relation (7.13). First, substitute the expansions (7.4) and (7.8) into the original equation (7.1) and equate the coefficients of the basis functions $\varphi^{(j)}_{m,n}$. As a result one obtains

$$\dot{\boldsymbol{C}}_{m,n} = \boldsymbol{V}_{m,n}\boldsymbol{C}_{m,n} + \sigma_{m,n}\boldsymbol{S}_{m,n}(t), \tag{7.18}$$

where the dot means the time differentiation, $\boldsymbol{S}_{m,n}(t) = (S^{(1)}_{m,n}(t), \ldots, S^{(4)}_{m,n}(t))$, and matrix $\boldsymbol{V}$ is given by

$$\boldsymbol{V} = \begin{bmatrix} -r & q & p & 0 \\ -q & -r & 0 & p \\ -p & 0 & -r & q \\ 0 & -p & -q & -r \end{bmatrix} \tag{7.19}$$

with

$$p = \frac{mu\pi}{a}, \quad q = \frac{nv\pi}{b}, \tag{7.20}$$

and r given by (7.10).

Let us integrate the system (7.18) through the interval $[(i-1)\Delta t, i\Delta t]$ where Δt is a fixed time interval and $i = 1, 2, \ldots, I$. Since (7.18) is an equation with constant coefficients, we arrive at the following formula

$$\begin{aligned} \boldsymbol{C}_{m,n}(i\Delta t) &= \exp(\Delta t \boldsymbol{V}_{m,n})\boldsymbol{C}_{m,n}((i-1)\Delta t) + \\ &+ \exp(i\Delta t \boldsymbol{V}_{m,n}) \int_{(i-1)\Delta t}^{i\Delta t} \exp(-s\boldsymbol{V}_{m,n})\boldsymbol{S}_{m,n}(s)ds. \end{aligned} \tag{7.21}$$

The matrix $\boldsymbol{V}$ is represented as

$$\boldsymbol{V} = -r\boldsymbol{I} + p\boldsymbol{I}_1 + q\boldsymbol{I}_2, \tag{7.22}$$

where $\boldsymbol{I}$ is the 4×4 identity matrix, $\boldsymbol{I}_1$ and $\boldsymbol{I}_2$ are 4×4 commuting matrices with entries 0 and 1 satisfying

$$\boldsymbol{I}_1^2 = \boldsymbol{I}_2^2 = -\boldsymbol{I}. \tag{7.23}$$

From the commutation property, for any real h

$$\exp(h\boldsymbol{V}) = \exp(hr)\exp(h\boldsymbol{I}_1)\exp(h\boldsymbol{I}_2) \tag{7.24}$$

By using (7.23) one can easily compute the exponents on the right-hand side of (7.24) and get

$$\exp(h\boldsymbol{V}) = \exp(-hr)\boldsymbol{\Phi}, \tag{7.25}$$

where $\boldsymbol{\Phi}$ is given by (7.15). After substituting $h = \Delta t$ one finds that the covariance matrix of the second term on the right hand side of (7.21) is equal to

$$\frac{\sigma^2_{m,n}}{2r_{m,n}}(\exp(2F_{m,n}) - 1)\boldsymbol{I}, \tag{7.26}$$

where the use of the orthogonal basis (7.5) is crucial and the orthogonal property of $\boldsymbol{V}$ has been applied. The formulas (7.21), (7.25), and (7.26) lead to (7.13).

Let M_1 and M_2 be some prescribed maximal wavenumbers corresponding to the x and y directions respectively. Then from (7.13) we readily obtain the likelihood function proper for $M = (2M_1 + 1)(2M_2 + 1)$ highest modes

$$\begin{aligned} L_M = \sum_{m=0}^{M_1} \sum_{n=0}^{M_2} \{ 4 \ln \tilde{\sigma}_{m,n} + \frac{4}{I} \ln \frac{\tilde{\sigma}_{m,n}}{\sqrt{1 - \exp(-2F_{m,n})}} \\ + \frac{1}{\tilde{\sigma}^2_{m,n}} [A_{m,n} + B_{m,n} \exp(-2F_{m,n}) - 2\exp(-F_{m,n}) \\ \times (E^{(1)}_{m,n}\varphi^{(1)}_{m,n} + E^{(2)}_{m,n}\varphi^{(2)}_{m,n} + E^{(3)}_{m,n}\varphi^{(3)}_{m,n} + E^{(4)}_{m,n}\varphi^{(4)}_{m,n})]\}. \end{aligned} \tag{7.27}$$

In the likelihood function (7.27),

$$\begin{aligned} A &= \hat{\rho}_0(1,1) + \hat{\rho}_0(2,2) + \hat{\rho}_0(3,3) + \hat{\rho}_0(4,4), \\ B &= A - \tfrac{1}{I}(\boldsymbol{C}(0)^2 + \boldsymbol{C}(I)^2), \\ E^{(1)} &= \hat{\rho}_1(1,1) + \hat{\rho}_1(2,2) + \hat{\rho}_1(3,3) + \hat{\rho}_1(4,4), \\ E^{(2)} &= \hat{\rho}_1(2,1) + \hat{\rho}_1(4,3) - \hat{\rho}_1(1,2) - \hat{\rho}_1(3,4), \\ E^{(3)} &= \hat{\rho}_1(3,1) + \hat{\rho}_1(4,2) - \hat{\rho}_1(1,3) - \hat{\rho}_1(2,4), \\ E^{(4)} &= \hat{\rho}_1(4,1) + \hat{\rho}_1(1,4) - \hat{\rho}_1(2,3) - \hat{\rho}_1(3,2), \end{aligned} \tag{7.28}$$

where the sample correlations are given by

$$\hat{\rho}_j(k,l) = \frac{1}{I} \sum_{i=0}^{I-j} C^{(k)}_{m,n}(i) C^{(l)}_{m,n}(i+j). \tag{7.29}$$

The indices m, n were dropped in formulas (7.28) for the sake of simplicity.

Let us assume that $D_{yx} = D_{xy}$. Differentiating of the likelihood function (7.27) with respect to the unknown parameters $u, v, D_{xx}, D_{xy}, D_{yy}$, and λ gives us the following system of six equations

$$\sum_{m,n=0}^{M_1,M_2} \begin{bmatrix} m^2 \\ mn \\ n^2 \\ 1 \end{bmatrix} \left\{ \frac{2(z_{m,n}^2 - 1 + 2F_{m,n}z_{m,n}^2)}{F_{m,n}(1 - z_{m,n}^2)} - \frac{2}{IF_{m,n}} + \right.$$

$$+ \frac{1}{\Delta t\sigma_{m,n}^2}[\frac{1 - z_{m,n}^2 - 2F_{m,n}z_{m,n}^2}{(1 - z_{m,n}^2)^2}(A_{m,n} + B_{m,n}z_{m,n}^2 -$$

$$\left. -2z_{m,n}G_{m,n}) + \frac{2F_{m,n}}{1 - z_{m,n}^2}(-B_{m,n}z_{m,n}^2 + z_{m,n}G_{m,n})] \right\} = 0, \qquad (7.30)$$

$$\sum_{m,n=0}^{M_1,M_2} \frac{F_{m,n}z_{m,n}^2}{1 - z_{m,n}^2} \frac{\partial G_{m,n}}{\partial u} = 0,$$

$$\sum_{m,n=0}^{M_1,M_2} \frac{F_{m,n}z_{m,n}^2}{1 - z_{m,n}^2} \frac{\partial G_{m,n}}{\partial v} = 0,$$

where $z_{m,n} = \exp(-2F_{m,n})$, and

$$G_{m,n} = E_{m,n}^{(1)}\varphi_{m,n}^{(1)} + E_{m,n}^{(2)}\varphi_{m,n}^{(2)} + E_{m,n}^{(3)}\varphi_{m,n}^{(3)} + E_{m,n}^{(4)}\varphi_{m,n}^{(4)},$$

$$\frac{\partial G_{m,n}}{\partial u} = m(-E_{m,n}^{(1)}\varphi_{m,n}^{(3)} - E_{m,n}^{(2)}\varphi_{m,n}^{(4)} + E_{m,n}^{(3)}\varphi_{m,n}^{(1)} + E_{m,n}^{(4)}\varphi_{m,n}^{(2)}), \qquad (7.31)$$

$$\frac{\partial G_{m,n}}{\partial v} = n(-E_{m,n}^{(1)}\varphi_{m,n}^{(2)} + E_{m,n}^{(2)}\varphi_{m,n}^{(1)} - E_{m,n}^{(3)}\varphi_{m,n}^{(4)} + E_{m,n}^{(4)}\varphi_{m,n}^{(3)}).$$

Now one can estimate the parameters $u, v, D_{xx}, D_{xy}, D_{yy}$, and λ by solving the system of the nonlinear equations (7.30).

7.2 SIMULATING THE TRACER FIELD

The numerical procedure consists of two parts: the forward and inverse problems. In the forward problem, the time evolution of tracer is simulated as a response of the conservation equation (7.1) to the forcing $S(t, \boldsymbol{r})$. We start with the velocity vector $\boldsymbol{u}$, the diffusion coefficients D_{xx}, D_{xy}, D_{yy}, and the feedback rate λ, which are fixed over some area in space. In order to test the ML estimates for different values of the model parameters, numerical experiments are carried out over a wide range of the dimensionless variables

$$\alpha = \lambda \Delta t, \quad \beta = D\Delta t/(\Delta l)^2, \quad \gamma = u\Delta t/\Delta l, \tag{7.32}$$

where Δl is the grid spacing, D and u are typical values of diffusivity and velocity, respectively.

Since our interest is in the application of ML techniques to climatic records for the ocean, we choose the grid spacing Δl and the time interval Δt to be 110 km and 10 days, respectively. These values correspond to the resolution of the sea surface temperature time series which will be analyzed in Chapter 10.

The forward problem involves the following four steps.

(1) Choose the values of the parameters $u, v, D_{xx}, D_{xy}, D_{yy}$, and λ.

(2) Generate the noise sequences $\varepsilon_{m,n}^{(j)}(i), \quad j = 1, \ldots, 4$ appearing in (7.13) and compute the initial values of the amplitudes $C_{m,n}^{(j)}(0)$ in accordance with (7.9) under the assumption

$$\sigma_{m,n} = \frac{1}{(m^2 + n^2 + 1)^{\mu}}. \tag{7.33}$$

In (7.33) μ is constant; it was chosen to be $\mu = 2$. While $\tilde{\sigma}_{m,n}$ depend upon the time interval Δt as in (7.16), it is reasonable to scale $\sigma_{m,n}$ by $\sqrt{\Delta t}$ in order to ensure that $\tilde{\sigma}_{m,n}$ does not turn out to be unrealistically large. An example of $\tilde{\sigma}_{m,n}$ is shown in Fig. 7.1.

The amplitudes $C_{m,n}^{(j)}(0)$ are simulated at the points of the regular grid $m = 0, 1, \ldots, 2M_1, \ n = 0, 1, \ldots, 2M_2$.

(3) Compute the array $\{C_{m,n}^{(j)}(i)\}, \quad m = 0, 1, \ldots, 2M_1, \ n = 0, 1, \ldots 2M_2, \quad j = 1, \ldots, 4$ at times $i = 1, \ldots, I$ by using (7.13) which is reproduced here

$$\boldsymbol{C}_{m,n}(i) = \exp\{-F_{m,n}\}\boldsymbol{\Phi}_{m,n}\boldsymbol{C}_{m,n}(i-1) + \tilde{\sigma}_{m,n}\boldsymbol{\varepsilon}_{m,n}(i), \tag{7.34}$$

Fig. 7.2 demonstrates an example of the function $\exp(-F_{m,n})$.

Fig. 7.3 displays the computational algorithm for the amplitudes $\boldsymbol{C}_{m,n}(t)$ schematically in the particular case of the highest mode $m, n = 0$. The other

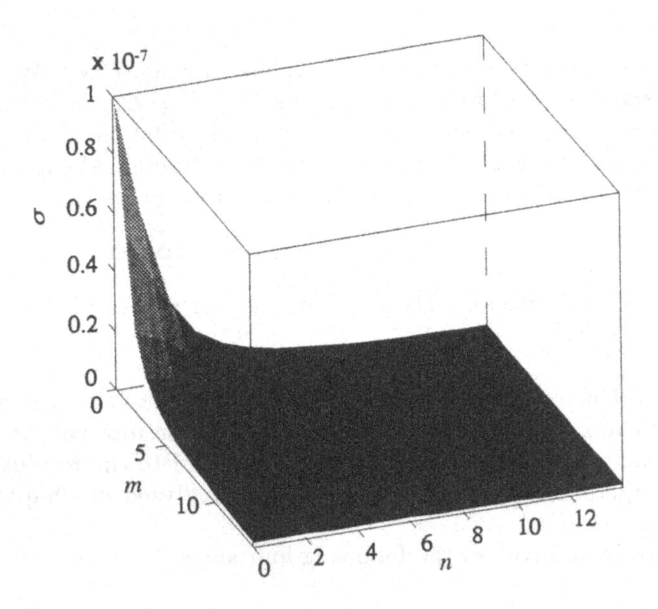

Figure 7.1 Example of the function $\tilde{\sigma}_{m,n}$. Units are arbitrary.

modes are characterized by the products $\exp(-F_{m,n})\mathbf{\Phi}_{m,n}$ some of which look like those below

$$\exp(-F_{0,1})\mathbf{\Phi}_{0,1} = \begin{bmatrix} 0.623 & 0.000 & 0.497 & 0.000 \\ 0.000 & 0.623 & 0.000 & 0.497 \\ -0.497 & 0.000 & 0.623 & 0.000 \\ 0.000 & -0.497 & 0.000 & 0.623 \end{bmatrix},$$

$$\exp(-F_{1,0})\mathbf{\Phi}_{1,0} = \begin{bmatrix} 0.623 & 0.497 & 0.000 & 0.000 \\ -0.497 & 0.623 & 0.000 & 0.000 \\ 0.000 & 0.000 & 0.623 & 0.497 \\ 0.000 & 0.000 & -0.497 & 0.623 \end{bmatrix}, \quad (7.35)$$

$$\exp(-F_{1,1})\mathbf{\Phi}_{1,1} = \begin{bmatrix} 0.463 & 0.369 & 0.369 & 0.294 \\ -0.369 & 0.463 & -0.294 & 0.369 \\ -0.369 & -0.294 & 0.463 & 0.369 \\ 0.294 & -0.369 & -0.369 & 0.463 \end{bmatrix}.$$

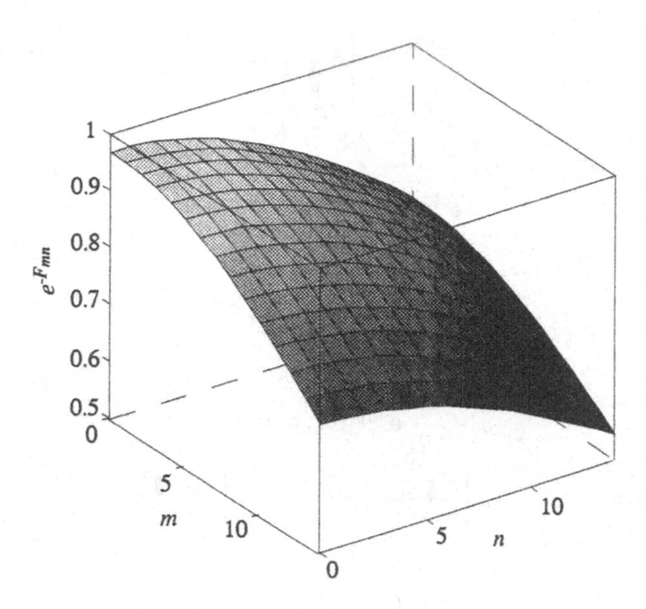

Figure 7.2 An example of the function $\exp(-F_{m,n})$.

The amplitudes $C_{m,n}^{(j)}(i)$ are used to compute the distribution of the tracer $c(t, \boldsymbol{r})$ itself in accordance with

$$c(t, x, y) = \sum_{m=0}^{2M_1} \sum_{n=0}^{2M_2} \sum_{j=1}^{4} C_{m,n}^{(j)}(t) \varphi_{m,n}^{(j)}(x, y). \tag{7.36}$$

This formula is obtained by shrinking the infinite series in (7.8). Fig. 7.4 demonstrates $c(t, \boldsymbol{r})$ calculated for the different values of the parameters $u, v, D = D_{xx} = D_{yy}, \lambda$ for $\mu = 2$ and $D_{xy} = 0$. Apparent in Fig. 7.4 is difference in propagation speed of the tracer disturbances: the higher the velocity u and v, the faster moving are the disturbances. Also noticeable is enhancement of dissipation due to the growth of the mixing coefficients D and/or the decay rate parameter λ. Abrupt changes in the tracer field occur due to occasional extreme forcing. Such fluctuations may introduce uncertainties in the solutions of the inverse problems.

(4) Compute the sample correlations $\hat{\rho}_s(k, l)$ of the series $C_{m,n}^{(j)}(i)$ in accordance with (7.29) for $s = 0, 1;\ \ k, l = 1 \ldots 4$.

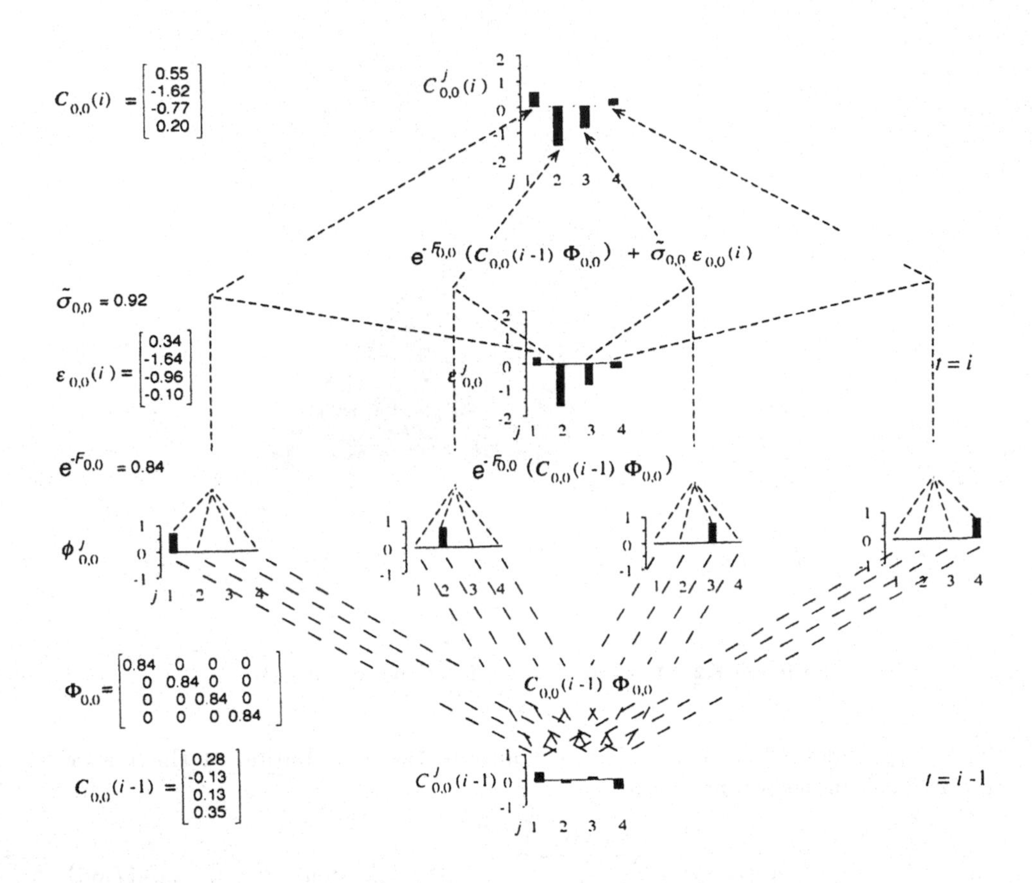

Figure 7.3 Schematic chart of the numerical scheme for simulation of the tracer field.

Thus in our simulations we start from the given amplitudes. The amplitudes of the first $(2M_1 + 1) \times (2M_2 + 1)$ modes in expansion (7.8) can be retrieved from the tracer field given on the grid

$$c_{k,l} \equiv c_{k,l}(t) = c(t, (k - M_1)\Delta x, (l - M_2)\Delta y), \tag{7.37}$$

where $k = 0, 1, \ldots, 2M_1$, $l = 0, 1, \ldots, 2M_2$, and

$$\Delta x = \frac{2a}{2M_1 + 1}, \quad \Delta y = \frac{2b}{2M_2 + 1}. \tag{7.38}$$

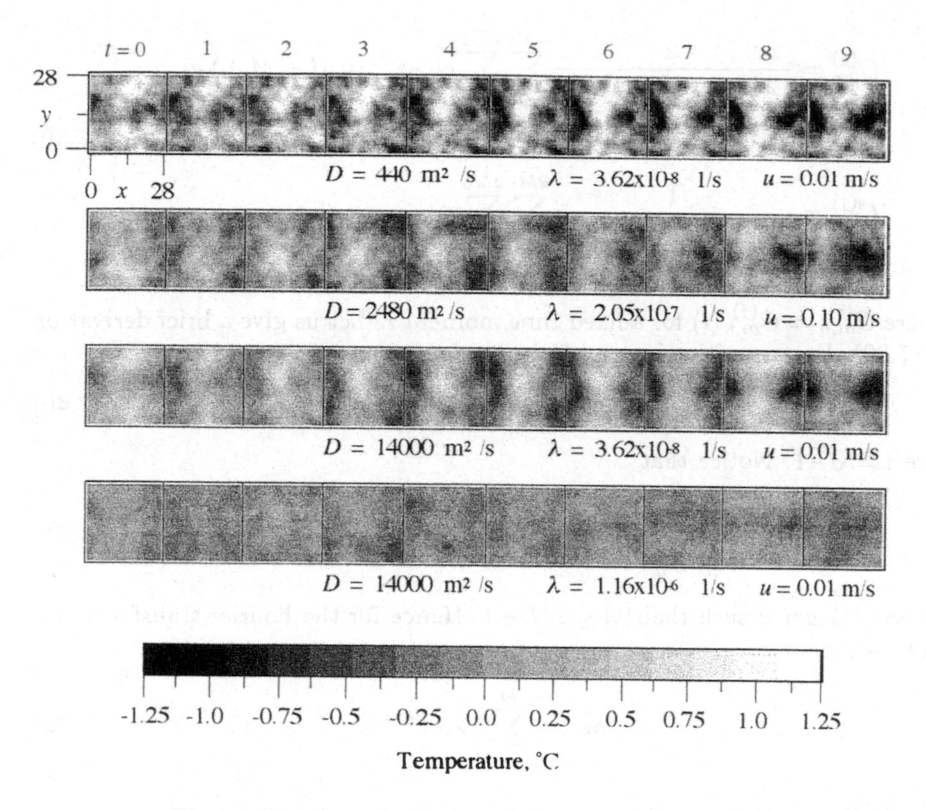

Figure 7.4 **Some simulations of the sequential tracer fields.**

The amplitudes can be computed by

$$C_{m,n}^{(j)} = \frac{4}{(2M_1+1)(2M_2+1)} \sum_{k=0}^{2M_1} \sum_{l=0}^{2M_2} c_{k,l} \varphi_{m,n}^{(j)}((k-M_1)\Delta x, (l-M_2)\Delta y),$$
$$m, n \neq 0, \quad j = 1, 2, 3, 4,$$

$$C_{m,0}^{(j)} = \frac{2}{(2M_1+1)(2M_2+1)} \sum_{k=0}^{2M_1} \sum_{l=0}^{2M_2} c_{k,l} \varphi_{m,0}^{(j)}((k-M_1)\Delta x, 0),$$
$$m \neq 0, \quad j = 1, 3$$

$$C_{0,n}^{(j)} = \frac{2}{(2M_1+1)(2M_2+1)} \sum_{k=0}^{2M_1} \sum_{l=0}^{2M_2} c_{k,l} \varphi_{0,n}^{(j)}(0, (l-M_2)\Delta y),$$
$$n \neq 0, \quad j = 1, 2, \tag{7.39}$$

$$C_{0,0}^{(j)} = \frac{1}{(2M_1+1)(2M_2+1)} \sum_{k=0}^{2M_1} \sum_{l=0}^{2M_2} c_{k,l}, \quad j = 1,$$

where $C_{m,n}^{(j)} = C_{m,n}^{(j)}(i)$ for a fixed time moment i. Let us give a brief derivation of (7.39). For a positive integer M let us denote

$$z = \exp(2\pi i/(2M+1)), \tag{7.40}$$

here $i = \sqrt{-1}$. Notice that

$$\sum_{m=-M}^{M} z^{km} = \begin{cases} 2M+1, & k = 0 \\ 0, & \text{otherwise} \end{cases} \tag{7.41}$$

for any integer k such that $|k| \leq 2M+1$. Hence for the Fourier transform of a finite sequence $(x_0, \ldots, x_M)$

$$a_k = \sum_{m=-M}^{M} x_m z^{km} \tag{7.42}$$

we obtain

$$x_m = \frac{1}{2M+1} \sum_{m=-M}^{M} a_k z^{-km}. \tag{7.43}$$

It follows from (7.43) that for the Fourier expansion

$$c_k = a_0 + \sum_{m=1}^{M} a_m \cos\frac{2km\pi}{2M+1} + \sum_{m=1}^{M} b_m \sin\frac{2km\pi}{2M+1} \tag{7.44}$$

we have the following inverse formulas

$$a_m = \frac{2}{2M+1} \sum_{k=-M}^{M} c_k \cos\frac{2km\pi}{2M+1},$$
$$\tag{7.45}$$
$$b_m = \frac{2}{2M+1} \sum_{k=-M}^{M} c_k \sin\frac{2km\pi}{2M+1}.$$

Let us set $\tilde{c}_{kl} = c(t, k\Delta x, l\Delta y) = c_{k+M_1, l+M_2}$, $k = -M_1, \ldots, M_1$; $l = -M_2, \ldots, M_2$. Expansion (7.36) after substituting (7.5) for a fixed time i, implies

$$\tilde{c}_{k,l} = \sum_{n=0}^{M_2} (C^{(1)}_{0,n} \cos \frac{2ln\pi}{2M_2+1} + C^{(2)}_{0,n} \sin \frac{2ln\pi}{2M_2+1}) +$$
$$\sum_{m=1}^{M_1} \cos \frac{2km\pi}{2M_1+1} [\sum_{n=0}^{M_2} (C^{(1)}_{m,n} \cos \frac{2ln\pi}{2M_2+1} + C^{(2)}_{m,n} \sin \frac{2ln\pi}{2M_2+1})] + \tag{7.46}$$
$$+ \sum_{m=1}^{M_1} \sin \frac{2km\pi}{2M_1+1} [\sum_{n=0}^{M_2} (C^{(3)}_{m,n} \cos \frac{2ln\pi}{2M_2+1} + C^{(4)}_{m,n} \sin \frac{2ln\pi}{2M_2+1})].$$

Using (7.44) one can invert (7.46) in the following way

$$\sum_{n=0}^{M_2} (C^{(1)}_{0,n} \cos \frac{2ln\pi}{2M_2+1} + C^{(2)}_{0,n} \sin \frac{2ln\pi}{2M_2+1}) =$$
$$\frac{1}{2M_1+1} \sum_{k=-M_1}^{M_1} \tilde{c}_{k,l},$$

$$\sum_{n=0}^{M_2} (C^{(1)}_{m,n} \cos \frac{2ln\pi}{2M_2+1} + C^{(2)}_{m,n} \sin \frac{2ln\pi}{2M_2+1}) =$$
$$\frac{2}{2M_1+1} \sum_{k=-M_1}^{M_1} \tilde{c}_{k,l} \cos \frac{2km\pi}{2M_1+1}, \quad m > 0, \tag{7.47}$$

$$\sum_{n=0}^{M_2} (C^{(3)}_{m,n} \cos \frac{2ln\pi}{2M_2+1} + C^{(4)}_{m,n} \sin \frac{2ln\pi}{2M_2+1}) =$$
$$\frac{2}{2M_1+1} \sum_{k=-M_1}^{M_1} \tilde{c}_{k,l} \sin \frac{2km\pi}{2M_1+1}, \quad m > 0.$$

Then by inverting (7.47) once again with respect to index n we arrive at (7.39).

7.3 MAXIMIZING THE LIKELIHOOD

In the inverse problem the advection-diffusion equation (7.1) is fitted to the data by means of ML estimators. The search for the unknowns $\boldsymbol{u}, \boldsymbol{D}$, and λ can be done either by finding the minimum of the function (7.27) or by solving the system of the nonlinear equations (7.30). Here, in order to test the method we compute the likelihood functions but the parameter estimation will be done via minimizing the sum of squares of the functions (7.30). Given the six functions in (7.30) of the variables $\theta = (u, v, D_{xx}, D_{xy}, D_{yy}, \lambda)$ and an initial vector θ_0 we obtain a final vector θ_f, that gives a local minimum of the sum of squares of the functions in (7.30) by using the Levenberg-Marquardt-Morrison method (Marquardt 1963, Osborne 1976, Dennis and Schnabel 1983). The numerical implementation of this method was accomplished by using the code NOLF1 in the SSLII library of mathematical subroutines (SSLII 1987).

The system (7.30) is fitted to the sample correlations $\hat{\rho}_s$ of the amplitudes $C_{m,n}^{(j)}(i)$ of the tracer fields simulated in the forward problem (see section 7.2). The adjustable parameters $\boldsymbol{u}, \boldsymbol{D}$, and λ are assumed to be constants over the same grid as in the forward problem. The source term $S(t, \boldsymbol{r})$ is taken to be either known as in the forward problem or to be noise with the standard deviation σ_{mn} as in the formula (7.33) where the parameter μ can be arbitrarily set. In the latter case, we are tempted to guess the slope of the spectrum of the forcing $S(t, \boldsymbol{r})$. This trick gives us surprising results, which will be discussed later.

It is noteworthy that this inversion can be constrained by minimizing certain norms. For example, it is easy to minimize the horizontal divergence, vorticity and kinetic energy norms for the advection velocities. This minimization is aimed at reducing solution uncertainties. Kelly (1989) and Kelly and Strub (1992) combined minimization of the heat equation and the horizontal divergence norm to get more realistic velocities from satellite infrared images. Here we do not discuss norm minimization, because our purpose is to understand the nature of the ML estimates themselves.

The nonlinear equations (7.30) can be scaled before minimization. The scaling factors can be chosen so as to find a family of solutions, which provides the best fit to the model (7.1) in terms of the desired degree of accuracy for the parameters $u, v, D_{xx}, D_{xy}, D_{yy}$, and λ. After a series of experiments we decided to scale only the last two equations in (7.30) by a factor of the order $M_1 M_2$.

Our numerical experiments involved simulation and fitting the modes $C^{(j)}_{m,n}(i)$ to the model (7.1). The time series $C^{(j)}_{m,n}(i)$ of the length 2, 3,..., 128 were generated and the inverse problems solved on mesh sizes containing 3×3, 4×4,..., and 15×15 points. Below we consider mainly results obtained for a limited number observations in time over a small spatial mesh. This allows us to test the feasibility of ML estimators in fairly extreme conditions of minimum observations in both time and space. Such conditions are realistic for practical applications because usually very limited sequences of sea tracer fields are available from satellite observations.

The spatial mesh can be further considered to be a subset of points in inversion from real data. For example, a satellite tracer image can be divided into subsets of pixels so the inversion problem can be solved on a grid with each point centered on a subset of pixels. The unknown parameters are assumed to be constants only within each subset. Thus the numerical experiments on spatial meshes will serve as a basic test for the inversion.

Noteworthy, Kelly (1989) considered advection between the subsets of points in the inversion of the differential form of the advection-diffusion equation. It was pointed out that the minimum subset size Δx is related to the maximum velocity, u, by a Courant-Friedrichs-Lewy (CFL) condition

$$\Delta l > u\Delta t. \tag{7.48}$$

Taking $\Delta t = 1$ day and the sea surface current velocity component $u \approx 18$ cm s^{-1} gives $\Delta l > 15$ km. Kelly (1989) chose a subset size of 16×16 points so that $\Delta l > 18$ km. This subset size allowed neglecting the diffusion. Kelly (1989) applied this technique to satellite radiometry images of the coastal zone off California. In this dynamically active area, the sea surface diffusion coefficient is as high as $D \approx 10^2 - 10^3$ m^2 s^{-1} on horizontal scales of 10 km (Davis 1985). So the diffusion term can be smaller than the advection term by at least one order of magnitude if the velocity is 0.1 m s^{-1}. The near surface diffusivity can be much smaller in other oceanic regions. The analysis by Okubo (1971) indicated that the diffusion coefficient D is in the range of 1 to 10 m^2 s^{-1} on the scales of 10 km.

As to the diffusivity, there is no such physical constraint. If one needs the velocity estimate alone then, following Kelly (1989), an easy solution is to exclude diffusion from consideration so long as the size of the diffusion term is much smaller than that of the advection term. However the problem turns out to be non-trivial if the relative sizes of the terms prevent us from neglecting the diffusion or if we have to estimate the diffusion coefficients as well. When the current velocity is very small, diffusion is the only contributor to tracer transport. Therefore in general it is not proper to neglect diffusion from consideration of tracer transport in the upper ocean. ML inversion provides a method for making general diffusivity estimates.

As it will be discussed below, results of this inversion do not depend directly upon the CFL condition. In this study we consider the transport within a subset of 7×7 points assuming that velocity and diffusivity are fixed over this subset.

In our numerical experiments, the time interval, $\Delta t = 10$ days, and the grid resolution, $\Delta l = 110$ km, are the same as in the forward problem. Thus we assume that the passive tracer can be sea surface temperature (SST) and the main forcing mechanism is air-sea heat flux fluctuations.

As was mentioned above, in order to test the ML method over a large range of the parameters $u, v, D_{xx}, D_{xy}, D_{yy}$, and λ, the numerical experiments were carried out using different values of the dimensionless variables given by (7.32). Though large parameters ranges were tested, we will focus our discussion on the conditions typical of the upper ocean.

We conducted two large groups of the numerical experiments. The groups differ in assumptions about the forcing term in the inverse problem. In Group I the velocity vector $\boldsymbol{u}$, the mixing coefficients D_{xx}, D_{xy}, D_{yy}, and the decay rate λ are assumed to be unknown constants but the forcing $S(t, \boldsymbol{r})$ is taken to be known as in the forward problem. These experiments correspond to the case when both tracer and forcing can be obtained experimentally. We ran numerical experiments with different values of the forcing parameter μ such as ($\mu = 0.5, 1.0$, and 2.0). The conclusions derived from these tests are basically identical. Below we focus on experiments when ($\mu = 0.5$). Group II represents the particular situation when only tracer data are available and possibly the spectral slope of the forcing. Thus here we do not attempt to identify the source term but, instead, we simulate it according to the formula (7.33) with an arbitrary value of μ. Each group includes an experimental series for the absolute values of the parameter $|\gamma| = 0.1, 0.15, \ldots, 1.55$. Each series consisted of about 1600 experiments with different values of the parameters α and β. These values change from one experiment to another so that their full ranges are three orders of magnitude in each series of the numerical experiments.

The proposed estimator is characterized by the likelihood function that can be computed from (7.27). Likelihood function minimums are of interest to us because the inverse problem solution we seek is in one of the minimums.

Minimizing (7.27) is equivalent to maximizing the probability that the data set $C^{(j)}_{m,n}(i)$ is governed by the model (7.1). In doing so we expect that the obtained values of $u, v, D_{xx}, D_{xy}, D_{yy}$, and λ will be close to the true values $u^*, v^*, D^*_{xx}, D^*_{xy}, D^*_{yy}$, and λ^*.

Minimizing (7.27) occurs in at least six dimensions if the source term is given. According to Theorem 6.2 (see Chapter 6) a unique solution can be expected only for the coefficients D_{xx}, D_{xy}, and D_{yy} due to the consistency of the ML estimator for the diffusivity. Much less is known from the theory about the behavior of ML estimates for the velocity and feedback when the spatial modes $C^{(j)}_{m,n}(i)$ are limited only to the highest ones and the number of observations is small in time.

Since it is impossible to illustrate the ML function (7.27) in six dimensions, in this section we restrict our attention to the cases when the diffusion is isotropic, $D_{xx} = D_{yy} = D$, the components of the velocity in x and y directions are equal $u = v$, and either the velocity u or the feedback rate λ are known. In other words, we consider cases with two unknown parameters, λ and D or u and D.

Let us consider results of the ML function calculations in the experiments with the shortest ($I = 1$) time series $C^{(j)}_{m,n}(i)$ over the small grid of 5×5 points, $M_1 = M_2 = 2$. Then we will turn to the case of the long ($I = 100$) time series over an enlarged grid of 17×17 points.

Fig. 7.5 illustrates the ML function for the case when, in the forward problem, the true values of the parameters are $\alpha^* = 2^{-5}$, and $\gamma^* = -1$. In the calculations (7.27), the feedback rate is taken to be known, $\lambda = \lambda^*$. The figure shows the ML estimates from two sequential fields $C^{(j)}_{m,n}(i)$, $i = 0, 1$.

In the case when $\beta^* = 2^{-5}$ (Fig. 7.5, left panel), the ML function forms a street of depressions with minimums at $\beta = 2^{-5}$ that exactly correspond to the true value $\beta^* = 2^{-5}$. Similar behavior of the ML estimator for the diffusion coefficient is found in the numerical experiments with $\beta^* = 1$ (Fig. 7.5, central panel). The ML function is characterized by a steady trough of local minimums. The minimums are situated at $\beta = 1$, which is correct solution for the diffusivity. In the case of $\beta^* = 2^5$ shown in Fig. 7.5 (right panel), local minimums merge into one deep steady trough corresponding to the true diffusivity. The ML estimator easily converges to the correct solution for diffusivity.

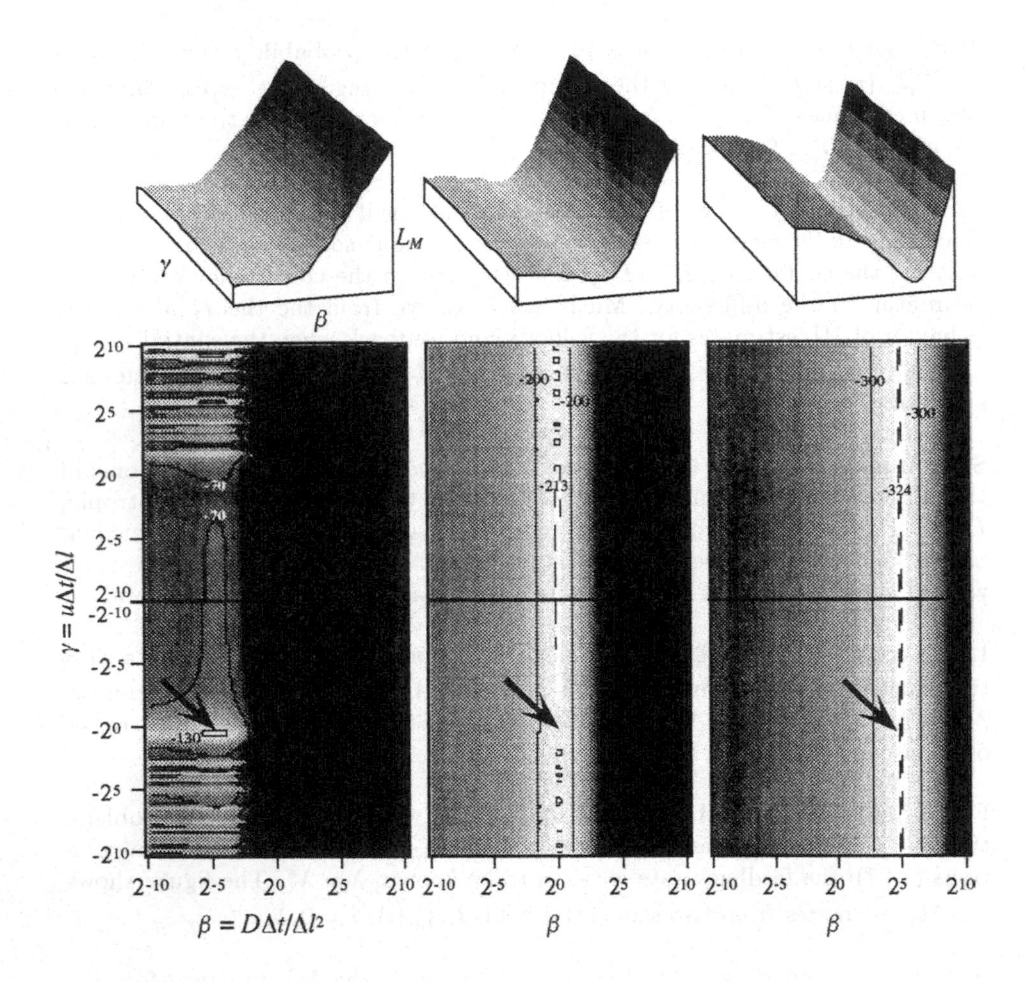

Figure 7.5 Maximum Likelihood functions computed in the domain of the dimensionless variables for velocity and diffusivity, γ and β, with a fixed value of the dimensionless feedback variable, α. The solutions of the inverse problems are the ML function minimums. The locations of the true solutions are indicated by arrows. The true value of the dimensionless velocity parameter is $\gamma^* = -1$. On the left, central, and right panels, the true values β^* are 2^{-5}, 1, and 2^5, respectively. Only two sequential tracer fields are considered. The mesh size is 5 $\times$ 5 points. Notice that the ML function minimums are always located along the troughs of the true values of the diffusivity variable β.

The ML estimator for velocity behaves differently. As it was shown above, while both the diffusivity and feedback are small the ML function is characterized by the local minima. For example, in Fig. 7.5 (left panel), the minimums are extended in β-direction. The appearance of local minima indicates, on the one hand, a degradation of the diffusivity estimates as the diffusion decreases and, on the other hand, the possibility of inferring the velocity. Since the data are limited by a short time series and a small observational grid, among the minimums there might be no correct estimate for the velocity, as it can be seen in Fig. 7.5 (central panel). The ML estimator is confused about the absolute value and direction of propagation for the tracer disturbances. Only by increasing the length of the observation period and/or the number of the spatial modes $C_{m,n}^{(j)}$ we can improve the estimate. Fig. 7.6 is the same as Fig. 7.5 except that the ML functions in the numerical experiments use long time series, $I = 100$, on a 17×17 grid of points, $M = N = 8$. As the number of observations increases, minimums emerge at the right locations as in the case of the small (Fig. 7.6, left panel) and moderate (Fig. 7.6, central panel) values of β. It is important that the true minima are nearest to the origin, $\gamma \approx 0$. Therefore starting from small initial values of γ the minimization will eventually lead us to the true minimum. If initially γ is large, the minimization might converge to the wrong minimum. The true minimums are usually the deepest so the right solution corresponds to the lowest minimum. That is the nature of ML estimators when the feedback rate is small, $\alpha < 1$, and diffusivity is small or moderate, $\beta < 1$.

When the diffusivity is large, $\beta > 1$, there is no hope of estimating the velocity. Instead of a local minimum, there is a steady trough. The ML estimator for the velocity does not converge to the true solution even when increasing both the length of the observation period and the number of the spatial modes $C_{m,n}^{(j)}$ (Fig. 7.6, right panel). In general, as the diffusivity grows, the velocity estimate becomes worse.

The results of the ML function evaluation for diffusivity vs feedback are shown in Fig. 7.7. The advection velocity is assumed to be known so $\gamma^* = 1$ and the ML function (7.27) was calculated over a mesh of the dimensionless variables α and β. The ML functions shown in Fig. 7.7a, b, c, d, and e were obtained, respectively, for the following pairs of the parameters: $\alpha^* = 2^5$, $\beta^* = 2^{-5}$; $\alpha^* = 2^{-5}$, $\beta^* = 2^{-5}$; $\alpha^* = 2^5$, $\beta^* = 2^5$; $\alpha^* = 2^{-5}$, $\beta^* = 2^5$; $\alpha^* = 1$, $\beta^* = 1$. Fig. 7b-d illustrate that usually there is only one minimum in ML function (7.27) for the estimator of both diffusivity and feedback. The presence of this unique minimum enables us to obtain feasible estimates. However, in the domain where the feedback rate is large $\alpha > 1$ and the diffusion is small $\beta < 2^{-5}$,

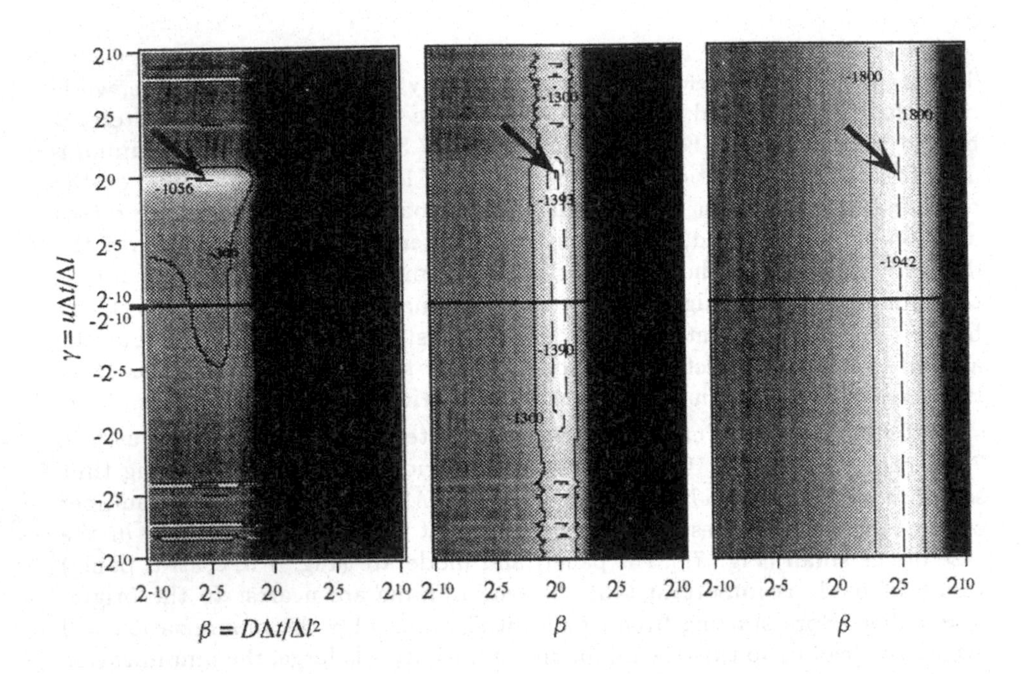

Figure 7.6 The same as Fig. 7.5 but for a long time series of tracer fields, $I = 100$, over a 17×17 mesh of points. In these numerical experiments the true velocity is fixed as $\gamma^* = 1$.

the ML estimator has no local minima but a half-trough extending along the β-axis (Fig. 7.7a). Such half-troughs appear even if we run the numerical experiments with long time series, $I = 100$, and a large number of the spatial modes $C_{m,n}^{(j)}$, $M = N = 8$ (Fig. 7.8). Notice, however, the trough location at $\alpha = 2^5$ indicates the correct solution for the feedback rate.

Summarizing the results of the ML function evaluation we notice the following. (1) The numerical results agree broadly with theory in that the ML estimator is consistent for the diffusivity. There is an exception in the region of small diffusivity $\beta < 2^{-5}$ and large feedback rate $\lambda > 1$. However this range of large feedback rate, $\lambda > 1$, is beyond the scope of practical applications. (2) The ML estimator easily converges to the correct solution for the diffusivity. This simplifies the search for the numerical solution in (7.30). Starting from any point within the range of the parameter $\beta > 2^{-5}$, the solution converges to the true value D^*. When solving the system (7.30), it is not necessary to start the search for the numerical solution with different initial values D_0 and λ_0. (3) The calculations give us conditions favorable for feasible velocity estimates.

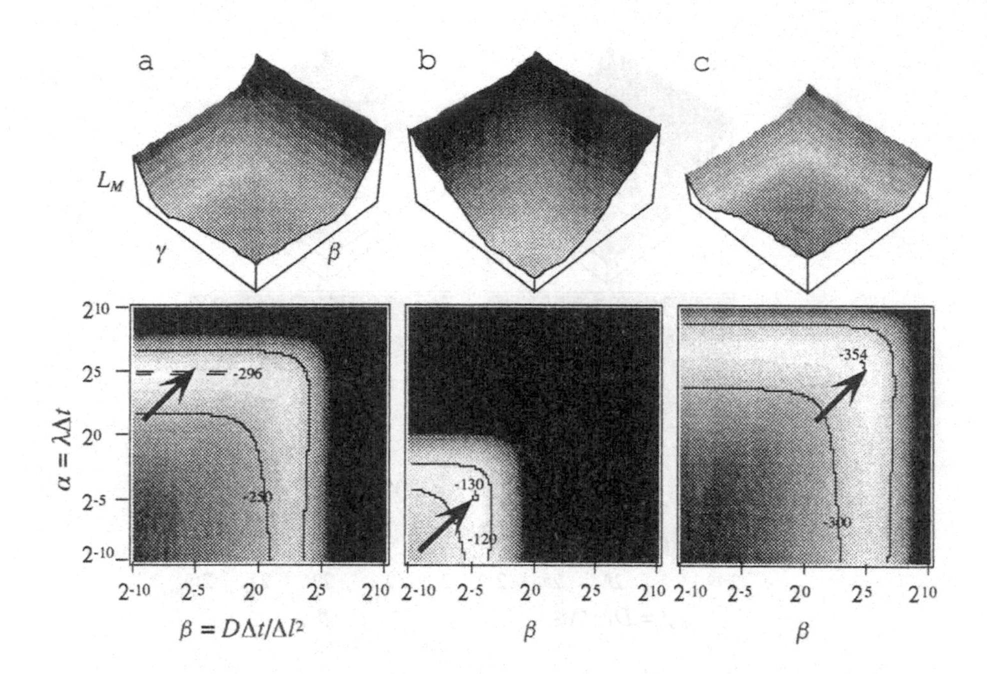

Figure 7.7 Maximum Likelihood functions computed in the domain of the dimensionless variables for feedback and diffusivity, α and β , with the fixed value of the variable for the velocity, γ. The solutions of the inverse problems are at the ML function minimums. The true solutions are shown by arrows. Only two sequential tracer fields are considered. The grid size is 5×5 points.

In the search for velocity, the initial value $\boldsymbol{u}_0$ should be small. Otherwise, starting from a large $\boldsymbol{u}_0$ the minimization finds some nearby minimum that is not necessarily the root of the nonlinear system (7.30). Since the true minimum is usually the deepest one, it is worthwhile, also, to try different initial directions of the velocity vector $\boldsymbol{u}_0$.

Finally it should be noted, that the ML function calculations were performed for a limited number of the unknown parameters. In practice, the minimization of (7.27) must be done in six dimensions, at least. Fortunately, however, as it will be shown below, the general numerical solution of the system (7.30) basically confirms the above findings.

To find the roots of the nonlinear system (7.30), it is not necessary solve the system explicitly but, instead, we minimize the sum of squares of the functions in (7.30). The search for the solution starts with different initial vectors $\theta_0 =$

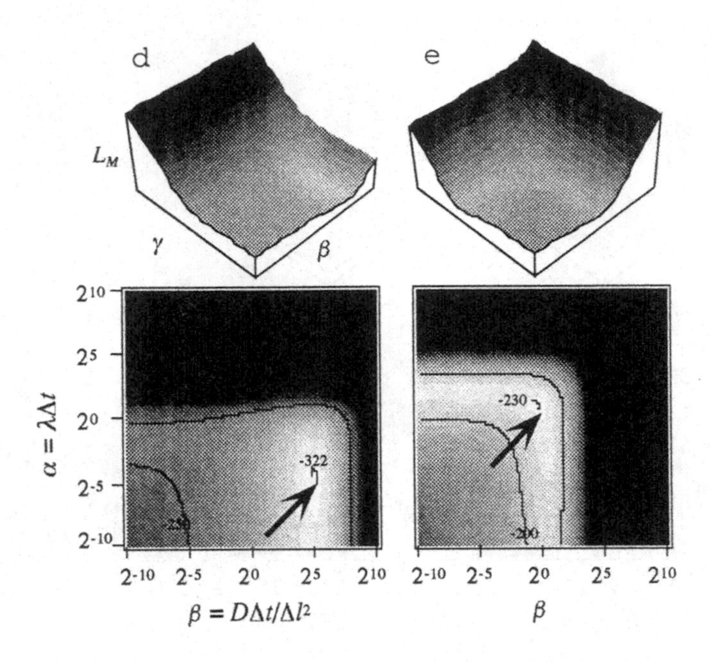

Figure 7.7 *(continued)*

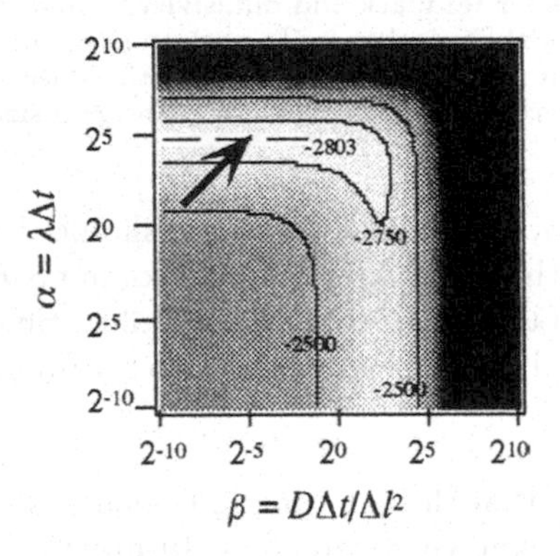

Figure 7.8 The same as Fig. 7.7a but for a long time series of tracer fields, $I = 100$, over a spatial grid of 17×17 points.

$(u, v, D_{xx}, D_{xy}, D_{yy}, \lambda)_0$ and the solution is chosen to be that vector θ_f, which gives the smallest value of the sum of squares in the minimizations. Due to a trade-off between computational efficiency and the maximum number of the correct solutions obtained, we adopted to start the minimizations from 100 initial vectors θ_0 in each numerical experiment. These vectors contain the same values of the diffusion coefficients $\boldsymbol{D}_0$ and the feedback rate λ_0 but different values of the velocity components u_0 and v_0. For example, among the velocity components we used are the following pairs (0.00, 0.00); (0.05, 0.05); (0.05, -0.05); (-0.05, 0.05); (-0.05, -0.05), m s^{-1}. The squares of functions in (7.30) are minimized iteratively until a predetermined convergence criteria is satisfied. Then one vector among the one hundred θ_f's is selected in accordance with the smallest misfit value in (7.30). This solution is to be compared with the true vector θ^*.

Preliminary experiments showed that the estimation of the set of the diffusion coefficients D_{xx}, D_{xy}, and D_{yy} leads to solution uncertainties. So, in this section, we will consider the system (7.30) with only two unknown diffusion coefficients D_{xx} and D_{yy}. Correspondingly, the system (7.30) is reduced to five equations with five unknowns u, v, D_{xx}, D_{yy}, and λ. Omitting the off-diagonal elements of the matrix $\boldsymbol{D}$ is conventional for many oceanographic applications.

Here we show the results of the ML estimates for data on a 7×7 grid of points. First, we will discuss the estimates for moderately short time series, $C^{(j)}_{m,n}(i), i = 0, 1, ..., 13$. The analysis is restricted to the case when the forcing term $S(t, \boldsymbol{r})$ is known as in the forward problem. These conditions are well suited to the practical application that will be discussed in Chapter 10. Second, we will consider the case of unknown forcing with an arbitrary spectral slope. In both these cases we assume that the diffusion is isotropic. Finally, we will give results for time series $C^{(j)}_{m,n}(i)$ of different lengths for both isotropic, $D_{xx} = D_{yy}$, and anisotropic diffusion, $D_{xx} \neq D_{yy}$.

As was emphasized above, in the Group I numerical experiments, the forcing field is assumed to be known as in the forward problem. Group I consists of 30 series of numerical experiments. The absolute value $|\gamma^*|$ increases from one series to another as $|\gamma^*| = 0.1, 0.15, ..., 1.55$, so the absolute value of the velocity vector $|\boldsymbol{u}^*|$ grows from about 0.01 m s^{-1} up to 0.20 m s^{-1}. Though $|\boldsymbol{u}^*|$ is fixed in each experimental series, its direction varies as

$$u^* = |\boldsymbol{u}^*| \sin l\pi/20 \quad and \quad v^* = |\boldsymbol{u}^*| \cos l\pi/20,$$

$l = 1, 2, ..., 41$. The other parameters $D^* = D^*_{xx} = D^*_{yy}$, and λ^* are subject to change from one experiment to another so as to cover a mesh of 41×41 points for the dimensionless variables $\alpha^* = 2^m$ and $\beta^* = 2^n$, where $m = -10.0, -9.75, -9.5, ..., 0.0$ and $n = -8.0, -7.75, -7.5, ..., 2.0$. Therefore the diffusion coefficient and the feedback rate lay in the ranges $55 < D^* < 56 \times 10^3$

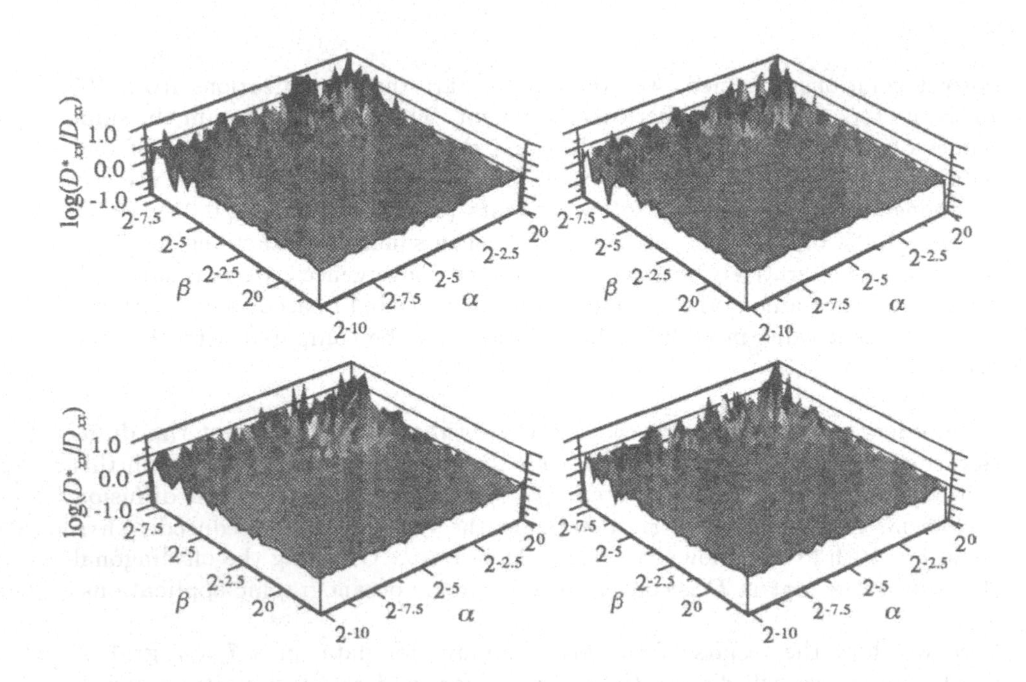

Figure 7.9 The ratios of the estimates D_{xx} obtained as solutions of the inverse problem to the true values D^*_{xx}. In the inversions, the advection-diffusion equation was fitted to the sequential tracer fields. The tracer distributions were simulated in the forward problems during 12 time intervals over elementary arrays of 7 × 7 points. Notice that if $|\log(D_{xx}/D^*_{xx})| < 0.3$ then the estimate D_{xx} differs from the true value by less than a factor of two. The horizontal axes of the plots specify the domain of the dimensionless parameters α and β. Clockwise from the upper left are the results from numerical experiments with the velocity parameter fixed at $|\gamma^*| = 0.1, 0.5, 1$, and 1.5.

m^2 s^{-1} and $4.5 \times 10^{-9} < \lambda^* < 5 \times 10^{-7} s^{-1}$. This domain of α^* and β^* represents variability covering the range of oceanic conditions of interest to us (see chapters 9-10). Fig. 7.9 demonstrates the results of minimizing in the system of equations in (7.30) for the diffusion coefficient D_{xx} in the experimental series with $|\gamma^*| = 0.1, 0.5, 1.0, 1.5$. The surface plots show the ratios of ML estimates for D_{xx} to the true values of D^*_{xx}. Similar results are obtained for the diffusivity component D_{yy}.

Apparent in Fig. 7.9 is a large region where the ratio $D/D^* \approx 1$. In this region, ML estimation of the diffusivity is feasible. The estimates of the diffusion coefficients are within a factor of 2 of the true values for 82% of the points over

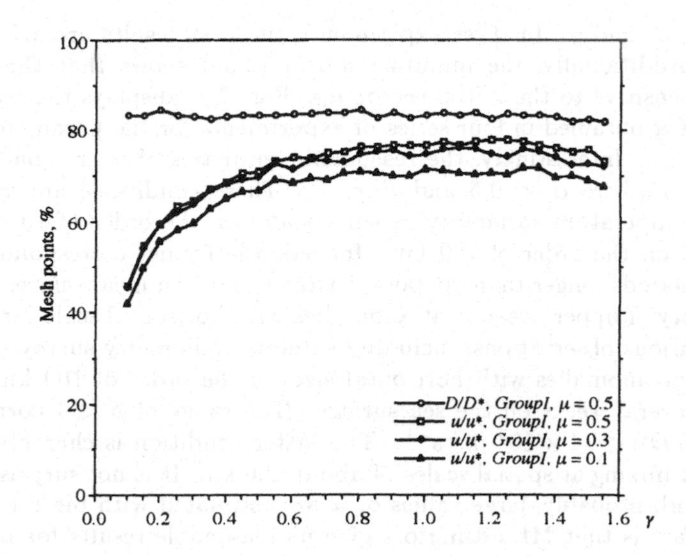

Figure 7.10 Numbers of reasonable estimates (see text for explanation) of diffusivity and velocity obtained as the inverse problem solutions under known (Group I) and unknown (Group II) forcing. The solutions were sought in the experimental series each for fixed true value of the dimensionless parameter for the velocity, γ. Each series consisted of 41×41 experiments with different true values of the dimensionless parameters for feedback and diffusivity, α^* and β^*. Each experiment is represented by one point in the mesh (α, β). The percentage is defined by the ratio of the points where the reasonable estimates were obtained to the total amount of 1684 points over the mesh.

the mesh (α, β). However, the results of the minimization become worse in the region for small $\beta < 2^{-6}$.

It is important that the ML estimator for the diffusivity is stable under changes of the parameter γ. Fig. 7.10 displays the number of the feasible solutions for diffusivity in each of 30 experimental series. This number represents the amount of points over the mesh (α, β) where the ML estimator provides a solution that satisfies the condition $0.5 < D/D^* < 2.0$. So, reasonable estimates for diffusivity are found in 82% of the inversions throughout the experimental series.

The number of reasonable estimates for velocity are also shown in Fig. 7.10. We regarded the estimate of $\boldsymbol{u}$ as reasonable if its error is less than 30° in direction and a factor of 2 in absolute value. Admittedly, ML estimates for velocity are not consistent, i.e. they are not correct for all values of the dimensionless variables α, β, and γ. In these experiments the best results are achieved for $\gamma > 0.6$. Additionally, the minimization of (7.30) shows that the velocity solution is sensitive to the initial vector $\boldsymbol{u}_0$. Fig. 7.11 displays the reasonable estimates of $\boldsymbol{u}$ obtained in four series of experiments for the parameter $|\gamma^*| = 0.1, 0.5, 1.0, 1.5$. Interestingly, the reasonable estimates of $\boldsymbol{u}$ are concentrated in the region where $\alpha < 0.5$ and $\beta < 1$. These conditions are typical of upper sea temperature variability at time scales on the order of 10 days and space scales on the order of 100 km. Indeed, the former corresponds to the relaxation periods longer than 20 days. Faster relaxation means unrealistically short memory of upper ocean heat anomalies with horizontal scale larger than 100 km. Various observations, including satellite radiometry surveys, suggest that the heat anomalies with horizontal size on the order of 100 km can be traced for several weeks on the sea surface. The range of $\beta < 1$ corresponds to values of $D^* < 14 \times 10^3$ m^2 s^{-1}. The latter condition is characteristic of upper ocean mixing at spatial scales of about 100 km. It is not surprising that the small and, probably, large values of $\boldsymbol{u}$ are estimated with big errors. The important fact is that ML estimators give us reasonable results for moderate velocities 0.08 - 0.2 m s^{-1}, which are typical of the upper ocean.

As for the feedback estimation in the parameter domain concerned, the ML estimator usually does not provide the proper solution. The feedback estimates are improved only for $\alpha > 0.5$. This is in agreement with the above ML function evaluation.

Summarizing the numerical experiments on the minimization of the nonlinear system (7.30) in the case of 13 sequential tracer fields limited in space by 7×7 points and forced by a known process, we conclude that the above results support the theory by demonstrating the following features.

(1) The ML estimator for the diffusivity is robust in the range of the dimensionless variable $\beta > 2^{-6}$ under realistic conditions on the feedback $\alpha < 0.5$. The ML estimator for diffusivity is stable under variations in the advection velocity. (2) The ML estimates for the velocity $\boldsymbol{u}$ are not reliable except for those values of the dimensionless variables for feedback and diffusivity in the ranges $\alpha < 0.5$, and $\beta < 1$. The feasibility of making velocity estimates depends upon the initial conditions in the minimization and the absolute value of the true velocity; the best results are obtained for $\gamma > 0.6$. (3) The ML estimates of the feedback λ are not feasible when the diffusivity estimates are consistent.

Group II numerical experiments concern the particular case of unknown forcing. It is interesting to investigate the ML estimator behavior in the situation when forcing fields are not available and attempt to guess the slope of the forcing

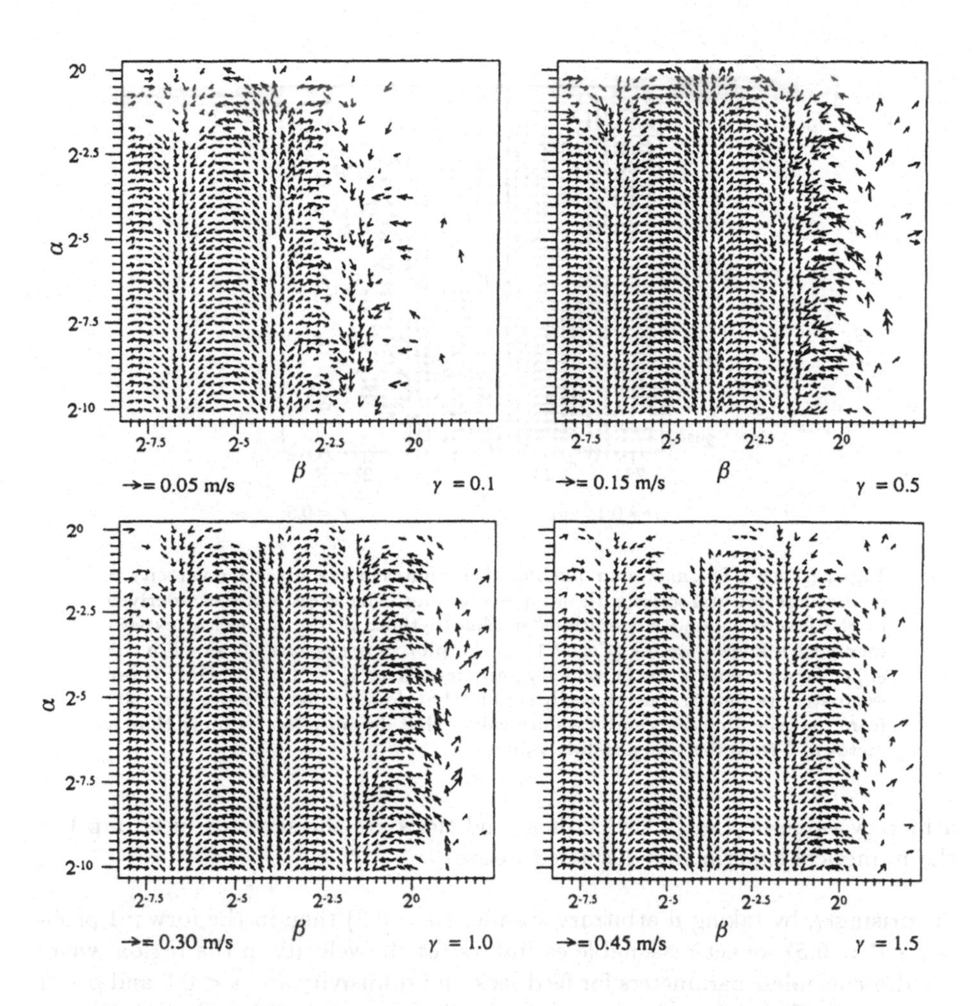

Figure 7.11 The maximum likelihood estimates of the advection velocity $\boldsymbol{u}$ in the inversion of the advection-diffusion equation from sequential tracer fields. The tracer distributions were simulated during 12 time intervals over elementary arrays of 7 × 7 points. The figures represent the results obtained in experimental series with different absolute values of the dimensionless velocity parameter, γ. Clockwise from upper left are the results from the experimental series with the velocity parameter fixed at $|\gamma^*| = 0.1, 0.5, 1$, and 1.5. Only those estimates $\boldsymbol{u}$ are shown, which have directions that do not differ by more than 30° from the true directions and which have absolute values that do not differ more than a factor of 2 from the true absolute values.

spectrum. Below we consider results of the minimization in (7.30) with an

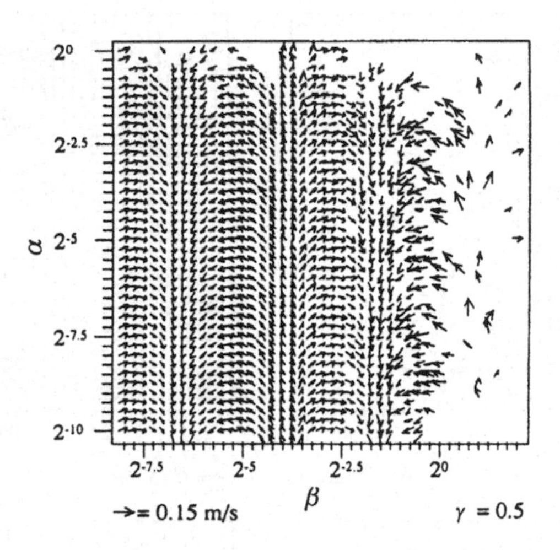

Figure 7.12 The maximum likelihood estimates of the advection velocity $\boldsymbol{u}$ in the experimental series with unknown forcing. The $\boldsymbol{u}$ estimates are derived in the inversions under conditions similar to those of the numerical experiments with known forcing (Fig. 7.11). The plot shows results obtained in the experimental series with the velocity parameter fixed at $|\gamma^*| = 0.5$. Only those estimates are shown, which have directions that do not differ by more than 30° from the true directions and absolute values that do not differ by more than a factor of 2 from the true absolute values.

arbitrary value of μ in (7.33). Other conditions are the same as in Group I of the numerical experiments discussed above.

Surprisingly, by taking μ arbitrary smaller ($\mu = 0.3$) than in the forward problem ($\mu^* = 0.5$) we get reasonable estimates for the velocity in the region where the dimensionless parameters for feedback and diffusivity are $\alpha < 0.5$ and $\beta < 1$ (Fig. 7.12). The region of reasonable estimates of $\boldsymbol{u}$ (Fig. 7.12) almost overlaps that of the Group I numerical experiments (Fig. 7.11). The price for this stability of the ML estimator for the velocity is an overestimation of the diffusion coefficient (Fig. 7.13). Calculating the ML function (7.27) under the value of $\mu = 0.1$ (other conditions are similar to those used to get Fig. 7.5) we find (Fig. 7.14) that the ML function minimum pinpoints diffusivity that is higher than the true one.

The quality of the $\boldsymbol{u}$ estimates derived in the Group II numerical experiments is shown in Fig. 7.10. The quality is characterized by the percentage of reasonable estimates of $\boldsymbol{u}$ in the inverse problems with the forcing parameter fixed at

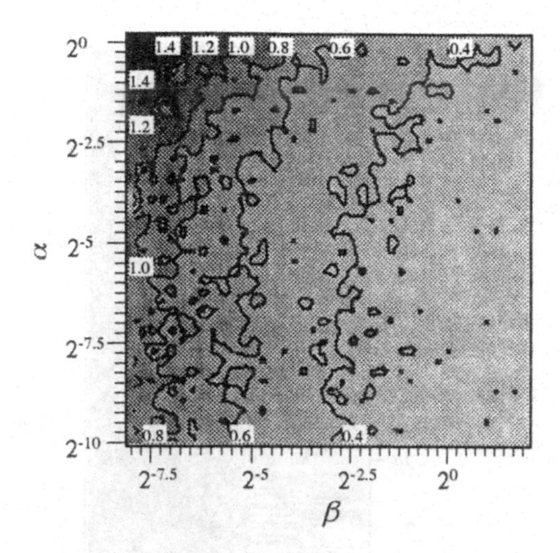

Figure 7.13 The distribution of $\log D_{xx}/D^*_{xx}$, where D_{xx} and D^*_{xx} are, respectively, the estimate and true value of the diffusivity in the experimental series with unknown forcing. The estimates D_{xx} are derived in the inversions under conditions similar to those of the numerical experiments with known forcing (Fig. 7.9). The plot demonstrates results obtained in the experimental series with the velocity parameter fixed at $|\gamma^*| = 0.5$.

$\mu = 0.1$ or $\mu = 0.3$ instead of $\mu^* = 0.5$ as in the forward problem. These results suggest that in practice when the forcing can not be measured, it might be possible to estimate the velocity by minimizing the system (7.30) for small values of μ.

Above we considered numerical experiments with a moderately short tracer time series. Now let us consider how the ML estimator approaches its asymptotic regime when the observation period tends to infinity. A key question is how fast does the ML estimator converge with increasing observations. In other words, how many observations in time are needed to get feasible ML estimates.

Fig. 7.15 shows the number of reasonable estimates for the diffusivity and velocity. These numbers are computed over the mesh (α, β), $\alpha = 2^m$ and $\beta = 2^n$, where $m = -10.0, -9.75, -9.5, ..., 0.0$ and $n = -8.0, -7.75, -7.5, ..., 2.0$, in a series of numerical experiments with $\gamma^* = 0.1, 0.2, ..., 1.6$ for the time series lengths $I = 1, 2, 4, 8, 16, 32, 64, 128$.

It is evident from Fig. 7.15 that the ML estimation of diffusivity is feasible even for the shortest time series. If one has only two sequential tracer fields then

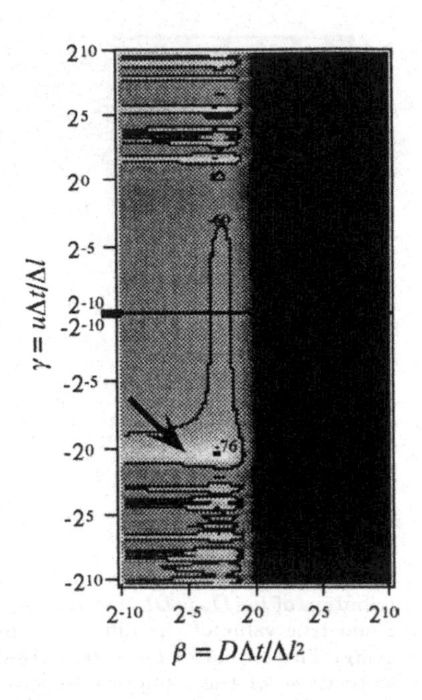

Figure 7.14 Maximum Likelihood function computed in the domain of the dimensionless variables for velocity and diffusivity, γ and β , with a fixed value of the feedback variable, α. While computing the ML function the forcing parameter μ was deliberately set to 0.1 instead of 0.5 as in the forward problem. The solution of the inverse problem is at the ML function minimum. The true solution shown by an arrow corresponds to the value $\beta^* = 2^{-5}$. Notice that unlike Fig. 7.5 the ML function minimum is shifted to a higher value of β. In these numerical experiments the dimensionless velocity parameter is fixed as $\gamma = -1$. Moderately short time series of tracer fields, $I = 13$, are considered. The mesh size is 5 × 5 points.

about 70% of the diffusivity estimates can be deduced correctly. Meanwhile, the estimation error is not homogeneous over the mesh but depends on the absolute value of the unknown diffusivity: the larger the diffusivity the better ML estimate (see Fig. 7.9). From $I = 1$ to $I = 128$ the number of reasonable estimates grows up to 94%. This growth can be approximated by the following formula

$$Number\ of\ Reasonable\ Estimates = 68.5 + 12.4 \log I, \ \text{r.m.s.} = 0.99. \quad (7.49)$$

The outstanding feature of the plot in Fig. 7.15 is the narrow scatter in the number of reasonable estimates for the diffusivity obtained with different values

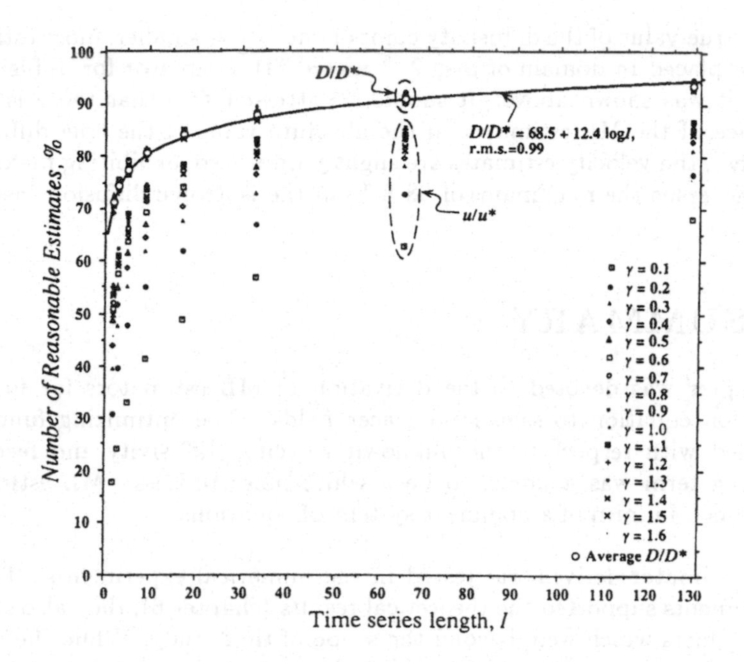

Figure 7.15 Number of reasonable estimates of diffusivity and velocity as functions of the tracer time series length. The solid line is a logarithmic approximation of the average number of reasonable estimates for diffusivity for the experiments with $0.1 \leq \gamma^* \leq 1.6$.

of γ at each fixed value of I. This demonstrates that diffusivity estimates are not affected by velocity variance for $0.1 < \gamma < 1.6$.

The ML estimator for velocity improves drastically with each new sequential data field until $I \approx 10$ (Fig. 7.15). At $I = 16$ it is possible to derive reasonably well 49-78% of the velocities. At the time asymptote, this amount rises to 68-87%. The quality of the velocity estimates depends drastically on the absolute value of the true velocity $\boldsymbol{u}^*$. The best estimates are obtained for $\gamma^* = 0.9$ - 1.2.

Similar experiments were carried out for anisotropic diffusion, $D_{xx} \neq D_{yy}$. The results, in general, are about the same as those of the case of the isotropic diffusion. For example, the number of reasonable estimates for anisotropic diffusivity, $D^*_{yy} = 0.5 D^*_{xx}$, is only 5% smaller with a maximum of 89%. This number can be approximated by the logarithmic function as follows

$$Number\ of\ Reasonable\ Estimates = 64.1 + 12.1 \log I, \text{ r.m.s.} = 0.99. \quad (7.50)$$

Since the true value of the diffusivity component D^*_{yy} is smaller, more estimates should be placed in domain of $\beta < 2^{-5}$ where ML estimator for diffusivity is worse as it was shown above. It should be stressed also that there is still a dependence of the ML estimates on the absolute value of the true diffusivity. By contrast, the velocity estimates are slightly improved, exhibiting maximums of 72-90% versus the maximums of 68-87% in the isotropic diffusion case.

7.4 SUMMARY

This Chapter was devoted to the derivation of ML estimators for fitting a conservation equation to sequential tracer fields. The optimizing functional was defined with respect to the unknown velocity, diffusivity and feedback. The forcing term was assumed to be a white noise process. ML estimators were obtained in form of a nonlinear system of equations.

The ML estimator derived was tested in the numerical experiments. Though the experiments supported the theoretical results (chapter 6), they also showed interesting facts which were beyond the scope of that study. While the theory concerned the asymptotic case of an infinitely large observational domain, here we considered a limited number of observations in space.

The numerical experiments involved the forward and inverse problems. In the forward problem the time evolution of tracer was simulated as a response of the conservation equation to the given forcing. The ML function was calculated from the simulated data in a huge domain of unknown parameters which allowed the diffusion coefficients and feedback rate to vary over six orders of magnitude and the velocity change in two orders of magnitude. These calculations pointed out the main advantages and limitations of ML estimators. Particularly, it was found that the ML estimator is usually consistent for diffusivity; the estimates of the diffusion coefficient were usually consistent over wide ranges of the unknown parameters even when the number of observations in time was limited to two over a spatial grid of 5×5 points. The calculations, also, give us the parameter ranges favorable for feasible velocity estimation and suggested an optimal strategy for searching for the velocity by minimizing the sum of squares of functions in a nonlinear system.

In the inverse problem, the conservation equation was fitted by means of ML estimation to simulated tracer data. The results obtained are akin to the theory and show the consistency of the diffusivity estimates under conditions on the dimensionless variables for feedback and diffusivity typical of upper ocean processes:

$$\alpha < 0.5, \quad \beta > 2^{-6}.$$

The amount of reasonable ML estimates grows as the logarithm of the number of observations in time, mainly due to improvements for the range of small β's while for the larger β's the diffusivity estimates are basically consistent even in the case of the shortest times series. The diffusivity estimator is feasible under the moderately anisotropic diffusion.

The ML estimates for velocity are not reliable except for those values of the variables α and β in the ranges

$$\alpha < 0.5, \quad \beta < 1.$$

The feasibility of making velocity estimates essentially depends upon the initial conditions in the minimization and the absolute value of the true velocity.

The numerical experiments also showed that the ML estimates for the feedback rate λ were not feasible under the conditions favorable for consistent diffusivity estimates. In other words it is impossible to correctly estimate both the diffusivity and the feedback using the ML inversion.

The ML estimator can be widely used for the derivation of the diffusion coefficients of transient flows. For example, one application of the ML technique is the quantitative analysis of satellite radiometry images of the sea. The ML estimator can be used as a new and useful tool in monitoring ocean turbulence.

8

THE INVERSE PROBLEM: AUTOREGRESSIVE ESTIMATORS

Roughly speaking, all numerical solutions of linear partial differential equations can be broken down into two groups: (1) the expansion of the solution in terms of some basis (Galerkin method), and (2) the approximation of derivatives by finite differences. The same is relevant to the inverse problem. While the two previous chapters addressed the Galerkin approach, here we discuss a method based on a finite difference approximation for the same transport equation

$$\frac{\partial c}{\partial t} + \boldsymbol{u} \cdot \nabla c + \lambda c = D\nabla^2 c + S(t, \boldsymbol{r}), \tag{8.1}$$

where the diffusion tensor is assumed to be isotropic and only the 2-dimensional case will be considered. One should be extremely careful when dealing with the discretization of stochastic partial differential equation containing white noise process since the solution in this case may not be differentiable. Fortunately for equation (8.1) with constant coefficients the discretization in time can be done exactly. It results in a recursive integral equation which can be approximated by a multidimensional autoregressive model. The crucial point is that the coefficients of this model are expressed in term of the unknown parameters $\boldsymbol{u}, D$, and λ via simple formulas. Therefore by estimating the autoregressive model via some standard method (least square, for instance) one can retrieve the parameters of interest. Unfortunately the procedure of discretization in space is not exact. For this reason we analyze the accuracy of this procedure. It should be stressed that we do not give any rigorous results for this method. Hopefully, properties of the obtained estimators can be derived from the well known properties of autoregressive models and the convergence properties of difference schemes for PDE's.

8.1 TIME-SPACE DISCRETIZATION OF THE ADVECTION-DIFFUSION EQUATION

The solution of (8.1) given the initial value $c(t_0, \boldsymbol{r}) = c_0(\boldsymbol{r})$ can be written as follows

$$c(t, \boldsymbol{r}) = \exp(-\lambda(t - t_0)) \int_E \Phi_{\boldsymbol{u},D}(t - t_0, \boldsymbol{r} - \boldsymbol{r}')c_0(\boldsymbol{r}')d\boldsymbol{r}' \tag{8.2}$$

$$+\exp(-\lambda t) \int_{t_0}^{t} \int_E \exp(\lambda s)\Phi_{\boldsymbol{u},D}(t - s, \boldsymbol{r} - \boldsymbol{r}')S(s, \boldsymbol{r}')d\boldsymbol{r}'ds,$$

where $E = E^2$ is the 2-dimensional Euclidean space, and

$$\Phi_{\boldsymbol{u},D}(t, \boldsymbol{r}) = \frac{1}{4\pi Dt} \exp(-\frac{(\boldsymbol{r} - t\boldsymbol{u})^2}{4Dt}) \tag{8.3}$$

is the Gaussian kernel (Fig. 8.1). The representation (8.2) follows from the fact that the Gaussian kernel satisfies the equation

$$\frac{\partial \Phi_{\boldsymbol{u},D}}{\partial t} + \boldsymbol{u} \cdot \nabla \Phi_{\boldsymbol{u},D} = D\nabla^2 \Phi_{\boldsymbol{u},D}, \tag{8.4}$$

with the initial condition

$$\Phi_{\boldsymbol{u},D}(t_0, \boldsymbol{r}) = \delta(\boldsymbol{r}), \tag{8.5}$$

where $\delta(\boldsymbol{r})$ is the Dirac delta function.

Fig. 8.1 illustrates the behavior of the solution (8.2) with respect to different values of both the advection velocity and diffusion coefficient in four idealized cases. Notice that the location and height of the Gaussian kernel are modified owing to specific advection and diffusion.

For time discretization, let us set

$$c_n(\boldsymbol{r}) = c(n\Delta t, \boldsymbol{r}), \tag{8.6}$$

where Δt is a prescribed time step and $n\Delta t$ denotes temporal position, $n = 0, 1, 2, \ldots$.

Substituting $t_0 = (n-1)\Delta t,\ \ t = n\Delta t$, and $c_0(\boldsymbol{r}) = c_{n-1}(\boldsymbol{r})$ into (8.2), we get the following recursive equation for any $n = 1, 2, \ldots$

$$c_n(\boldsymbol{r}) = \nu \int_E \Phi_{\boldsymbol{u},D}(\Delta t, \boldsymbol{r} - \boldsymbol{r}')c_{n-1}(\boldsymbol{r}')d\boldsymbol{r}' + \varepsilon_n(\boldsymbol{r}), \tag{8.7}$$

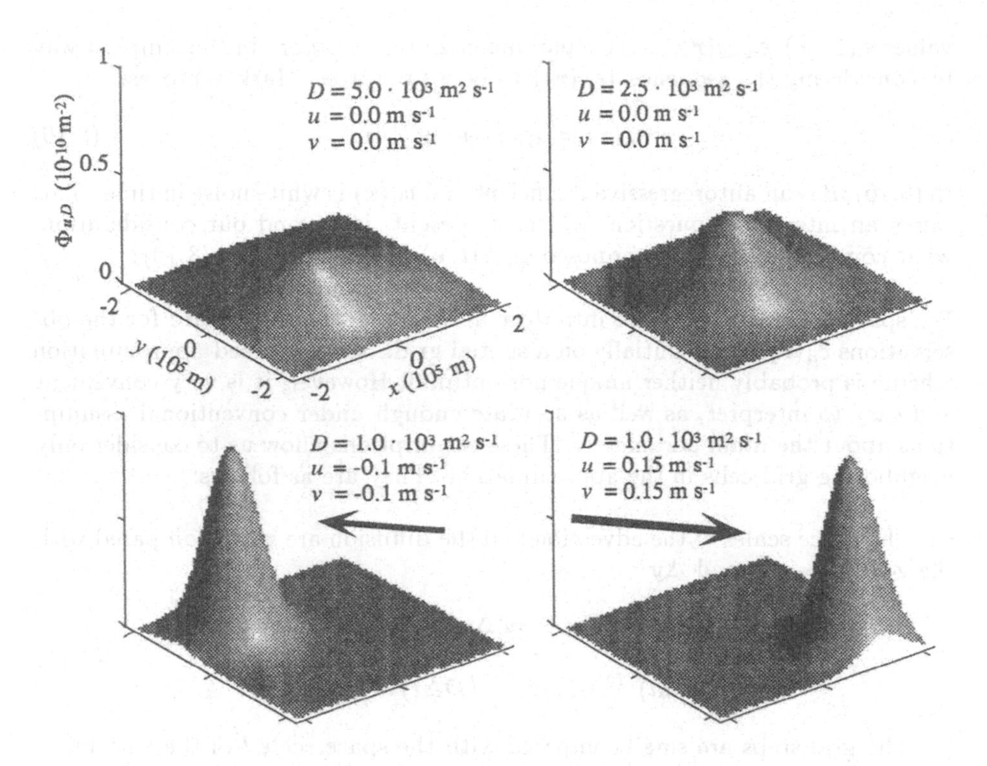

Figure 8.1 The Gaussian kernel for the solution of the advection-diffusion equation. Notice that the location and height of the function maximum are modified owing to the specific velocity and diffusivity in the Gaussian kernel. The arrows indicate advection.

where

$$\nu = \exp(-\lambda\Delta t), \tag{8.8}$$

and

$$\varepsilon_n(\boldsymbol{r}) = \exp(-\lambda n\Delta t)\int_{(n-1)\Delta t}^{n\Delta t}\int_E \exp(\lambda s)\Phi_{\boldsymbol{u},D}(n\Delta t - s, \boldsymbol{r} - \boldsymbol{r}') \tag{8.9}$$

$$\times S(s,\boldsymbol{r}')d\boldsymbol{r}'ds$$

is a stationary sequence of homogeneous random fields.

In this section we do not suppose that $S(t,\boldsymbol{r})$ is white noise in t, instead we assume that the forcing field $\varepsilon_n(\boldsymbol{r})$ appearing in (8.7) depends on the previous values $\varepsilon_{n-1}(\boldsymbol{r}), \varepsilon_{n-2}(\boldsymbol{r}), \ldots$. We parameterize this memory in the simplest way by considering the sequence $\{\varepsilon_n(\boldsymbol{r})\}$ to be a first-order Markov process

$$\varepsilon_n(\boldsymbol{r}) = \beta\varepsilon_{n-1}(\boldsymbol{r}) + b_n(\boldsymbol{r}). \tag{8.10}$$

In (8.10), β is an autoregressive coefficient and $b_n(\boldsymbol{r})$ is white noise in time. This raises an interesting question, which at present, is beyond our consideration: what conditions should be imposed on $S(t, \boldsymbol{r})$ in order to have (8.10)?

For spatial discretization, we introduce an interpolation procedure for the observations $c_n(\boldsymbol{r})$, given initially on a spatial grid. The proposed approximation scheme is probably neither unique nor optimal. However, it is very convenient and easy to interpret, as well as accurate enough under conventional assumptions about the main parameters. These assumptions allow us to consider only neighboring grid cells in the approximation. They are as follows:

(1) The space scales of the advection and the diffusion are small compared with the grid steps Δx and Δy

$$\begin{aligned} |u|\Delta t &\ll \Delta x, \qquad |v|\Delta t \ll \Delta y, \\ (D\Delta t)^{1/2} &\ll \Delta x, \qquad (D\Delta t)^{1/2} \ll \Delta y. \end{aligned} \tag{8.11}$$

(2) The grid steps are small compared with the space scale l of the variability of $c_n(\boldsymbol{r})$ (and, evidently, $\varepsilon_n(\boldsymbol{r})$)

$$\Delta x \ll l, \qquad \Delta y \ll l. \tag{8.12}$$

The scale l can be thought as in Chapter 1.

Now let us proceed from (8.7) to its discrete space form. Our purpose is to derive a relation between the observations of $c_n(\boldsymbol{r})$ and $\varepsilon_n(\boldsymbol{r})$ on a discrete grid with the specific steps Δx and Δy. Unlike the procedure of the time discretization, here it is impossible to derive an exact formula. We can get only an approximate expression that connects the observations of $c_n(\boldsymbol{r}), c_{n-1}(\boldsymbol{r})$, and $\varepsilon_n(\boldsymbol{r})$ on the grid.

The idea behind this procedure is to replace the field $c_{n-1}(\boldsymbol{r}')$ in (8.7) by its approximating polynomial. The latter has coefficients, which are expressed as functions of the values $c_{n-1}(\boldsymbol{r}')$ taken on the grid.

Now let us omit the index $n-1$ of $c_{n-1}(\boldsymbol{r})$ for the sake of clarity. We approximate $c(\boldsymbol{r}) = c(x, y)$ in the rectangle $G = \{(x, y) : |x - x_0| < \Delta x, |y - y_0| < \Delta y\}$

with the sides $2\Delta x$ and $2\Delta y$ centered at the point $\boldsymbol{r}_0 = (x_0, y_0)$ by the quadratic function

$$g(x,y) = A + B(x-x_0) + C(y-y_0) + D(x-x_0)^2 + E(y-y_0)^2 + F(x-x_0)(y-y_0). \tag{8.13}$$

We seek the coefficients A, B, C, D, E, and F, which minimize the sum of squared errors

$$\sum_{m=0}^{8} [c(\boldsymbol{r}_m) - g(\boldsymbol{r}_m)]^2 \to \min, \tag{8.14}$$

where $\boldsymbol{r}_1 = (x_0, y_0 + \Delta y)$, $\boldsymbol{r}_2 = (x_0 + \Delta x, y_0)$, $\boldsymbol{r}_3 = (x_0, y_0 - \Delta y)$, $\boldsymbol{r}_4 = (x_0 - \Delta x, y_0)$, $\boldsymbol{r}_5 = (x_0 - \Delta x, y_0 + \Delta y)$, $\boldsymbol{r}_6 = (x_0 + \Delta x, y_0 + \Delta y)$, $\boldsymbol{r}_7 = (x_0 + \Delta x, y_0 - \Delta y)$, $\boldsymbol{r}_8 = (x_0 - \Delta x, y_0 - \Delta y)$ (see Fig. 8.2). Differentiating (8.14) with respect to A, B, C, D, E, and F, one can show that the minimum is attained when

$$\begin{aligned}
A &= \frac{1}{9}[5c(\boldsymbol{r}_0) + 2c(\boldsymbol{r}_1) + 2c(\boldsymbol{r}_2) + 2c(\boldsymbol{r}_3) + 2c(\boldsymbol{r}_4) \\
&\quad - c(\boldsymbol{r}_5) - c(\boldsymbol{r}_6) - c(\boldsymbol{r}_7) - c(\boldsymbol{r}_8)], \\
B &= \frac{1}{6\Delta x}[c(\boldsymbol{r}_1) - c(\boldsymbol{r}_3) + c(\boldsymbol{r}_5) + c(\boldsymbol{r}_6) - c(\boldsymbol{r}_7) - c(\boldsymbol{r}_8)], \\
C &= \frac{1}{6\Delta y}[c(\boldsymbol{r}_2) - c(\boldsymbol{r}_4) - c(\boldsymbol{r}_5) + c(\boldsymbol{r}_6) + c(\boldsymbol{r}_7) - c(\boldsymbol{r}_8)], \\
D &= \frac{1}{6\Delta x^2}[-2c(\boldsymbol{r}_0) - 2c(\boldsymbol{r}_1) + c(\boldsymbol{r}_2) - 2c(\boldsymbol{r}_3) + c(\boldsymbol{r}_4) \\
&\quad + c(\boldsymbol{r}_5) + c(\boldsymbol{r}_6) + c(\boldsymbol{r}_7) + c(\boldsymbol{r}_8)], \\
E &= \frac{1}{6\Delta y^2}[-2c(\boldsymbol{r}_0) + c(\boldsymbol{r}_1) - 2c(\boldsymbol{r}_2) + c(\boldsymbol{r}_3) - 2c(\boldsymbol{r}_4) \\
&\quad + c(\boldsymbol{r}_5) + c(\boldsymbol{r}_6) + c(\boldsymbol{r}_7) + c(\boldsymbol{r}_8)], \\
F &= \frac{1}{4\Delta x \Delta y}[-c(\boldsymbol{r}_5) + c(\boldsymbol{r}_6) - c(\boldsymbol{r}_7) + c(\boldsymbol{r}_8)],
\end{aligned} \tag{8.15}$$

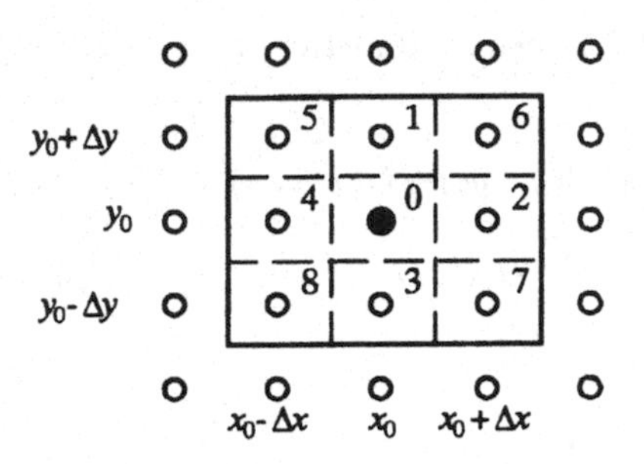

Figure 8.2 The set of the grid cells for the autoregressive model.

In the integral (8.7), instead of $c_{n-1}(\boldsymbol{r})$, we write the above described approximation for the SST anomaly field

$$c_{n-1}(\boldsymbol{r}) \approx A_{n-1} + B_{n-1}(x - x_0) + C_{n-1}(y - y_0) + D_{n-1}(x - x_0)^2 + \\ + E_{n-1}(y - y_0)^2 + F_{n-1}(x - x_0)(y - y_0), \tag{8.16}$$

where the coefficients $A_{n-1}, B_{n-1}, C_{n-1}, D_{n-1}, E_{n-1}, F_{n-1}$ are given by the formulas (8.15), and where $c(\boldsymbol{r}_m) = c_{n-1}(\boldsymbol{r}_m)$ represents the values of $c_{n-1}(\boldsymbol{r})$ at the set of the grid cells $m = 0, \ldots, 8$.

By computing the obtained integrals of the form

$$\int_E (x - x_0)\Phi_{\boldsymbol{u},D}(\Delta t, \boldsymbol{r} - \boldsymbol{r}')d\boldsymbol{r}', \ \ldots \ , \int_E (x - x_0)(y - y_0)\Phi_{\boldsymbol{u},D}(\Delta t, \boldsymbol{r} - \boldsymbol{r}')d\boldsymbol{r}' \tag{8.17}$$

from (8.7) we get

$$c_n(\boldsymbol{r}_0) = \nu\{\frac{1}{9}[5c_{n-1}(\boldsymbol{r}_0) + 2c_{n-1}(\boldsymbol{r}_1) + 2c_{n-1}(\boldsymbol{r}_2) + 2c_{n-1}(\boldsymbol{r}_3)$$

$$+2c_{n-1}(\boldsymbol{r}_4) - c_{n-1}(\boldsymbol{r}_5) - c_{n-1}(\boldsymbol{r}_6) - c_{n-1}(\boldsymbol{r}_7) - c_{n-1}(\boldsymbol{r}_8)]$$

$$-\frac{u\Delta t}{6\Delta x}[c_{n-1}(\boldsymbol{r}_1) - c_{n-1}(\boldsymbol{r}_3) + c_{n-1}(\boldsymbol{r}_5)$$

$$+c_{n-1}(\boldsymbol{r}_6) - c_{n-1}(\boldsymbol{r}_7) - c_{n-1}(\boldsymbol{r}_8)]$$

$$-\frac{v\Delta t}{6\Delta y}[c_{n-1}(\boldsymbol{r}_2) - c_{n-1}(\boldsymbol{r}_4) - c_{n-1}(\boldsymbol{r}_5) + c_{n-1}(\boldsymbol{r}_6)$$

$$+c_{n-1}(\boldsymbol{r}_7) - c_{n-1}(\boldsymbol{r}_8)] \tag{8.18}$$

$$+\frac{u^2\Delta t^2 + 2D\Delta t}{6\Delta x^2}[-2c_{n-1}(r0) - 2c_{n-1}(\boldsymbol{r}_1) + c_{n-1}(\boldsymbol{r}_2) - 2c_{n-1}(\boldsymbol{r}_3)$$

$$+c_{n-1}(\boldsymbol{r}_4) + c_{n-1}(\boldsymbol{r}_5) + c_{n-1}(\boldsymbol{r}_6) + c_{n-1}(\boldsymbol{r}_7) + c_{n-1}(\boldsymbol{r}_8)]$$

$$+\frac{v^2\Delta t^2 + 2D\Delta t}{6\Delta y^2}[-2c_{n-1}(r0) + c_{n-1}(\boldsymbol{r}_1) - 2c_{n-1}(\boldsymbol{r}_2) + c_{n-1}(\boldsymbol{r}_3)$$

$$-2c_{n-1}(\boldsymbol{r}_4) + c_{n-1}(\boldsymbol{r}_5) + c_{n-1}(\boldsymbol{r}_6) + c_{n-1}(\boldsymbol{r}_7) + c_{n-1}(\boldsymbol{r}_8)]$$

$$+\frac{uv\Delta t^2}{4\Delta x\Delta y}[-c_{n-1}(\boldsymbol{r}_5) + c_{n-1}(\boldsymbol{r}_6) - c_{n-1}(\boldsymbol{r}_7) + c_{n-1}(\boldsymbol{r}_8)]\} + \varepsilon_n(\boldsymbol{r}_0).$$

After some algebra, we obtain the following model for the random fields $c(t, \boldsymbol{r})$ and $\varepsilon(t, \boldsymbol{r})$ taken at the specific time intervals over the discrete grid

$$c_n(\boldsymbol{r}_0) = \sum_{m=0}^{8} \alpha_m c_{n-1}(\boldsymbol{r}_m) + \varepsilon_n(\boldsymbol{r}_0) \tag{8.19}$$

where

$$\alpha_0 = \nu[\frac{5}{9} - \frac{1}{3}(\frac{u^2\Delta t^2}{\Delta x^2} + \frac{v^2\Delta t^2}{\Delta y^2} + \frac{2D\Delta t}{\Delta x^2} + \frac{2D\Delta t}{\Delta y^2})],$$

$$\alpha_1 = \nu[\frac{2}{9} - \frac{v\Delta t}{6\Delta y} - \frac{1}{3\Delta x^2}(u^2\Delta t^2 + 2D\Delta t) + \frac{1}{6\Delta y^2}(v^2\Delta t^2 + 2D\Delta t)],$$

$$\alpha_2 = \nu[\frac{2}{9} - \frac{u\Delta t}{6\Delta x} - \frac{1}{3\Delta y^2}(v^2\Delta t^2 + 2D\Delta t) + \frac{1}{6\Delta x^2}(u^2\Delta t^2 + 2D\Delta t)],$$

$$\alpha_3 = \nu[\frac{2}{9} + \frac{v\Delta t}{6\Delta y} - \frac{1}{3\Delta x^2}(u^2\Delta t^2 + 2D\Delta t) + \frac{1}{6\Delta y^2}(v^2\Delta t^2 + 2D\Delta t)],$$

$$\alpha_4 = \nu[\frac{2}{9} + \frac{u\Delta t}{6\Delta x} - \frac{1}{3\Delta y^2}(v^2\Delta t^2 + 2D\Delta t) + \frac{1}{6\Delta x^2}(u^2\Delta t^2 + 2D\Delta t)],$$

$$\alpha_5 = \nu[-\frac{1}{9} + \frac{u\Delta t}{6\Delta x} - \frac{v\Delta t}{6\Delta y} + \frac{1}{6}(\frac{u^2\Delta t^2}{\Delta x^2} + \frac{v^2\Delta t^2}{\Delta y^2} + \frac{2D\Delta t}{\Delta x^2} + \frac{2D\Delta t}{\Delta y^2}) - \frac{uv\Delta t^2}{4\Delta x\Delta y}],$$

$$\alpha_6 = \nu[-\frac{1}{9} - \frac{u\Delta t}{6\Delta x} - \frac{v\Delta t}{6\Delta y} + \frac{1}{6}(\frac{u^2\Delta t^2}{\Delta x^2} + \frac{v^2\Delta t^2}{\Delta y^2} + \frac{2D\Delta t}{\Delta x^2} + \frac{2D\Delta t}{\Delta y^2}) + \frac{uv\Delta t^2}{4\Delta x\Delta y}],$$

$$\alpha_7 = \nu[-\frac{1}{9} - \frac{u\Delta t}{6\Delta x} + \frac{v\Delta t}{6\Delta y} + \frac{1}{6}(\frac{u^2\Delta t^2}{\Delta x^2} + \frac{v^2\Delta t^2}{\Delta y^2} + \frac{2D\Delta t}{\Delta x^2} + \frac{2D\Delta t}{\Delta y^2}) - \frac{uv\Delta t^2}{4\Delta x\Delta y}],$$

$$\alpha_8 = \nu[-\frac{1}{9} + \frac{u\Delta t}{6\Delta x} + \frac{v\Delta t}{6\Delta y} + \frac{1}{6}(\frac{u^2\Delta t^2}{\Delta x^2} + \frac{v^2\Delta t^2}{\Delta y^2} + \frac{2D\Delta t}{\Delta x^2} + \frac{2D\Delta t}{\Delta y^2}) + \frac{uv\Delta t^2}{4\Delta x\Delta y}]. \tag{8.20}$$

Taking into account (8.10) we get from (8.19) the following model:

$$c_n(\boldsymbol{r}_0) = \sum_{m=0}^{8} \alpha_m[c_{n-1}(\boldsymbol{r}_m) - \beta c_{n-2}(\boldsymbol{r}_m)] + \beta c_{n-1}(\boldsymbol{r}_0) + b_n(\boldsymbol{r}_0) \tag{8.21}$$

which is the discrete time-space form of the stochastic partial differential equation (8.1). Multiplying (8.21) by $c_{n-1}(\boldsymbol{r}_0)$ and taking sums over the time series, one obtains the following relation:

$$\gamma_1^{00} = \sum_{m=0}^{8} \alpha_m(\gamma_0^{0m} - \beta\gamma_1^{0m}) + \beta\gamma_0^{00}, \tag{8.22}$$

where

$$\gamma_s^{ij} = \langle c_n(\boldsymbol{r}_i)c_{n-s}(\boldsymbol{r}_j)\rangle \tag{8.23}$$

is the covariance between the time series $c_n(\boldsymbol{r}_i)$ and $c_{n-s}(\boldsymbol{r}_j)$ with a time lag s. Note the covariance

$$\langle b_n(\boldsymbol{r}_i)c_{n-s}(\boldsymbol{r}_j)\rangle = 0, \quad s \geq 1 \tag{8.24}$$

because $\{b_n(\boldsymbol{r})\}$ is white noise in time.

Repeating the same procedure for $c_{n-1}(\boldsymbol{r}_m)$, $m = 1, 2, \ldots, 8$, we get a system of nine nonlinear simultaneous equations of the form (8.22).

Let us estimate the error of the space approximation, assuming for simplicity that $u = v = 0$ and $\Delta x = \Delta y = h$.

Let

$$\tilde{c}_n(\boldsymbol{r}) = \nu \int_E \Phi(\Delta t, \boldsymbol{r} - \boldsymbol{r}')g_{n-1}(\boldsymbol{r}')d\boldsymbol{r}' + \varepsilon_n(\boldsymbol{r}), \tag{8.25}$$

where $g_{n-1}(\boldsymbol{r})$ is given by (8.13) with coefficients from (8.15) in which $c = c_{n-1}$. For the sake of clarity we omitted the subscripts $(\boldsymbol{u}, D)$ of the function $\Phi_{\boldsymbol{u},D}$. The difference between the exact and approximate solutions is

$$\begin{aligned}
&\Delta c = c_n(\boldsymbol{r}) - \tilde{c}_n(\boldsymbol{r}) = \\
&\textstyle\int_E \Phi(\Delta t, \boldsymbol{r} - \boldsymbol{r}')c_{n-1}(\boldsymbol{r}')d\boldsymbol{r}' - \int_E \Phi(\Delta t, \boldsymbol{r} - \boldsymbol{r}')g_{n-1}(\boldsymbol{r}')d\boldsymbol{r}' = \\
&\textstyle\int_{E\setminus G_h} \Phi(\Delta t, \boldsymbol{r} - \boldsymbol{r}')c_{n-1}(\boldsymbol{r}')d\boldsymbol{r}' + \int_{G_h} \Phi(\Delta t, \boldsymbol{r} - \boldsymbol{r}')[c_{n-1}(\boldsymbol{r}') \\
&-g_{n-1}(\boldsymbol{r}')]d\boldsymbol{r}' + \int_{E\setminus G_h} \Phi(\Delta t, \boldsymbol{r} - \boldsymbol{r}')g_{n-1}(\boldsymbol{r}')d\boldsymbol{r}',
\end{aligned} \tag{8.26}$$

where G_h is the square with the side $2h$ centered at $\boldsymbol{r}$. Let us denote the first, second, and third terms on the right-hand side of (8.26) by I_1, I_2, and I_3, respectively. Notice,

$$\int_{D_l} \Phi(\Delta t, \boldsymbol{r} - \boldsymbol{r}')\boldsymbol{r}' \le \int_E \Phi(\Delta t, \boldsymbol{r} - \boldsymbol{r}')d\boldsymbol{r}' = 1 \tag{8.27}$$

so for the first integral in (8.26) we have

$$|I_1| \le \max|c(\boldsymbol{r}') - g(\boldsymbol{r}')|, \quad \boldsymbol{r}' \in G_h, \tag{8.28}$$

where the subscripts have been omitted. Let us estimate the right-hand side of (8.28). First, we expand $c(\boldsymbol{r}_m)$, $m = 1, \ldots, 8$ up to third-order terms in a power series centered at the point $\boldsymbol{r}_0$, for example,

$$c(\boldsymbol{r}_1) = c(\boldsymbol{r}_0) + h\frac{\partial c(\boldsymbol{r}_0)}{\partial y} + \frac{h^2}{2}\frac{\partial^2 c(\boldsymbol{r}_0)}{\partial y^2} + \frac{h^3}{6}\frac{\partial^3 c(\boldsymbol{r}_*)}{\partial y^3}, \tag{8.29}$$

where $\boldsymbol{r}_*$ is a point which belongs to G_h.

Then, by substituting the obtained results into the expressions for the coefficients in (8.15) and taking into account $\Delta x = \Delta y = h$, we get

$$\begin{gathered} |A - c(\boldsymbol{r}_0)| \le \frac{4}{3}h^3 M_3 \quad |B - \frac{\partial c(\boldsymbol{r}_0)}{\partial x}| \le \frac{8}{9}h^2 M_3, \\ |C - \frac{\partial c(\boldsymbol{r}_0)}{\partial y}| \le \frac{8}{9}h^2 M_3, \quad |D - \frac{1}{2}\frac{\partial^2 c(\boldsymbol{r}_0)}{\partial x^2}| \le \frac{19}{8}hM_3, \\ |E - \frac{1}{2}\frac{\partial^2 c(\boldsymbol{r}_0)}{\partial y^2}| \le \frac{19}{8}hM_3, \quad |F - \frac{1}{2}\frac{\partial^2 c(\boldsymbol{r}_0)}{\partial x \partial y}| \le \frac{4}{3}hM_3, \end{gathered} \tag{8.30}$$

where

$$M_3 = \max|\frac{\partial^3 c(\boldsymbol{r}_*)}{\partial^i x \partial^j y}|, \quad \boldsymbol{r}_* \in G_h, \quad i + j = 3. \tag{8.31}$$

Finally, let us expand $c(\boldsymbol{r})$ in the power series as well. Taking into account (8.30), we obtain that

$$|c(\boldsymbol{r}') - g(\boldsymbol{r}')| \le \frac{22}{3}h^3 M_3, \quad \boldsymbol{r}' \in G_h. \tag{8.32}$$

Using the scale relation

$$M_3 \sim \frac{\sigma_c}{l^3} \tag{8.33}$$

where σ_c is a typical absolute value of the tracer concentration, for instance the standard deviation and l is its spatial scale, we derive from (8.28) and (8.30)

$$|I_1| \le \frac{22}{3}\sigma_c(\frac{h}{l})^3. \tag{8.34}$$

To estimate the second integral in (8.26), first, we find

$$\int_{|x'-x|>h,|y'-y|>h} \Phi(\Delta t, \boldsymbol{r}-\boldsymbol{r}')d\boldsymbol{r}' \leq$$

$$4\int_{h}^{\infty}\int_{-\infty}^{\infty}\Phi(\Delta t, x, y)dxdy < \frac{2\sigma}{\sqrt{\pi}h}\exp(-\frac{h^2}{4\sigma^2}), \tag{8.35}$$

where $\sigma = (D\Delta t)^{1/2}$ is the diffusion length scale.

Then one can write

$$|I_2| < \sigma_c \frac{2\sigma}{\sqrt{\pi}h}\exp(-\frac{h^2}{4\sigma^2}). \tag{8.36}$$

Note that

$$\int_E x\Phi(\Delta t, x, y)dxdy = \int_E y\Phi(\Delta t, x, y)dxdy =$$
$$\int_E xy\Phi(\Delta t, x, y)dxdy = 0. \tag{8.37}$$

So, the estimate of the third integral in (8.26) is given by

$$|I_3| < A\frac{2\sigma}{\sqrt{\pi}h}\exp(-\frac{h^2}{4\sigma^2}) + (D+E)\frac{4h\sigma^2}{\sqrt{\pi}\sigma}\exp(-\frac{h^2}{4\sigma^2}) \sim$$
$$\sigma_c\frac{2}{\sqrt{\pi}}\exp(-\frac{h^2}{4\sigma^2})(\frac{4\sigma}{3h}+\frac{4\sigma^2}{l^2}). \tag{8.38}$$

Finally, summarizing the results (8.34), (8.36) and (8.38), we find that the relative error of the approximation (8.25) can be estimated as follows:

$$\delta = \frac{|\Delta c|}{\sigma_c} \leq \frac{22}{3}\frac{h^3}{l^3} + \frac{2}{\sqrt{\pi}}\exp(-\frac{h^2}{4\sigma^2})(\frac{7}{3}\frac{\sqrt{D\Delta t}}{h} + \frac{4D\Delta t}{l^2}). \tag{8.39}$$

Using the dimensionless variables $\zeta = h/l$, $\eta = 2(D\Delta t)^{1/2}/h$, we write

$$\delta(\zeta,\eta) = \frac{22}{3}\zeta^3 + \frac{2}{\sqrt{\pi}}\exp(-\frac{1}{\eta^2})(\frac{7}{6}\eta + \eta^2\zeta^2). \tag{8.40}$$

A plot for $\delta(\zeta,\eta)$ is shown in Fig. 8.3. The expression for the relative error of the approximation (8.25) is more difficult to derive when $u, v \neq 0$, but it is clear that the order of the error remains the same as that of $\delta(\zeta,\eta)$ in (8.40).

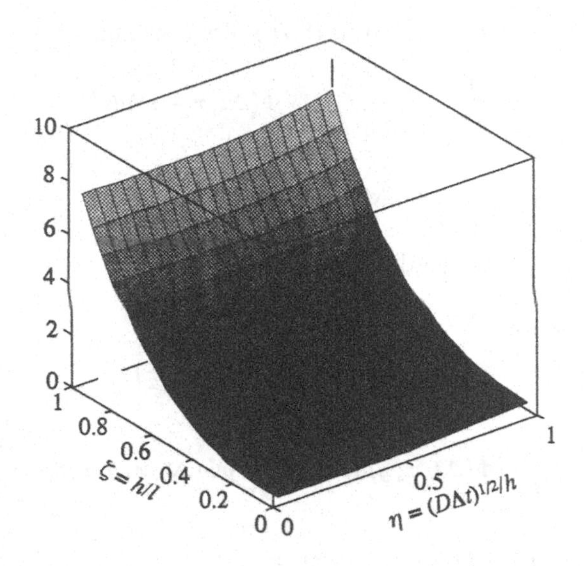

Figure 8.3 The relative error δ of the discrete time-space approximation of the advection-diffusion equation.

8.2 SUMMARY

An alternative method of estimating the velocity, diffusivity and feedback parameter in the transport equation (8.1) is discussed. This equation is reduced to a multivariate autoregressive model by discretization in time and space. The coefficients of the autoregressive model are non-linear functions of the unknown parameters. Estimating the coefficients via the least squares method and solving the mentioned non-linear equations, we get estimators of the parameters of interest. This is the essence of autoregressive inversion.

This approach does not require the assumption that the forcing is white noise. We give an estimate for the space discretization accuracy in (8.41).

9

A STOCHASTIC MODEL FOR SST ANOMALY TRANSPORT

Chapters 1-5 concerned the theory and numerical simulation of advection and diffusion. To delineate these processes we took advantage of tracers - Lagrangian particles and scalar concentrations. In practice, however, while using tracer models, we run into difficulties since experimental data constrain the theory.

On the one hand, the number of Lagrangian particles, for example drifters in the ocean, is often not enough to obtain statistically significant estimates because the particle density decreases when they spread from source making their number insufficiently large to obtain stable statistics (this is in stark contrast to numerical simulation where, at the limit of infinitely large number of particles, their density can be converted into a tracer concentration, for a discussion see Richards et al. 1995). It should be noted that since 1988 the Global Drifter Program has made a tremendous effort to correct the spatial imbalance in the distribution of drifters drogued in the upper ocean (Niiler 1995).

On the other hand, it is difficult to identify a real passive scalar in the environment because some measured quantities can be treated as passive scalars only at certain time-space scales. If the scales are selected properly then a data assimilation, including inverse modeling (e.g. Wunsch 1978, Wunsch and Minster 1982, among others), can operate with these quantities assumed to be tracers. In oceanic and atmospheric physics, data assimilation is emerging in response to the rapid growth of data collected via satellite remote sensing. Although climate studies are moving swiftly into the era of global observing systems (CLIVAR 1995), many observations of various parameters, which can be considered as proxies for tracers, are poorly analyzed so tremendous amounts of information remain hidden in data archives. Development of methods for tracer studies is a high priority task in meteorology and physical oceanography.

In chapters 6 through 8, we introduced estimators which are aimed at inferring flow dynamics from time dependent distributions of scalar concentration. Although these methods suit more or less the general case, they originated from our interest in geophysical flows. Application to geophysical data such as those collected in the ocean are of particular importance providing a sort of 'ground truth' for our methodology.

In the remaining two chapters we, firstly, apply the autoregressive and maximum likelihood estimators for a sea surface temperature (SST) variability and, secondly, turn estimate the near-surface ocean advection and diffusion from sequential SST fields.

In the late 1960s, SST variability caught the attention of oceanographers partly because it was realized that the SST field contains a wealth of information about the upper ocean. For the first time, perhaps, McLeish (1970) suggested that the temperature patterns represent the indirect effect of turbulence on the upper ocean. As a result, infrared radiometry was accepted as a method for the study of ocean dynamics. Since the middle 1970s, satellite infrared imagery has been designated to measure SST. The Very High Resolution Radiometer (VHRR) aboard the ATS-6 spacecraft in 1974 was followed by Advanced Very High Resolution Radiometer (AVHRR) which became the most popular operational remote-sensing instrument for physical oceanography after launch of the Tiros N satellite in 1978.

Now, satellite-born SST images are routinely used to trace oceanic currents, eddies, long waves, and fronts (for a review see Fedorov and Ginzburg 1988). A correlation method (Emery et al. 1986) and an inversion technique (Kelly 1989) have been developed to derive sea currents from sequential SST images. A spectral analysis (for review see Fedorov 1986), a fractal dimension estimate (Bunimovich et al. 1993), and a wavelet visualization (Ostrovskii 1995) were used to provide a qualitative description of upper ocean turbulence revealed by SST fields. However, in general, how best to extract information on ocean dynamics from remotely sensed SST data is still an open question. It should be noted that SST is both a tracer and a 'player' in the thermodynamics of the ocean.

The fact that SST is not an ideal tracer but a governing thermodynamic parameter gives us insight into the problem of heat transport in the upper ocean that is important for climate change (Hasselmann 1982). So the question can asked how can one extract information on upper ocean heat advection and diffusion from the SST time series? The answer can be found by considering the conservation equation (3.7) defined for SST anomaly (departure from climate mean or space-time average, e.g. Adem 1970, 1975). For certain time and space scales, SST anomalies are assumed to undergo the advection and diffusion, so they are analogous to scalar concentration. This definition is granted by Hasselmann's (1976) stochastic climate model theory.

Hasselmann (1976) paid attention to the fact that the spectra of the oceanic signals, including that of SST anomalies, exhibit a monotonous growth in energy with decreasing frequency for period bands lasting a dominant, synoptic, time scale of atmospheric processes. It was hypothesized that there is a two-scale separation of weather and climate variables so the long-period SST anomalies can be viewed as an integral response of the upper ocean to short period weather forcing (Hasselmann 1976, Frankignoul and Hasselmann 1977). According to that theory, slowly responding components of the climate system such as oceans and ice sheets act as integrators of random input by short time scale 'weather' components much in the same way as Brownian particles integrate the forces exerted on them by the light particles.

Numerical simulations and data analyses for midlatitude oceans agree broadly with Hasselmann's (1976) stochastic climate model theory (for a review see Frankignoul 1985 and Dobrovolski 1992; more recent results can be found in Miller 1992, Alexander 1992, Miller et al. 1994, GarciaOrtiz and RuizdeElvira 1995, Ostrovskii and Piterbarg 1995). Hasselmann's hypothesis (1976) and rigorous approach to averaging the heat balance equation (Piterbarg 1987) provide a rationale for embedding the results of the above chapters 6-8 in a stochastic model of the SST anomaly transport. As a result, the autoregressive and maximum likelihood estimators can be applied to the SST anomaly field in order to extract information on heat advection and diffusion in the upper ocean.

9.1 TEMPERATURE ANOMALIES IN THE UPPER OCEAN

Concerning the upper ocean, we must bear in mind that there is so-called 'mixed layer' that can penetrate to considerable depths. Indeed, in midlatitudes, the top part of the water column is usually characterized by a layer of almost uniform temperature over a stratified thermocline (Fig. 9.1). The uniform

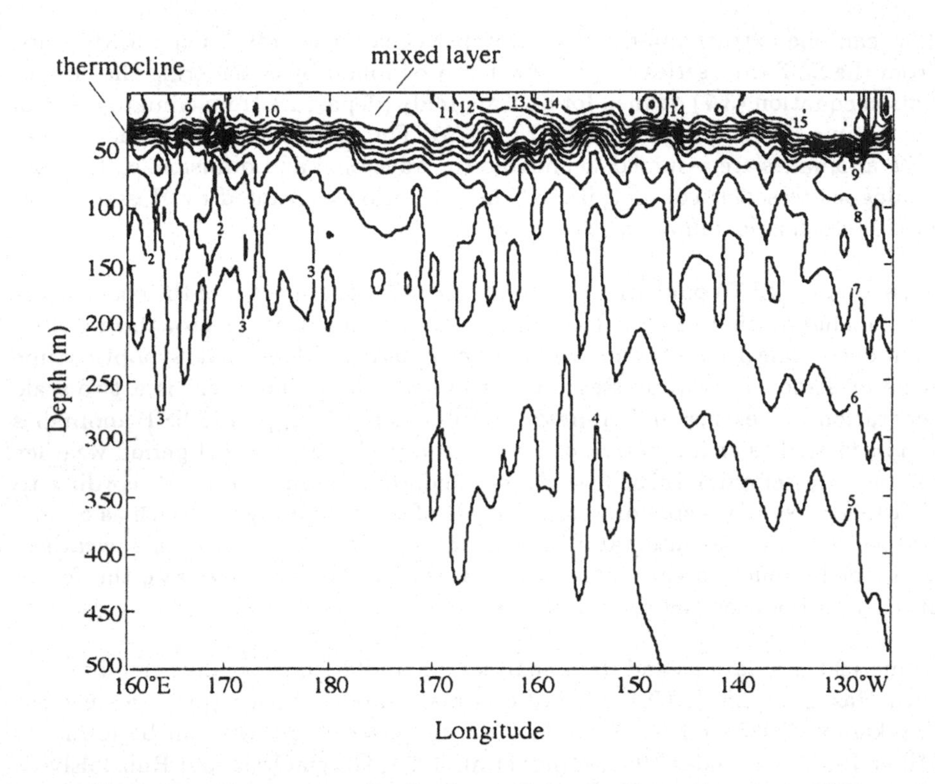

Figure 9.1 Temperature distribution at Trans Pacific section along 47°N done by R/V 'Thompson'. Drawn from the ATLAST data by Rhines (1992).

temperature layer is mainly generated by local turbulence in a stably stratified water (Turner and Kraus 1967). The upper mixed layer undergoes seasonal evolution in phase with the annual cycle of buoyancy and momentum fluxes across the air-sea boundary. According to Bathen (1972), in the central North Pacific, the upper mixed layer has a minimum depth in summer due to the response of solar heating. This shallow warm layer remains until October when it begins to deepen due to the fall cooling and the storms which mix the surface water down. By November the mixed layer deepens to 60-80 meters. During the period between December and April its depth generally exceeds 100 meters north of 30°N.

The upper mixed layer has been often treated as horizontally statistically homogeneous. This idea has led to the development of one-dimensional models (e.g. Kitaigorodskii 1960, Kraus and Rooth 1961). The latter were extended to 'bulk models' by Turner and Kraus (1967), Niiler (1975), Niiler and Kraus (1977), and Garwood (1977). In the bulk models, the conservation equations for momentum, temperature, salinity and buoyancy are integrated vertically within the mixed layer. The turbulence energy equation is used to derive an expression for the evolution of the layer depth that is needed for closure. Thus, the obtained system governs the momentum and heat balance of the entire mixed layer under the action of momentum and buoyancy fluxes at the sea surface while entrainment at the bottom of the mixed layer controls deepening.

In this approach, major complications are associated with the parameterization of the entrainment (for discussion and review see Martin 1985, Large et al. 1994, Kantha and Clayson 1994). Nevertheless the bulk model concept has proved to be a good approximation for the upper ocean. It is mostly suitable for applications during periods when the upper mixed layer depth is rather stable.

In this study we consider SST averaged over geographical boxes of about 100 km in space and over time intervals of 10 days as a proxy for the mixed layer temperature averaged throughout the same time-space grid. We neglect the surface 'skin-layer' associated with the heat transport through the top millimeter of the water column (for review see Fedorov and Ginzburg 1988).

More important could be the diurnal cycle of thermal structure in the upper mixed layer (e.g. Fedorov and Ginzburg 1988, Kudryavtsev and Soloviev 1990). This effect is greatest on the afternoons of relatively calm days. Since solar radiation is progressively absorbed as it penetrates the water column, the temperature rises most near the surface. During weak wind events, solar heating suppresses the turbulence and limits the downward penetration of the turbulent mixing so the heating is localized in a thin near-surface layer on the order of 1 m. The temperature differences can be about 1° C in the top 5 meters of the ocean (Kudryavtsev and Soloviev 1990). Therefore, analyzing satellite radiometer data, one has to take into account the effect of the diurnal thermocline. However it is possible to exclude this effect by using night-time infrared imagery because after sunset the convection generated by heat losses to the atmosphere destroys the diurnal thermocline. Therefore, the night time SST measurements should be free of the diurnal thermocline problem (Robinson 1985).

Concerning the vertical thermal anomaly structure, it is worth mentioning that the observations demonstrate SST anomalies do not only extend throughout the oceanic mixed layer but penetrate to the deeper ocean. However such phenomena have an interannual rather than a seasonal scale. Analysis of monthly mean bathythermograph data from the Ocean Weather Ships N (30°N, 140°W), P

(50°N, 145°W), and V (34°N, 164°E) in the North Pacific showed that anomalous subsurface thermal structure was generally deeper than 275 m, the maximum depth of observation (White and Walker 1974). At the two stations N and P, the subsurface anomaly was tilted with a depth indicating that the anomaly had formed initially near the surface and had penetrated deeper at an approximate rate of 100 m year^{-1}. Further analysis of observations at weather station N (Emery 1976) showed an appreciable interannual variability of slow vertical motions in upper ocean thermal variations. Airborne expendable bathythermograph measurements in the central North Pacific (Barnett 1981) indicated that spatially coherent features of the non-seasonal, approximately monthly, anomalies are confined largely to the upper 100 m, with small spatial scales dominating the variance below. Other curious data collected during 1956-1978 at the weather station P and along Line P (Tabata 1981) showed that while there were extreme temperature anomalies at depths of 200-1000 m nothing of this kind occurred at the surface. As to the North Atlantic, its SST anomalies exhibited the 20-25-year trends associated with the amounts of a heat absorbed by the ocean (Bunker 1980).

Now, let us consider the value of SST, $T(t, \boldsymbol{r})$, at a fixed geographical point $\boldsymbol{r}$. Assume that this value is obtained by averaging individual observations during a considerable time interval $(t - \tau/2, t + \tau/2)$, say 10 days, which is enough to filter out the diurnal cycle. Suppose that the climate mean or norm of $T(t, \boldsymbol{r})$

$$\overline{T}(t, \boldsymbol{r}) = \frac{1}{N} \sum_{n=1}^{N} T(t + (n-1)t_0, \boldsymbol{r}), \tag{9.1}$$

where N is a number of observational years and t_0 is the annual time period. Notice, that the averaging (9.1) gives us a quasiperiodic function with period t_0. The departure of the SST from the norm:

$$T'(t, \boldsymbol{r}) = T(t, \boldsymbol{r}) - \overline{T}(t, \boldsymbol{r}) \tag{9.2}$$

is called the SST anomaly.

The definitions (9.1)-(9.2) can be subject to objections. For example, in some calendar year the annual cycle of SST variability may not have a 1 year period but instead it can be modified by delay or early onset of any season (Lappo and Gulev 1984). Here, following tradition, we stick to the conventional average (9.1) and the simple removal of the annual cycle (9.2). It is hard to provide another formalism that would be satisfactory in different situations. However, in certain cases it might be useful to define the annual cycle as a superposition of the annual, semiannual, 4-months, 3-months and so on harmonics (Fig. 9.2,

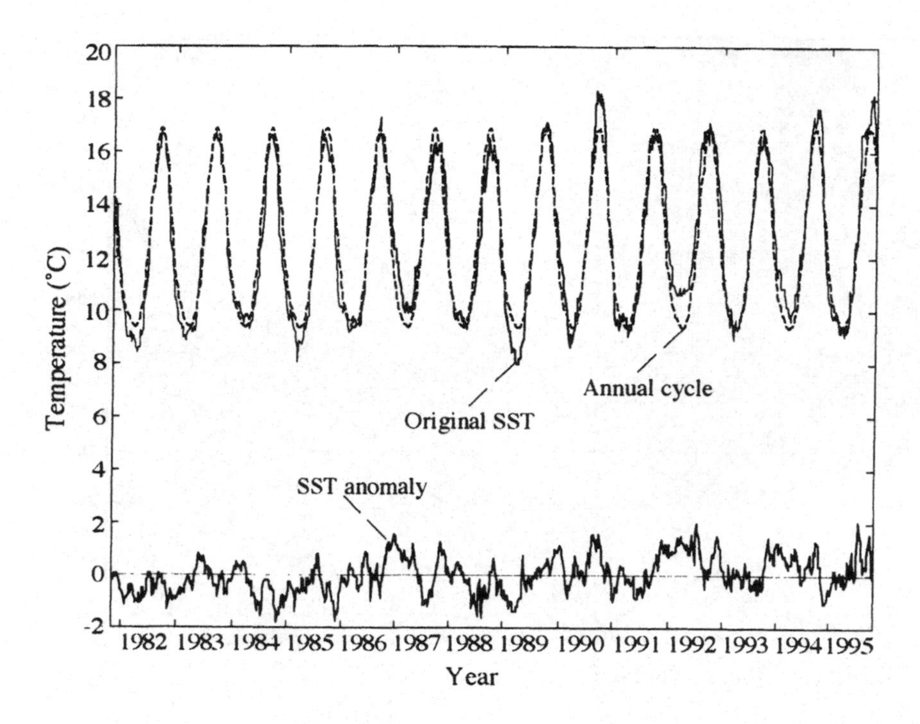

Figure 9.2 Examples of the SST time series observed at 46°N, 131°W, the SST annual cycle and the SST anomaly time series. Data courtesy R. Reynolds (Reynolds 1988, Reynolds and Smith 1994).

also see Horel 1982). Another point is that the number of observational years N_0, beyond which the SST statistics do not vary, may not be inferred from relatively short time series of ocean measurements. Here, as usual the average is computed throughout the whole observational period.

If we extend the considerations (9.1)-(9.2) to other geographical locations we find that significant departures (those exceeding a standard deviation, for example) from the norm occur over a considerable area (say, 10^6 km^2) during a long period (say, several months). This may be a climatically significant SST anomaly (Fedorov and Ostrovskii 1986) which results in a change of atmosphere circulation. A famous example of such an anomaly is the El-Nino/Southern Oscillation (ENSO) event in the equatorial Pacific (Philander 1990).

At fixed geographic coordinates $\boldsymbol{r}$, the function $T'_{\boldsymbol{r}}(t) = T'(t, \boldsymbol{r})$ comprises a time series of departures from the norms. An example of a SST anomaly time

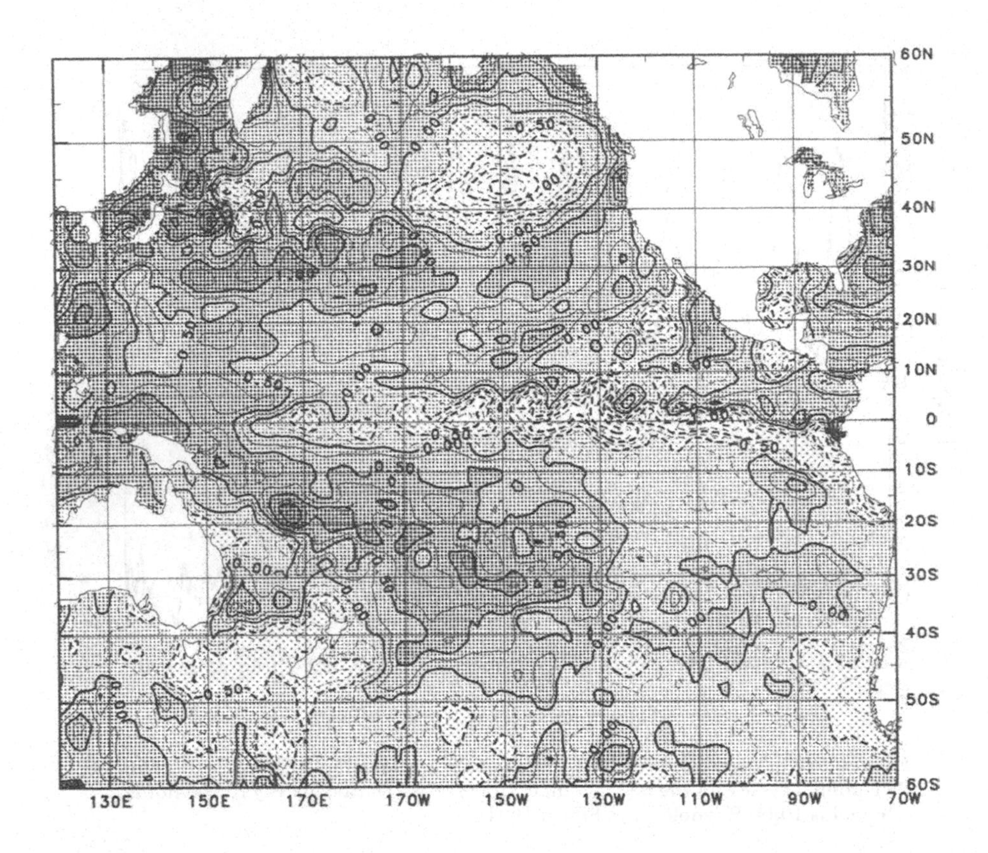

Figure 9.3 An example of SST anomalies in the Pacific Ocean on Oct. 15-21, 1995. Data courtesy R. Reynolds (Reynolds 1988, Reynolds and Smith 1994).

series is drawn in Fig. 9.2. Anomalies of the same sign observed continuously over sufficiently long time, say, more than one month, we call 'long-term SST anomaly'.

At a fixed time t, the 'frozen' field $T'_t(\boldsymbol{r}) = T(t, \boldsymbol{r})$ comprises a map of SST anomalies. One such map for the North Pacific is shown in Fig. 9.3. Apparent in Fig. 9.3 is the tendency for large SST anomaly maximums to be associated with large regions of the coherent SST anomaly variations. Spatial coherence with magnitudes higher than one half of the SST anomalies standard deviations can be observed in large areas on the order of 10^6 km^2. We call such phenomenon 'large-area SST anomalies'.

Actually, two formal definitions can be introduced for large-area SST anomalies. The first one is associated with our intuitive understanding; the second is more convenient for analytical study. According to the first definition (Piterbarg 1985, for more formalism see section 1.1), a large-area positive SST anomaly comprises the connected components of the set

$$\{(x, y) : T'(t, \boldsymbol{r}) > T_0\},$$

where $\boldsymbol{r} = (x, y)$ and T_0 is a threshold that can be taken either as a fixed temperature value, for example, 2°C or as a function of the standard deviation of SST anomalies, σ_T, for example, $0.5\sigma_T$. This choice for the threshold hinges on the researcher and the anomaly magnitude of interest. Similarly large-area negative SST anomalies can be defined. Fig. 9.4 demonstrates an example when the threshold is taken to be $T_0 = 0.5\sigma_T$.

Within the framework of the first definition, to investigate the lifetime of SST anomalies, it is worth considering space-time phenomenon. In other words, one can analyze the connected components of the three dimensional set

$$\{(t, x, y) : T'(t, \boldsymbol{r}) > T_0\}.$$

Long-term and large-area anomalies are represented as temporal and spatial sections of this set, respectively.

The second definition of large-area SST anomalies is as follows. The set of all points near the given location $\boldsymbol{r}_0$, for which it is known that $|T'(t, \boldsymbol{r}_0)| > T_0$, is considered to be a connected subset, i.e. part of the same anomaly if the correlation between $T'(t, \boldsymbol{r})$ and $T'(t, \boldsymbol{r}_0)$ is higher than a certain level σ_0. A criterion used to define the value σ_0 is admittedly ad hoc. The second definition incorporates the geographical coherence of anomalies and is often assumed implicitly while introducing a correlation radius for interpolation purposes. Nosko (1969, 1985) showed that asymptotically both definitions are equivalent for high excursions from norm. More specifically, high excursions in homogeneous random field yield the shape of the correlation function.

Midlatitude SST anomalies are characterized by an e-folding time scale. The bigger the horizontal size the larger the heat anomaly capacity and, consequently, the longer the decay time. In the midlatitude North Pacific, the e-folding time of monthly SST anomalies ranges between 5 months on basin scale to 1 month for the scale of order 500 km (Reynolds 1978). Large-scale SST anomalies have bigger horizontal sizes and longer lifetimes.

The large-scale SST anomalies interact with overlying atmospheric circulation. This was discovered by Namias (1959) and Bjerknes (1962) as early as at the

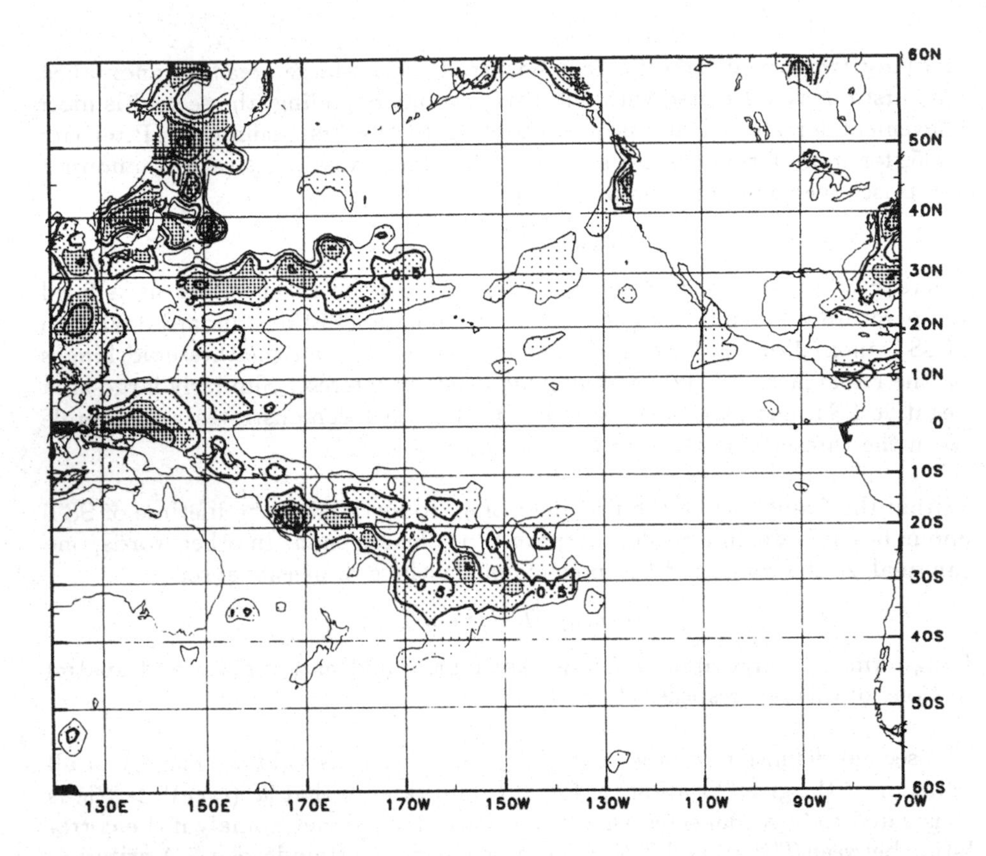

Figure 9.4 Large-area SST anomalies in the Pacific Ocean on Oct. 15-21, 1995. These anomalies are defined as departures from the threshold $T_0 = 0.5\sigma_T$.

end of the 1950's. In a series of studies Namias (1963, 1965, 1969, 1972, 1976) further described various interactions between SST anomalies and atmospheric circulation in midlatitudes (see also Namias and Born 1970, Namias and Cayan 1981). Bjerknes (1962, 1964) proposed physical links between the SST variations and an index for atmospheric circulation in the region between Iceland and the Azores. Later he focused attention on equatorial SST anomalies (Bjerknes 1966, 1969). Many references to the interactions between the midlatitude SST anomalies and the overlying atmosphere can be found in (Frankignoul 1985 and Dobrovolski 1992). Some processes in the climate system leading to the generation and evolution of SST anomalies will be discussed below.

9.2 SCALES OF SST AND ATMOSPHERIC VARIATIONS

Here we give an outline of the basic results obtained from statistical analyses of climate variable time series relevant to the problem discussed. Of particular importance for this study is the statistical relationship between SST and atmospheric pressure anomalies. The sea level pressure (SLP) determines the wind in the boundary layer above the sea. Surface wind stress is a major factor in the atmospheric forcing of the upper ocean. In accordance with the bulk parameterization (c.f. Kitaigorodskii 1970, Kraus 1972, Girdyuk and Malevskii-Malevich 1973, Businger 1975, Kondo 1975, Bunker 1976, Reed 1977, Isemer and Hasse 1987), turbulent flux across the air-sea interface relies on the wind stress both explicitly and implicitly. The latter is due to the dependence of the bulk exchange coefficients on wind speed. It is worth mentioning that turbulence generated in response to the wind stress is partly responsible for vertical heat transport in the upper ocean. It leads to entrainment and mixed layer deepening (c.f. Kraus and Turner 1967, Phillips 1969, Elsberry and Camp 1978b).

Let us focus on North Pacific SST anomalies which were studied rather intensively during the past three decades. The standard deviation of SST anomalies as a departure from the climate norm is less than 2.5°C in the midlatitude North Pacific. This holds for the anomalies deduced from different data sources with averaging periods from 1 week to 1 month over geographical boxes of the sizes from $1^\circ \times 1^\circ$ to $5^\circ \times 5^\circ$ (Frankignoul and Reynolds 1983, Dobrovolski 1992, Ostrovskii and Piterbarg 1985, 1995). The anomaly amplitude is characterized by an increase in summer and a decrease in winter. Fig. 9.5 demonstrates the standard deviation σ_T of the weekly SST anomalies in the Pacific Ocean. The winter and summer seasons are represented by the third weeks of January and July.

The patterns in Fig. 9.5 clearly show the seasonal nonstationarity of the second order moment σ_T (see also Saino 1992, Ostrovskii and Piterbarg 1995). For the midlatitudes it is typical that the SST anomaly variance is largest in the central North Pacific during the northern hemisphere summer. The maximum values of $\sigma_T \approx 2$°C occurred in a latitudinal belt between 35° and 45°N. By contrast, during the boreal winter, the SST anomaly variance is enchanced ($\sigma_T \approx 1.25$°C) in the southern hemisphere, especially in two areas, near New Zealand and at the region bounded approximately by 25°S and 40°S, 110°W and 150°W. This seasonal change seems to be associated with the annual march of the upper ocean mixed layer depth. While the mixed layer is thin in summer, its thermal

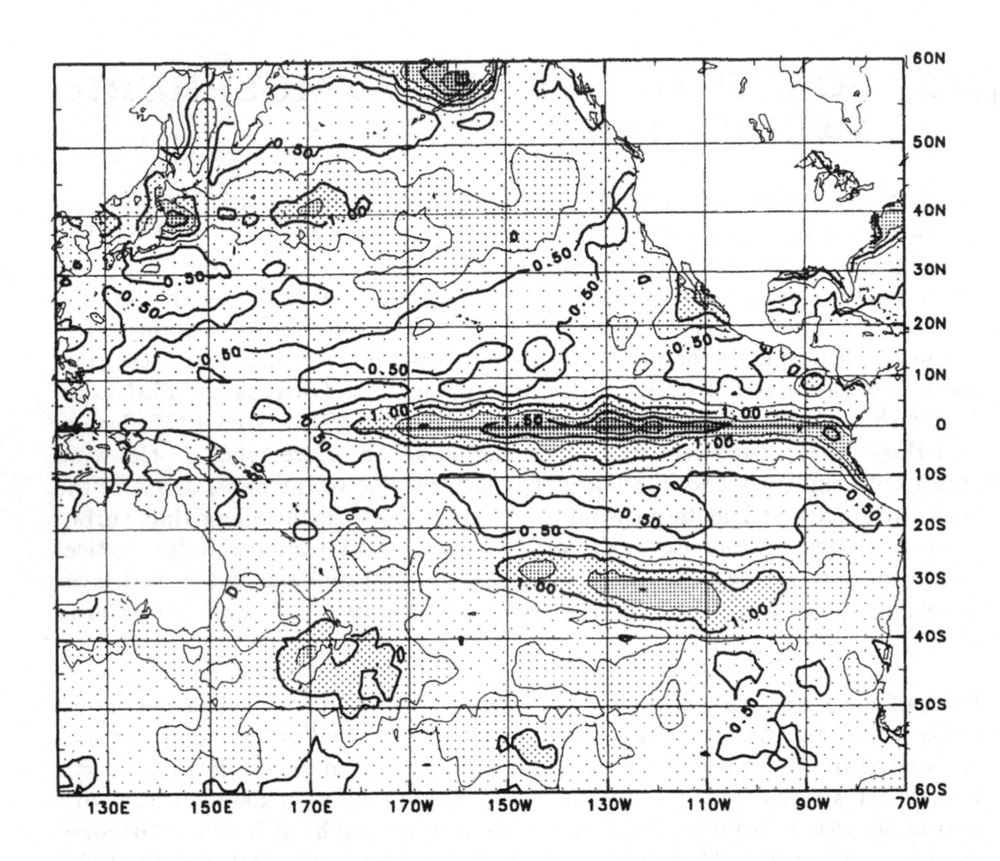

Figure 9.5 *a*) Distribution of the SST anomaly r.m.s. σ_T for the third week of January. Data courtesy R. Reynolds (Reynolds 1988, Reynolds and Smith 1994).

inertia is small compared to that of the cold season when the mixed layer is relatively deep. This meridional seesaw of the SST anomaly variance and the relationship between the air-sea heat flux and SST anomalies (Cayan 1992a, see below) implies that during the same season the upper layers of the North and South Pacific play different roles in the ocean-atmosphere interactions.

In the North Pacific, a lower SST anomaly variance ($\sigma_T < 0.5°$C) is observed in the western tropical Pacific far away from the equator. Another region of the low variance is located in the subpolar latitudes of 50° – 60°N between 170°E and 170°W. The variance is higher ($\sigma_T > 1°$C) in the vicinity of the western boundary along the convergence of the warm Kuroshio and cold Oyashio currents. In this region, the SST variability is associated with the frontal zone.

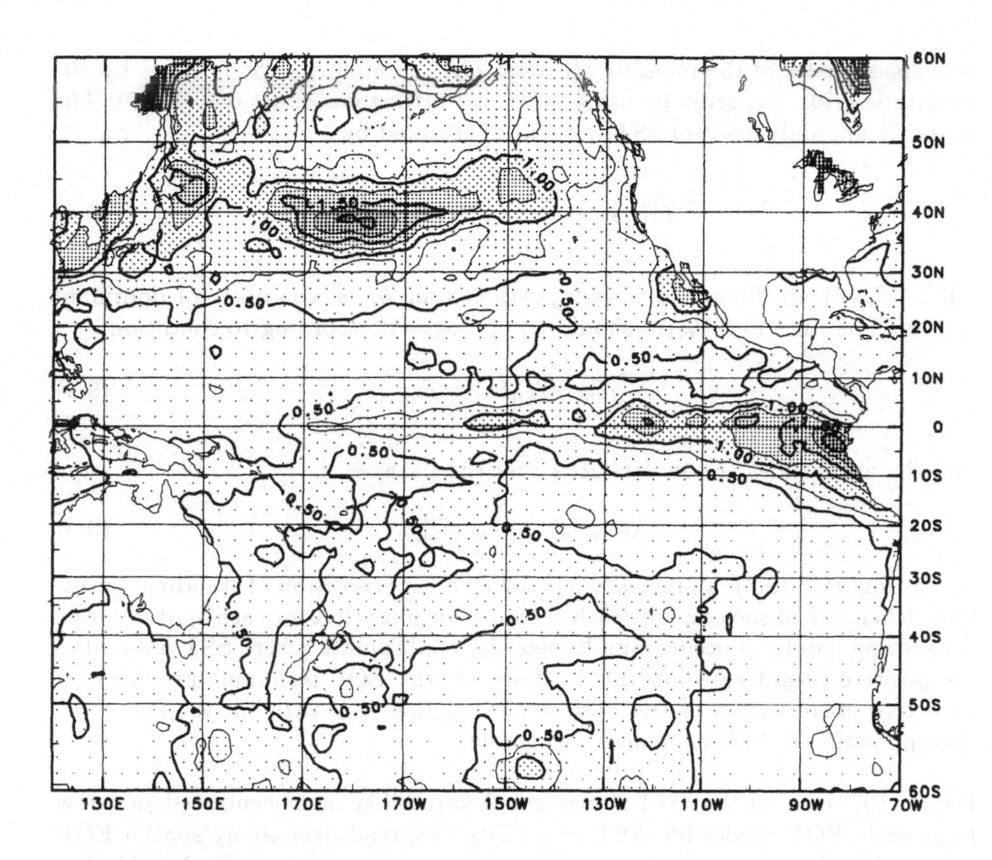

Figure 9.5 *(continued)*
b) Distribution of the SST anomaly r.m.s. σ_T for the third week of July. Data courtesy R. Reynolds (Reynolds 1988, Reynolds and Smith 1994).

The overall pattern of the SST anomaly variance in the North Pacific is similar to that in the North Atlantic (Ostrovskii and Piterbarg 1986) in that it shows an increase in σ_T near western boundary currents.

The SST anomaly variance is high ($\sigma_T > 0.75°C$) in the zonal belt of $35° - 50°N$ in October. There are two local centers of such a high activity at 40°N, 160°E and 45°N, 150°W. This intensification is well known from earlier observations (Elsberry and Camp 1978a, 1978b, Lomakin and Rogachev 1983, Nesterov 1983, Rogachev 1984).

Nonseasonal variability of both SLP and SST was examined by Davis (1976) for the central North Pacific (20° – 55°N). A basic description of the time and space scales of the monthly SST and SLP anomalies in 5° latitude by 10° longitude grids was given by using empirical orthogonal functions (EOF). The monthly anomaly fields of SST were approximated by

$$\tilde{T}'(t,\boldsymbol{r}) = \sum_{m=1}^{M} C_m(t) T_m(\boldsymbol{r}), \tag{9.3}$$

where $T_m(\boldsymbol{r})$ are functions of grid position and M is a number of dominant modes. The functions $T_m(\boldsymbol{r})$ should be orthogonal according to condition

$$\sum_{\boldsymbol{r}} T_m(\boldsymbol{r}) T_n(\boldsymbol{r}) = \delta_{nm}, \tag{9.4}$$

and the amplitudes are uncorrelated over the data set

$$\langle C_m C_n \rangle = \delta_{nm} \langle C_m^2 \rangle. \tag{9.5}$$

According with (9.3), a process is defined as a superposition of standing waves. One deficiency of such an approach is that it implies that for each spatial mode a warming (cooling) occurs simultaneously at all points where SST anomalies are positive (negative) (Gill 1983). However, the EOF technique provided an objective examination of the basic scales of the SST variability. The EOF decomposition of SLP anomalies was similar.

Davis (1976) found that SLP nonseasonal variability is concentrated in a few large-scale EOF modes but SST variability is spread over many spatial EOF modes. In other words, unlike the case of oceanic temperature, a relatively few large-scale patterns describe most of the month to month atmospheric pressure anomalies. Fig. 9.6 shows the fraction of variance associated with the first 20 EOF. As it can be seen in Fig. 9.6, to describe SST variability to the same accuracy as SLP variability many more empirical functions are needed. Thus the ten EOF modes for SST and the dominant five modes for SLP account for 73% and 84% of the respective total variability.

Davis (1976) emphasized that the first mode, P_1, so dominates the SLP variability that its pattern resembles the distribution of variance itself. In contrast, the highest modes, T_1 and T_2, control SST variability in the central part of the North Pacific and in the subarctic gyre, respectively, but the large variance associated with the active region of the Kuroshio - Oyashio frontal zone seems to be divided among several modes lower than T_2. Recent EOF also did not associate the highest modes in SST variability with the active region near western boundary currents (see below).

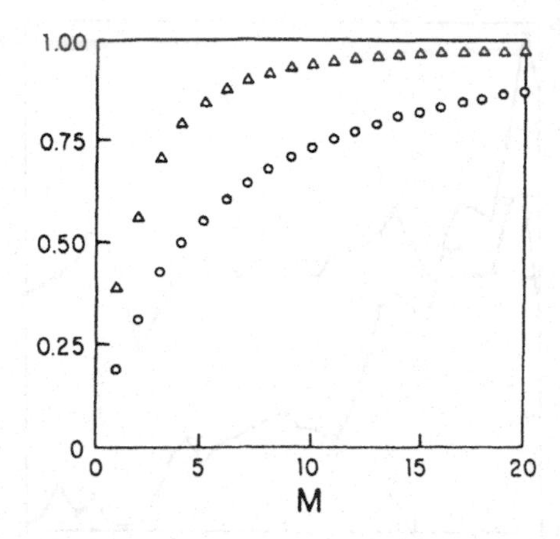

Figure 9.6 The fraction of total SST (circles) and SLP (triangles) anomaly variance accounted for the first M empirical orthogonal functions. Reprinted with permission (Davis 1976).

Computing the frequency spectra of the amplitudes C_1 and S_1 of the dominant SST and SLP modes, Davis (1976) found that the time scale of SST anomalies is significantly longer than that of SLP (Fig. 9.7). It is important that spectra of the SLP amplitudes $S_2 - S_5$ are similar to that of S_1. As to the SST variability, spectra of the amplitudes $C_2 - C_{10}$ are similar to that of C_1 except for a general increase in the relative importance of higher frequencies in the higher modes. Therefore, Davis (1976) concluded that 1) SST variability appears to have a continuous spectrum dominated by low frequencies; 2) SLP variability is characterized by a nearly uniform distribution over all time scales.

Using linear statistical predictors, Davis (1976) examined the predictability of monthly SST and SLP anomalies. The analysis showed that on time scales of a month to a year, previous SLP variability is better specified by SST data than by the instantaneous SLP variability. This fact indicates the insignificance of the ocean as a factor in determining atmospheric variability on a monthly scale in the middle latitudes. The observed connection between SST and SLP anomalies and the results of other empirical studies encouraged Davis (1976) to suggest that SST anomalies are the result of anomalous heat fluxes and wind driven advection of the meridional SST gradient. Several studies (e.g. Lanzante 1984, Iwasaka et al. 1987, Wallace et al. 1990 among others) of extratropical SST and atmospheric pressure variability confirmed the results of Davis (1976).

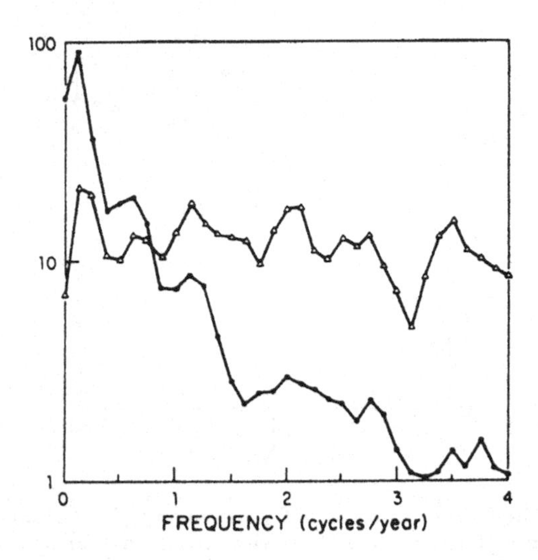

Figure 9.7 The frequency spectra of C_1, the amplitude of the dominant SST mode (circles), and S_1, the amplitude of the dominant SLP mode. Spectral units are arbitrary. Reprinted with permission (Davis 1976).

Global studies of SST anomaly fields (Hsuing and Newell 1983, Kawamura 1994, Tourre and White 1995) are mostly aimed at a description of the climatic impact of ENSO. Indeed, it has been demonstrated that the dominant mode of global SST variability is associated with ENSO. However, in the midlatitude North Pacific the ENSO related signal is much weaker or at least much less regular compared to ENSO in the equatorial Pacific.

Reexamining the spatio-temporal variability of the leading ENSO mode by means of rotated EOF analysis, Kawamura (1994) concluded that although remarkable loadings are observed in the equatorial Pacific SST anomaly field, we do not necessarily find significant loadings over the central North Pacific which were suggested by past similar studies (i. e. Weare et al. 1976, Hsuing and Newell 1983, Kawamura 1984). According to the most recent analysis in terms of the rotated EOF technique by Tourre and White (1995), the ENSO signal accounts for about 33% of the total SST anomaly variance in the whole Pacific Ocean. The overall pattern of the first global SST mode (Fig. 9.8) obtained by Tourre and White (1995) is similar to that presented by Kawamura (1994). However these patterns exhibit considerable differences locally in the midlatitudes. For example, from the analysis of Tourre and White (1995) it follows that there is significant negative loading in the eastern North Pacific in the region bounded approximately by 20° and 40°N, 130° and 170°W. A large region of negative loading also exists in the first mode pattern derived by Kawamura (1994) but its minimums are located in a different place. As it is obviously seen in Fig. 1 of Kawamura (1994) there is an elliptical pattern of relatively small negative loading in the area bounded by 30° to 50°N, 150°W and 150°E. In this respect the ENSO related pattern shown by Kawamura (1994) for the North Pacific, also, differs from the patterns obtained in other studies (Weare et al. 1976, Hsuing and Newell 1983, and Kawamura 1984). Unlike the results of Kawamura (1994), in the pattern obtained by Tourre and White (1995) there is no significant loading in the region centered at 20°N and 120°W, but there is a small region south-east of Japan where positive loading can be significant. Also, it is unfortunate that the second SST mode patterns derived by Kawamura (1994) and Tourre and White (1995) are totally different.

In summary, although there is general correspondence between the global ENSO related patterns obtained in different studies, problems remain to be resolved for the EOF description of SST variability in the middle latitudes. The ENSO mode is dominant near the equator. Its temporal variability has a quasiperiodicity of 2-5 years and corresponds basically to the occurrence of the ENSO events (Kawamura 1994, Tourre and White 1995). In midlatitudes, the SST variability is largely distributed over smaller space scales. Particularly, the conclusion that SST variability in the northwestern Pacific is of a different character than that in the equatorial Pacific is supported by the fact that the most active Kuroshio-Oyashio frontal zone does not define any of the leading modes. Within the midlatitude North Pacific, SST variability in the central basin differs from that of the western basin (Davis 1976). The role of regional processes is crucial to midlatitude SST anomalies. For time scales smaller than about 2 years, these anomalies can be largely explained as the integrated response of the oceanic mixed layer to the 'weather' stochastic forcing by the atmosphere (Herterich and Hasselmann 1987). Our study will focus on such the SST anomalies.

9.3 HASSLEMANN'S STOCHASTIC CLIMATE MODEL THEORY

Below we introduce stochastic climate model theory as it was stated by Hasselmann (1976). The theory provides a conceptual background for stochastic modeling of SST anomalies. According to the theory, the climate variability is accounted for as a reaction to internal random forcing without invoking vari-

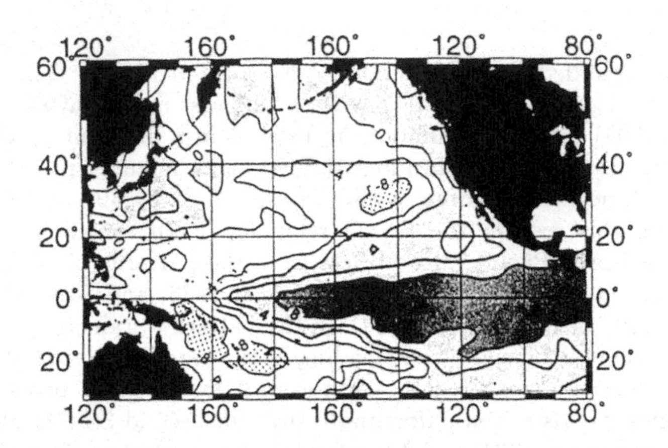

Figure 9.8 Spatial patterns (loadings × 10) of the first two EOF modes for SST over the Pacific Ocean for the 13-year period 1979-91. Areas with positive values are shaded. Extrema of the same order of magnitude are either dark shaded (positive) or stippled (negative). Reprinted with permission (Tourre and White 1995).

able external boundary conditions or internal instabilities in the coupled ocean-atmosphere-cryosphere-land system. The theory predicts the basic structure of climate spectra at frequencies lower than the high frequency limit corresponding approximately to 1 cycle per month. The climate variance spectra are theoretically proportional to the inverse frequency squared ('red noise' spectrum). Such spectral slopes are often found in the analysis of climate records (see for references Hasselmann 1976 and Dobrovolski 1992). Some of the ideas underlying Hasselmann's approach had been proposed by J. M. Mitchel as early as the middle of the 1960's (see Hasselmann 1976).

According to Hasselmann (1976), the state of the climate system is defined by a set of fields

$$Z = \{z_1(t, \boldsymbol{r}), z_2(t, \boldsymbol{r}), ...\}. \tag{9.6}$$

The set Z can be divided into two parts: subsets of the rapidly changing variables $X = x_1, x_2, ...$ and the slowly changing variables $Y = y_1, y_2, ...$. The variables X are basically attributed to weather forcing. The weather time scale τ_x is typically on the order of a few days. The subset X includes not only the atmospheric fields but certain oceanic fields with response times about the same as τ_x. The climate variables Y are associated with such fields as SST anomalies. They have response scales τ_y on the order of 1 month or larger.

The time scale separation can be defined as follows

$$O(x_i(\frac{\partial x_i}{\partial t})^{-1}) = \tau_x \ll \tau_y = O(y_i(\frac{\partial y_i}{\partial t})^{-1}). \tag{9.7}$$

Evolution in the ocean-atmosphere system is governed by the following equations

$$\frac{\partial x_i}{\partial t} = Q_i(X, Y), \tag{9.8}$$

$$\frac{\partial y_i}{\partial t} = P_i(X, Y), \tag{9.9}$$

where Q_i and P_i are in general nonlinear operators on the vector-functions X and Y. Equations (9.8)-(9.9) represents the atmospheric and oceanic components, respectively. The model (9.9) describes the SST variability as a response to stochastic forcing by weather variables.

Stochastic climate models are easier to understand if compared with statistical dynamical models (SDM's) which have been widely used in the past. SDM's are based on assumption that (9.9) can be averaged over some period τ such that for $\tau_x \ll \tau \ll \tau_y$

$$\frac{\partial \langle y_i \rangle}{\partial t} = \langle P_i(X, Y) \rangle, \tag{9.10}$$

so that the average rate of change $\langle P_i \rangle$ of $\langle y_i \rangle$ will depend on the statistical properties of X as well as on Y because the operator $\langle P_i \rangle$ is a nonlinear function of these variables. For the time scale τ_y the sense of (9.10) is the removal of weather fluctuations while treating the climate variables as constant on the right hand side of (9.9). If ergodicity holds then the ensemble average and time average are equivalent, so in (9.10), the averaging $\langle \cdot \rangle$ can be treated as an ensemble average over a set of realizations X for given Y. Since the average rate of change $\langle P_i \rangle$ of y_i depends on the statistical properties of X, Y, and their coupled statistical properties, the statistics of X must be expressed in terms of Y by means of a closure hypothesis.

The model (9.10) may be called 'statistical' because it involves the averaging and a statistical closure. It is, in fact, deterministic rather than statistical. "Most of the better known SDM's predict a unique, time-independent asymptotic state for any given initial state. These models appear inherently incapable of generating internally the time variable solutions with continuous variance spectra, as required by observations", to quote Hasselmann (1976). In conventional SDM simulations, climate variability is studied as a response to fluctuations due to external boundary conditions (see for discussion Dobrovolski 1992).

If the model (9.9) is not averaged but, instead, is considered to be a system of stochastic partial differential equations then the climate variability can be explained through internal interactions. The latter does not exclude the significance of external fluctuations in the climate change. In the framework of Hasselmann's approach, the climate variables are treated as random fields with a correlation time scale on the order of τ_y and longer. The red spectrum typical for climate variability can be deduced from (9.9) by the following consideration. The function P in (9.9) generally depends on all the past $x^{(t)} = \{x_s, \ s \leq t\}$, $y^{(t)} = \{y_s, \ s \leq t\}$. Assuming the departures from the norms, $x' = x - \langle x \rangle$ and $y' = y - \langle y \rangle$, to be small one can linearize (9.9) to obtain

$$\frac{\partial y'}{\partial t} = \frac{\delta P}{\delta x^{(t)}} x' + \frac{\delta P}{\delta y^{(t)}} y', \tag{9.11}$$

where $\delta P / \delta x^{(t)} x'$ and $\delta P / \delta y^{(t)} y'$ are the variational derivatives at the point $(\langle x^{(t)} \rangle, \langle y^{(t)} \rangle)$.

In the simplest case when the operators depend only upon the present, (9.11) becomes

$$\frac{\partial y'}{\partial t} = \sigma x'(t) - \lambda y'(t), \tag{9.12}$$

where λ and σ are constants. This equation is called the Langevin equation in the physical literature. Its solution, $y'(t)$, is called the Ornstein-Uhlenbeck process in the mathematical literature (Gardiner 1985). Taking the correlation time to be small for the fast process $x'(t)$, equation (9.12) implies a red spectrum for $y'(t)$. To obtain a red spectrum of SST variability from the heat balance equation it is not necessary to linearize (9.11). The short-correlation approximation (section 2.3) allows investigation of the statistical properties of the advection-diffusion equation because the equations for the first and second moments are linear and closed. Hasselmann (1976) termed the non-averaged model (9.9) a stochastic forcing model.

Following Hasselmann (1976), let us consider the relationship between atmospheric general circulation models (GCM's), SDM's, and stochastic forcing models by using an analogy from classical Brownian motion. Hasselmann (1976) drew an analogy between the climate variables y and weather variables x with the position and momentum of large and small particles in the Brownian motion. SDM simulations of climate variations without external forcing are then equivalent to modeling the interactions between large particles and the mean pressure and stress fields produced by small particles in motion. In terms of the Brownian problem, the atmospheric GCM's attempt to simulate all the trajectories of small particles for fixed positions of the large particles.

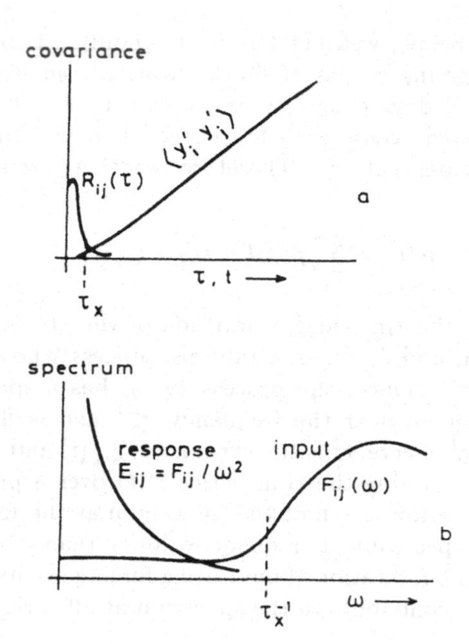

Figure 9.9 Input and response functions of a stochastically forced climate model without feedback: *a*) - covariances, *b*) - spectra. Reprinted with permission (Hasselmann 1976).

Finally, the stochastic forcing model outputs large particle dispersion by utilizing the statistics of small particles as input. In reality, the statistics for weather variables can be inferred from observational data.

Let us explain how the energy can be transferred from weather time scales to climate time scales in the coupled ocean-atmosphere system. Firstly, lets us concern ourselves briefly with the spectrum of atmospheric variability (for more detail see the next section). The basic structure of the input and output spectra for the stochastic climate model without feedback is shown in Fig. 9.9 from Hasselmann (1976). The input atmospheric forcing spectrum is similar to the familiar spectrum of the wind velocity (Van der Hoven 1957) for the time scales of 1 day - several months. The spectrum has a 'clear-cut' maximum at the period of a few days that represents synoptic atmospheric processes. Towards the lower frequencies the spectral density initially decreases and then flattens near the frequencies of 2×10^{-2} cph.

As will be discussed below, weather forcing is mainly attributed to intensive, but relatively rare, forcing events of short duration; the strongest wind forcing events last about 2 days (Elsberry and Camp 1978a, see also Large et al. 1986). These short-lived events are superposed on weak atmospheric fluctuations with a typical time scale τ_x. Therefore, weather forcing time series can be approximated by

$$p'(t) = \sum_k p_k \delta(t - t_k) + \xi_{\tau_x}(t), \tag{9.13}$$

where t_k and p_k are the time and amplitude of the strong forcing event, δ is the Dirac function, and $\xi_{\tau_x}(t)$ is a random process whose spectrum has a peak at the period τ_x. Hence, the process (9.13) has a spectrum (Fig. 9.9) that features a maximum near the frequency τ_x^{-1} and is flattening at lower frequencies to $\langle p_k^2 \rangle + \sigma_\xi^2$, where σ_ξ^2 is the variance of $\xi_{\tau_x}(t)$ and the angle brackets denote averaging. Smoothing the time series $p'(t)$ over a period $\tau \gg \tau_x$, say 1 month, leads to the elimination of the maximum at the frequency τ_x^{-1}, but does not change the spectrum at frequencies lower than τ^{-1}. The smoothed time series holds basic information about strong forcing events because synoptic atmospheric processes contribute to the spectrum at all periods longer than τ_x.

The model (9.13) serves to combine Hasselmann's hypothesis of energy transfer in the climate system with experimental fact that synoptic atmospheric forcing is main player in SST anomaly generation throughout midlatitudes. Additionally, the model supports conclusion of Davis (1976) by showing that it is not possible to infer effect of SST anomalies on synoptic atmospheric processes from the smoothed time series of observations.

Overall, the distinction between the basic time scales of the ocean and the atmosphere enables the time scale separation (9.7) in the coupled ocean-atmosphere system. This separation leads to statistical closure and, consequently, to the statistical reduction of the coupled ocean-atmosphere system. In the resulting model, the ocean component (9.12) includes the stochastic excitation determined by the second moments of the atmospheric input and the mean feedback from the ocean to the atmosphere.

In more general sense, Hasselmann's stochastic model can be viewed as an alternative concept to averaging which has dominated statistical physics. The averaging implies that large-scale processes can be described by deterministic equations while the smaller scale stochastic fluctuations are parameterized with efficient coefficients which hinge on the statistical properties of the smaller scale processes. In the late 1950's, Anderson and Mott (e.g. Anderson 1958) examined this approach critically. Instead they suggested the concept of localization in order to accommodate the effect of energy, momentum, mass and other quantities concentrating around some locations, even at individual points in inhomogeneous media. As to nonstationary random media, both the averaging concept and the localization concept are limited in application (Piterbarg 1989). The averaged equations appear inherently incapable of describing intermittency - extremely irregular and nonuniform distributions of substance in the nonstationary media. Localization is rather a property of stationary (steady) media than random flows. The above discussion implies that an adequate consideration should be based rather on the stochastic differential equations than on deterministic equations for averaged quantities. The simplest model of this type describes certain features of SST variability. As was shown by Frankgnoul and Hasselmann (1977), Reynolds (1978), Piterbarg and Ostrovskii (1984), OrtizBevia and RuizdeElvira (1985), Dobrovolski (1992) and other studies, much of the SST anomaly variance in midlatitudes can be explained simply by a local version of Hasselmann's model (see below).

Here, within the framework of Hasselmann's approach, we shall model the horizontal transport of SST anomalies. It should be noted that the ocean heat flux anomaly variations on a monthly scale may affect the thermohaline circulation. So, in ocean GCM, the variations of the heat and mass transport can not be meaningfully separated (for discussion see Hasselmann 1982). However, unlike the anomalous transport of heat, the advection and diffusion of heat anomalies relates basically to the near-equilibrium ocean circulation driven by the seasonally varying quasi-steady wind and density fields and, as it will be shown below, there are appropriate estimators for inferring the advection and diffusion of the heat anomalies from the time series of SST anomalies themselves.

9.4 ATMOSPHERIC FORCING

The spectral analysis of marine wind data is often reported in the literature. The relevant results published in the 1960-70's have been reviewed by Frankignoul and Müller (1979) (see also Reznik 1986). Though at that time little information was available on the wavenumber structure of the wind stress, which represent the atmospheric forcing better than the wind velocity itself, the general conclusion was that the frequency spectrum of the wind forcing is practically 'white' at the periods longer than 10-20 days. Among those studies it is worth mentioning the comprehensive analysis by Willebrand (1978) who evaluated the frequency spectra of quasi-geostrophic wind stress deduced from

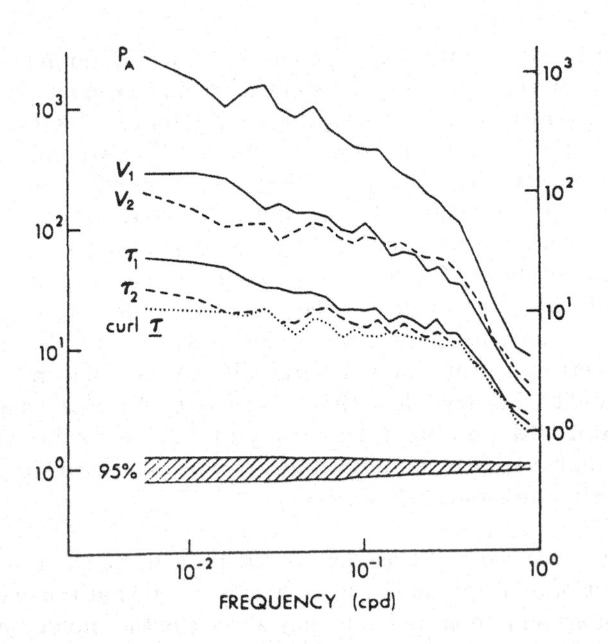

Figure 9.10 Longitudinally averaged autospectra of horizontal stress over the North Pacific at 43°N. P_a atmospheric pressure, v_1 and v_2 east and north surface wind components respectively, τ_1 and τ_2 wind stress components. Left scale: $(\mathrm{m\,s^{-1}})^2\mathrm{cpd}^{-1}$ and $(10^{-8}\mathrm{dyn\,cm^{-3}})^2\mathrm{cpd}^{-1}$; right scale: $\mathrm{mb^2cpd^{-1}}$ and $(\mathrm{dyn\,cm^{-2}})^2\mathrm{cpd}^{-1}$. Reprinted with permission (Willebrand 1978).

synoptic maps of surface pressure for the period of 1973-76 over the North Pacific and North Atlantic.

The typical spectrum was found to be flat at periods longer than 10 days (Fig. 9.10). At shorter time scales the spectrum decreases monotonously with increasing frequency. But the spectral slope changes markedly at frequency 0.3 cpd, which corresponds to the synoptic atmospheric time scale. Similar behavior was demonstrated by autospectra of surface pressure and wind stress curl. Additionally, comparison with weathership observations indicated that the geostrophic and in situ winds have identical spectra. Thus, the forcing fields were shown to be white in the low-frequency band due to the dominance of extension of high-frequency process with a short correlation time scale (for discussion see Frankignoul and Müller 1979). As to the seasonal variability, Willebrand (1978) found that only the energy levels are different in summer and winter seasons, whereas the spectral slope and propagation direction remained essentially unchanged.

There is a notion that some of the oceanic mesoscale variability can be associated with the response to wind stress fluctuations. Willebrand (1978) showed that the wind spectra are symmetric with respect to wavenumber at periods longer than 10 days, thereby indicating no preferred propagation direction. Such atmospheric fluctuations can resonantly excite westward traveling oceanic Rossby waves. At higher frequencies, 0.1-1 day^{-1}, eastward propagation is consistently favored. This asymmetry may lead to eastward propagation in the ocean, although with rather small amplitudes. The generation of oceanic geostrophic eddies in response to a realistic wind stress spectrum was studied by Müller and Frankignoul (1981). Lippert and Müller (1995) explored further consequences of the basic theory in order to accommodate the effects of observed significant nonlocal coherence between some oceanic and atmospheric variables (e.g. Chave et al. 1991, among others).

Chave et al. (1991) examined the spatial and temporal characteristics of the wind stress curl from the Fleet Numerical Oceanography Center data product over the North Pacific during 1985-88. It was found that the spectral density of the curl varies with season in an essentially frequency-independent manner. The seasonal range at a single location and frequency is typically a factor of 2-4 with maximum spectral density in winter and minimum in summer. Although the flatness dominates at low-frequencies, the curl spectra were shown not to be exactly white at periods 5-100 days. Chave et al. (1991) discussed a number of differences between their analysis of the curl wavenumber and both prior experimental and model results. Chave et al. (1991) also found a significant spatial variability of the curl. The spatial correlation structure of the wind stress curl was shown to have a main lobe of $\approx$ 1000 km and secondary peaks separated by $\geq$ 2000 km. Some regional features of the wind forcing in the North Pacific were further reported by Bond et al. (1994).

Perhaps, the main conclusion from the above findings concerns the flatness and wavenumber symmetry of low-frequency wind forcing spectra. However, the complex spatial structure of the wind curl and its seasonal fluctuations are due to an inhomogeneous and variable wavenumber spectrum. Thus simple parameterizations are often inadequate at reproducing the properties of the real wind stress. Therefore we should be aware of seasonally dependent model with spatially varying atmospheric forcing which is not necessary white noise in space. Nevertheless, in the frequency domain, a white noise process would serve as a first approximation for the wind forcing at periods longer than 10 days.

There are some spectral estimates for the air-sea heat fluxes (for references see Dobrovloskii 1992). M. Frost (see Frankignoul and Hasselmann 1978) obtained

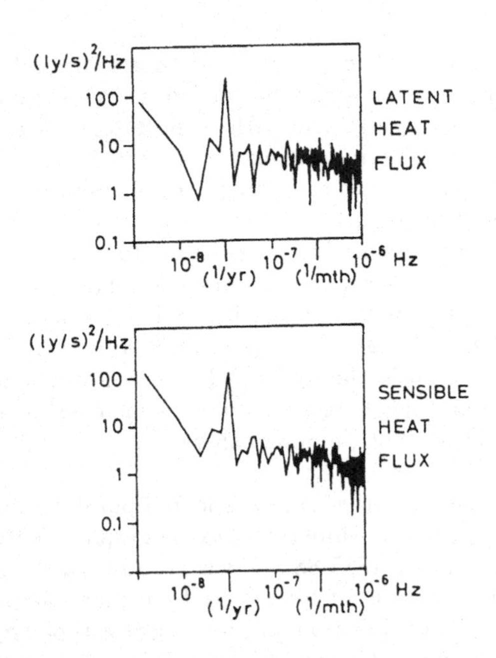

Figure 9.11 Spectra of latent and sensible heat flux at Ocean Weather Ship India for the period 1949-64. Reprinted with permission (Frankignoul and Hasselmann 1977).

spectra of the latent and sensible heat fluxes (Fig. 9.11) at Ocean Weather Ship I (59°N, 19°W). The heat fluxes were calculated from individual marine observations. The resulting time series comprised daily averaged heat fluxes yielding a Nyqvist frequency of about 1 cycle per 2 days. Although the seasonal signal has not been removed from the time series, the spectra were found to be approximately white at periods of a few days to several months.

The statistical relationship between atmospheric variability and SST anomalies in the middle latitudes has been considered in various studies (references can be found in review by Frankignoul 1985, see also Dobrovolski 1992). Unfortunately, the experimental research is still limited by data which are sparse and noisy. However, observational studies usually arrive at the conclusion that the atmosphere drives the ocean, at least in the winter season. Below we will briefly summarize some important experimental results which support this point of view.

Davis (1976) used linear statistical estimators to examine the relationship between SST and SLP nonseasonal (anomaly) variability in the central North Pacific. To accomplish this he tested two hypotheses (H_1) SLP anomalies can be specified from SST observations, i.e. with certain confidence level it is possible to reconstruct the SLP anomaly field at the current or previous months from the SST anomaly field of the current month; (H_2) SLP anomalies can be predicted from prior SST observations.

The procedure for testing the hypotheses was to examine the skill with which the optimal linear statistical predictor estimates the SLP variability forward or backward in time over the 28 years of available data. Davis (1976) took the SST field $\tilde{T}$ described by the ten EOF modes $T_1 - T_{10}$ as the predictor. The predictand considered representative of SLP was $\tilde{P}$, the field described by the five dominant modes $P_1, ..., P_5$. As was mentioned above, the spatially filtered fields, $\tilde{T}$ and $\tilde{P}$, account for the major portions of the SST and SLP variabilities. The skill of the SLP anomaly reconstruction and prediction was evaluated by means of the mean square misfit between the estimate $\hat{P}$ and the corresponding observation $\tilde{P}$. For each observed month the misfits were summed throughout the spatial grid and the global skill index SI was computed from the spatial mean values of the misfits averaged over the whole observational period

$$SI = 1 - \frac{\sum_{\boldsymbol{r}} \langle (\hat{P} - \tilde{P})^2 \rangle}{\sum_{\boldsymbol{r}} \langle \tilde{P}^2 \rangle}, \tag{9.14}$$

where angle brackets denote time average. This skill index is equivalent to the square of the correlation between the predicted and observed SLP fields

$$SI^{1/2} = \frac{\sum_{\boldsymbol{r}} \langle \hat{P}\tilde{P} \rangle}{[\sum_{\boldsymbol{r}} \langle \hat{P}^2 \rangle^{1/2}][\sum_{\boldsymbol{r}} \langle \tilde{P}^2 \rangle^{1/2}]}. \tag{9.15}$$

Therefore, large values SI indicate better predictability.

The skill index of estimating SLP variability from the ten SST modes is shown in Fig. 9.12. As seen in Fig. 9.12, the reconstruction of SLP anomalies is best when using the current and prior SST fields. Furthermore, the correlation estimate for the amplitudes S_1 and C_1 clearly demonstrates that the first mode of SST variability lags that of SLP variability. The atmospheric response to SST changes can not be verified on a monthly scale from the analysis by Davis (1976). Indeed, SST anomalies specify SLP anomalies three months in the past much better than one month in the future. The monthly predictability of SLP variability from prior SST fields is smaller than the predictability from the SLP field itself. Thus the hypothesis (H_1) is correct whereas the hypothesis (H_2) is

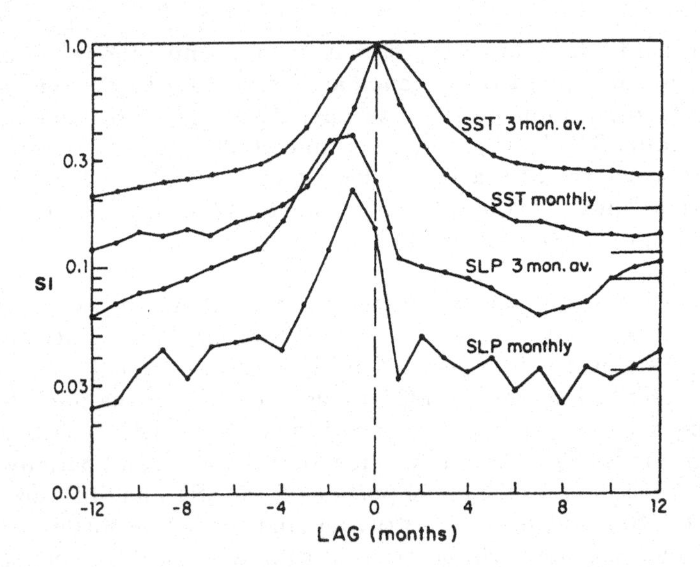

Figure 9.12 The skill of estimating SST and SLP in month t+lag using as data ten SST modes from month t. The skill index SI measures the fraction of the entire spatial field correctly estimated. The top two curves refer to estimating one and three month averages of SST; the lower two curves are SLP estimation skill. Reprinted with permission (Davis 1976).

not; the monthly SST anomaly field is driven by atmospheric variability rather than vise versa.

The conclusion about the leading role of the atmosphere is correct only for the certain time-space scales. For longer time scales, Davis (1978) showed a significant effect of SST anomalies on SLP variability. That result may demonstrate feedback in the ocean-atmosphere system, thereby supporting Hasselmann's stochastic climate model theory.

Apart from these findings, Davis (1976) made a comment which is worth mentioning. He found some autocorrelations in the first SLP mode at one month separations. This intrinsic predictability is weak but strong enough to lead to the conclusion that either the e-folding time of the first SLP mode is about twice as long as is generally accepted for midlatitude pressure patterns (3-4 days) or pressure development over the Pacific is not a Markov process.

The results of Davis (1976) about the ocean-atmosphere connections were confirmed later. Frankignoul and Reynolds (1983) computed the cross-correlations between the same monthly SST anomaly time series as in Davis (1976) on the one hand and the heat flux and Ekman advection anomalies deduced from sources such as ship-of-opportunity data on the other hand. As a result it was shown that SST anomalies are delayed compared with the heat flux anomalies so that the extreme values of the cross correlations appear at the lag of 1 month. The estimates also indicated that the heat flux forcing is more important for the generation of SST anomalies than advection by Ekman currents.

Although Hasselmann (1976) made use of limited information on synoptic atmospheric forcing, further observations confirmed his assumption that short-term forcing is integrated by the oceanic mixed layer. Analyzing the long-term observations at the Ocean Weather Ships N, P, and V, Elsberry and Camp (1978a) found that a significant oceanic thermal response occurs in association with limited periods of strong weather forcing. In September - December, the large, but relatively rare (13% of observations), forcing events define half of the wind forcing, thus contributing largely to the mechanical production of turbulent kinetic energy in the upper ocean. Large heat fluxes, which are also rare (26% of observations), account for half of the accumulated heat flux forcing. Synoptic event atmospheric forcing was defined by Elsberry and Camp (1978a) as the period during which the atmospheric forcing is greater than the long-term mean. Such events are basically short, taking only 2-4 days.

Other observations at Ocean Weather Ships in the North Atlantic support Hasselmann's assumption, too. For example, observations made at weather station E in 1958 and 1959 showed that a strong negative SST anomaly was formed in summer 1959 (Ostrovskii and Tuzhilkin 1982). While from the second half of June to first half of July the average wind stress was about the same in 1958 and 1959, its variance was higher in 1959 by factor 1.5-2.5, thereby affecting both the air-sea heat flux which was 25-30% larger and the entrainment of the cold water into the mixed layer from the interior which increased by a factor of 2-3 compared with that of the same period in the preceding year. As a result, by the middle of July 1959, the SST was 3.8°C lower than in July 1958. This anomaly evolved over three months until the middle of October 1959.

In midlatitudes, the SST anomaly variability should be considered with respect to season. The seasonal cycle in the climate system determines the conditions for SST anomaly generation and evolution. Cayan (1992a) examined atmospheric forcing of the upper ocean by relating monthly anomalies of latent and sensible heat fluxes to changes in SST anomalies over the North Atlantic and North Pacific. The focus of that study was on seasonal behavior. The air-sea turbulent heat flux and SST estimates were calculated from the Comprehensive Ocean - Atmosphere Data Set (COADS) for the period 1946-86. Of particular interest were cold seasons when the turbulent heat flux and its variance are large compared to radiative components of the surface heat budget and when the upper ocean releases heat to the atmosphere.

Cayan (1992a) computed local correlations between heat flux anomalies and temporal (month-to-month) changes in SST anomalies, $\partial T'/\partial t$. It was found that the two components are significantly correlated in midlatitudes, with anomalous positive (negative) flux associated with anomalous cooling (warming). These negative correlations are strongest during winter suggesting that the SST anomaly field is driven by heat flux anomalies. During winter, in the extratropical North Pacific, the correlations are usually less than -0.3 and can be as low as -0.7 to -0.8; the amount of variance in $\partial T'/\partial t$ accounted for by the flux anomalies mainly ranges from 10 to 30%. In contrast, during summer, the flux anomalies tend to be in phase with SST anomalies. Also, there are indications that the anomalous heat flux has greater impact on SST variability than does anomalous monthly wind speeds.

Additionally, Cayan (1992a) determined a link between flux anomalies and the tendency $\partial T'/\partial t$ under cases of strong anomalous atmospheric circulation (8-25% of all winter months). To accomplish this, winter months with extreme positive and negative EOF amplitudes were selected. Composites of the flux anomalies and the month-to-month changes in SST anomalies were formed by averaging the fields during the respective extreme months. It was shown that in the North Pacific, patterns of the SST anomaly tendency $\partial T'/\partial t$ are well aligned with those of the heat flux anomaly. Therefore, while the heat flux forcing of SST anomalies operates locally, the flux and SST tendency variability can be organized over large spatial scales when anomalous atmospheric forcing is extremely strong.

9.5 A LOCAL MODEL FOR SST VARIABILITY

The idea of proceeding from a stochastic differential equation to an autoregressive model first was realized in the simplest case of a local model. Following Frankignoul and Hasselmann (1978), let us consider the local (one-dimensional) model for temperature anomalies in the upper ocean mixed layer. Neglecting horizontal transport and heat exchange with the interior below, one can write a simple prognostic equation for the heat evolution as follows

$$\frac{\partial T}{\partial t} = \frac{H}{h\rho_w C_{pw}}, \tag{9.16}$$

where T is the uniform temperature of the upper mixed layer, $H = H_E + H_S$ is the surface heat flux which consists of the latent and sensible heat fluxes, ρ_w and C_{pw} are the density and specific heat of water. The upper mixed layer depth h is assumed to be constant, thereby ignoring upward entrainment flow when mixed layer deepens. Notice, however, that advection, diffusion, entrainment, and radiation flux are omitted temporally, but will be hold in the general case (see next section and Frankignoul 1985).

On the right hand side of (9.16), the heat flux can be treated by using conventional bulk formulae (Kondo 1975)

$$H_E + H_S = C_S(1+B)\rho_a C_{pa}(T_a - T)|\boldsymbol{w}_a|, \tag{9.17}$$

where T_a, ρ_a, and C_{pa} are the air temperature, density and specific heat, respectively, $\boldsymbol{w}_a$ is the wind speed, B the Bowen ratio, C_S the bulk transfer coefficient for sensible heat.

Separating the atmospheric and oceanic parts of the heat fluxes, one gets

$$\frac{\partial T}{\partial t} + \lambda T = S, \tag{9.18}$$

where

$$\lambda = C_S(1+B)\rho_a C_{pa}|\boldsymbol{w}_a|(h\rho_w C_{pw})^{-1}, \tag{9.19}$$

$$S = C_S(1+B)\rho_a C_{pa} T_a|\boldsymbol{w}_a|(h\rho_w C_{pw})^{-1}. \tag{9.20}$$

λ is called the feedback parameter. It is practical for the description of a tracer's decay time in the conservation equation; for example, it can be a radioactive decay coefficient (Lee and Veronis 1989).

Equation (9.18) is not yet a case of Hasselmann's stochastic forcing model. To obtain such a model one has to average (9.18) and subtract the averaged equation from the original one. Then, neglecting variations in λ, one gets the following equation for fluctuations (anomalies)

$$\frac{\partial T'}{\partial t} + \lambda T' = S', \tag{9.21}$$

where the prime denotes anomaly or departure from the norm. Notice that in (9.21) the feedback factor describes the speed of increase (decrease) in the incoming heat flux with declining (growing) SST as follows

$$\lambda = \frac{\partial}{\partial T'} \langle \frac{H'}{\rho_w C_{pw} h} \rangle, \tag{9.22}$$

where angle brackets denote the ensemble mean for a given T'. Therefore its dimension is reciprocal time. The feedback factor is useful when an explicit description of the heat flux becomes difficult. In general, a negative stabilizing feedback is needed to maintain a statistically stationary response to atmospheric forcing.

As was discussed above, the correlation time τ_a of the process S' is much smaller that the response time of the process T'. In the limiting case of $\tau_a \to 0$ one has a δ-correlated process with correlation function given by

$$\langle S'(t_1)S'(t_2)\rangle = F\delta(t_1 - t_2), \tag{9.23}$$

where F is a parameter characterizing the value of the spectrum of S. The δ-correlation assumption results in a solution for (9.21) in the form of a Gaussian stationary process with correlation function given by

$$R(t) = \sigma_T^2 e^{-\lambda t}, \tag{9.24}$$

where $\sigma_T^2 = \langle T'^2 \rangle$. The corresponding spectral function of the process T' is as follows

$$E_T(\omega) = \frac{\lambda \sigma_T^2}{\pi} \frac{1}{\lambda^2 + \omega^2}, \tag{9.25}$$

where ω is frequency.

The relation (9.25) first was found by Frankignoul and Hasselmann (1978) in the form

$$E_T(\omega) = \frac{F_L(0) + F_S(0)}{(h\rho_w C_{pw})^2(\lambda^2 + \omega^2)}, \tag{9.26}$$

where $F_L(0)$ and $F_S(0)$ are the low frequency levels of the sensible and latent heat flux spectra. Frankignoul and Hasselmann (1978) compared the model spectrum (9.26) with an experimental one calculated from the data of the Ocean Weather Ship I (59°N, 19°W). The hypothetical and experimental spectra are redrawn in Fig. 9.11. Notice, the annual cycle has been subtracted from the SST time series used to compute the experimental spectrum.

The average duration of T' fluctuations above the level σ_T is known to be equal to $(2\pi)^{1/2}\lambda^{-1}$ (Gardiner 1985). Therefore the parameter λ^{-1} characterizes the decay time of SST anomalies at given location. The value $\lambda^{-1} \approx 4.5$ months was derived by Frankignoul and Hasselmann (1978) from the SST anomaly time series at weather station I. This estimate is in accord with common notions about SST anomalies life time (see Dobrovolski 1992). Formula (9.19) provides the same order of the decay time $\lambda^{-1} = 1$ month if the upper mixed layer depth, h, is 50 m and the weather conditions are usual. Therefore the simple model (9.21) mostly yields SST variability in terms of the spectrum and the response time, though such factors as horizontal advection and diffusion are omitted. Frankignoul and Hasselmann (1978) mentioned that equation (9.21) corresponds to a first-order autoregressive (Markov) process. Indeed, this process given by

$$T'_n = \alpha T'_{n-1} + \epsilon_n, \tag{9.27}$$

where $T'_n = T'(n\Delta t)$ is a discrete signal, $\{\epsilon_n\}$ is a sequence of uncorrelated random variables with zero mean and variance σ_ϵ^2. Equation (9.27) can be related to the continuous model (9.21) through the connection between the parameters (Box and Jenkins 1976, Reynolds 1978, Ostrovskii and Piterbarg 1985):

$$\lambda = -\frac{1}{\Delta t}\log\alpha, \quad F = \frac{2\lambda\sigma_\epsilon^2}{1 - e^{-2\lambda\Delta t}}. \tag{9.28}$$

Applying familiar algorithms due to Box and Jenkins (1976), one can compute the statistical parameters α and σ_ϵ^2 from the time series T'_n. Then by using (9.28), it is easy to get the physical parameters, namely the feedback factor and the spectral level of the heat flux anomaly variance. Notice, the model (9.27) does not imply that the current value of the signal depends only on the prior value and is independent of observations lagged further behind. It rather states that all information about the past is contained in the prior value, i.e. the optimal prediction of the value T'_n given the past is completely determined by the knowledge of T'_{n-1}.

Reynolds (1978) further verified the local model (9.21) by fitting the power spectra of observed SST anomalies to that of the hypothetical autoregressive process. Three types of spectra generated by autoregressive models of the first and second orders were tested: (i) conventional Markov process, i.e. the response of a system with negative linear feedback to white atmospheric forcing given by (9.25), (ii) first-order autoregression with an extra noise term, (iii) second-order autoregression. A 95% confidence limit was chosen for statistical significance of the fit.

For analysis Reynolds (1978) used a monthly SST anomaly time series over 5° × 5° geographical grid in the midlatitude North Pacific (20°-55°N). The Markov process was found to be a good approximation for a little over 50% of the total five-degree squares. The models of the second and third types were jointly valid over 14% of the investigated regions. In the remaining regions, no model among (i) - (iii) fitted the true spectrum. Noticeably, the regions where the Markov process was found not valid were western boundaries, tropics and relatively high latitudes.

Later a family of univariate autoregressive models for SST anomalies was tested with the Maximum Entropy Method (MEM) (see for review Dobrovolski 1992). The MEM spectral density estimate for autoregression (Box and Jenkins 1970)

$$T'(t) = \alpha_1 T'(t - \Delta t) + \alpha_2 T'(t - 2\Delta t) + \ldots + \alpha_M T'(t - M\Delta t) + \epsilon(t) \tag{9.29}$$

is given by the following formula

$$S(\omega) = \frac{2\sigma_\epsilon^2}{\omega_N |1 - \sum_{n=1}^{M} \alpha_n \exp(-\imath 2\pi\omega n\Delta t)|^2}, \tag{9.30}$$

where $T'(n\Delta t)$ are descrete samples of the stationary signal $T'(t)$, $\epsilon(t)$ is a zero mean white noise with variance σ_ϵ^2, α_n the autoregression coefficients, ω_N the Nyquist frequency, and M the order of the autoregressive process. According with model (9.29) each SST anomaly observed during the time interval $n\Delta t$ contains unique information about future SST anomalies. MEM power spectrum estimation reduces to (i) fitting the autoregressive model to the time series, thereby identifying the order M and the coefficients α_n in (9.29) and (ii) calculation of the spectral density (9.30).

The analysis indicated that among the models (9.29) the first-order autoregressive process usually was the best approximation for monthly SST anomaly time series at a fixed geographical location (Dobrovolski 1992). Thus, in addition to the theoretical arguments, the statistical test also pointed to the Markov process as a model for nonseasonal SST variability.

In summary, this section concentrated on the simplest case that, perhaps, is too idealistic. The transition from the continuous model (9.21) to the statistical parametric model (9.27) through the relationships (9.28) only illustrated connection between the stochastic forcing model and the statistics of the climate variables. Further to model SST variability, we must consider Markov processes with respect to a sequence of fields instead of a single time series at a fixed location. The spatial extension of the one-dimensional model (Ostrovskii 1983, Piterbarg and Ostrovskii 1984) incorporates the fundamental effects of advection and diffusion into stochastic SST anomaly modelling.

In conclusion, let us propose an approach that we call 'objectively consistent estimation' (Fig. 9.13). Here, our aim is towards a solution of the inverse

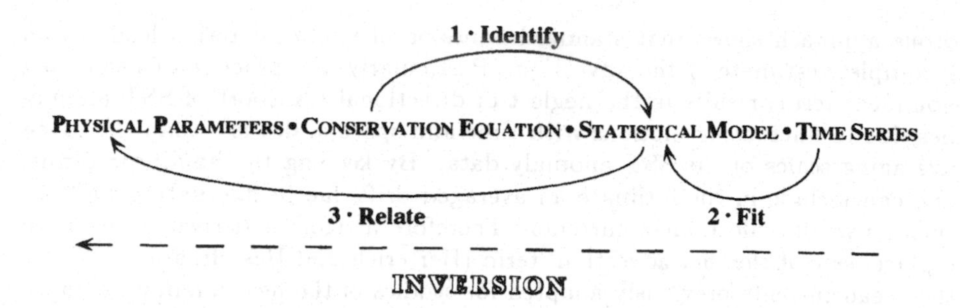

Figure 9.13 An explanation of objectively consistent estimation.

problem that serves to obtain the physical parameters, such as velocity and diffusivity, from a time series of tracer measurements. Generally speaking, starting from the data, we find their most suitable approximation in terms of a statistical model. This model is not any model but must be derived from the conservation equation which governs the underlying process recorded in the time series. If the conservation equation is defined in form of a stochastic partial differential equation then the relationship between the statistical model and the governing equation is a mathematically tractable function. Thus, the parameters in the conservation equation can be computed from the coefficients of the statistical model fitted to the data. The point of objectively consistent estimation is that it takes advantage of identifying the statistical model from the governing equation, instead of identification from the data, which is often treated as an art rather than a technique. The role of the statistical model as a mediator between the observations and the conservation equation is vital for both compressing available information and filtering out noise.

9.6 AN EQUATION FOR HEAT ANOMALY BALANCE IN THE UPPER OCEAN MIXED LAYER

This section addresses the heat anomaly balance equation that describes the generation and evolution of SST anomalies lasting the dominant (synoptic) time scale of atmospheric forcing. In this equation, derived rigorously by Piterbarg (1987), the advection is basically attributed to all motions with time and space scales exceeding the averaging scales of SST anomaly data, for example, longer than 10 days and larger than 1°latitude × 2°longitude, respectively. This rigorous approach shows that standard omission of nonlinear terms leads to an incomplete estimate of the advection. Particularly, the practice of neglecting nonlinear terms results in the neglect of directional transport of SST anomalies due to fluctuating motions with time and space scales less than the given averaging scales of the SST anomaly data. By keeping the nonlinear terms, one can include in the estimate an averaged drift due to fluctuating motions such as small scale Ekman currents. Therefore a rigorous derivation leads to replacement of the 'net advection' term (Herterich and Hasselmann 1987) for the mean currents previously adopted for studies of the heat anomaly balance (Frankignoul and Reynolds 1983).

As was mentioned by Herterich and Hasselmann (1987) the net advection velocities contain an unknown contribution from wind-driven currents. So the net velocities derived from SST anomaly data themselves may provide a more appropriate effective advection field than the near surface current measurements. In this experimental study we shall infer the heat anomaly transport from the covariances of SST anomaly time series. In addition to information on the mean current, the covariances contain information about other processes that lead to directional transport of the SST anomalies.

Now we will derive a stochastic partial differential equation which describes SST variability. Further this equation will be used for constructing a parametric statistical model of SST.

The advection-diffusion equation for the upper ocean mixed layer heat balance averaged over small time and space scales, say, 1 day and 1 km, respectively, can be written as follows (Frankignoul 1985):

$$\frac{\partial T}{\partial t} + \boldsymbol{u} \cdot \nabla T = \kappa \nabla^2 T - \lambda T + E, \tag{9.31}$$

where $T = T(t, \boldsymbol{r})$ is the temperature of the upper ocean mixed layer at time t and location $\boldsymbol{r}$, $\boldsymbol{u}$ is the velocity, and κ is the coefficient for small-scale diffusion processes. The second and third terms on the right-hand side of (9.31) represent the heat fluxes into the mixed layer $Q/(\rho_w C_{pw} h) = E - \lambda T$, where λT is the part of the flux, which depends on T. Note that $\boldsymbol{u} = \boldsymbol{u}(t, \boldsymbol{r})$, $\lambda = \lambda(t, \boldsymbol{r})$, and $E = E(t, \boldsymbol{r})$ are stochastic fields with a full range of variations with periods longer than 1 day and time scales bigger than 1 km.

For the sake of clarity we did not explicitly include the mixed layer depth, h, in (9.31). These effects seems to be most important during the transient seasons, spring and autumn, when mixed layer depth changes drastically (see section 9.1). In order to avoid uncertainties related to the depth variations, in this analysis we focus on the SST anomaly variability in the winter season when the upper mixed layer is deep and rather stable.

The relationship between the entrainment process and SST anomaly variability was thoroughly considered by Frankignoul (1985). Since by definition, entrainment operates during the deepening of the mixed layer, its role is crucial during fall. In winter, the mixed layer does not deepens much. Instead its depth may fluctuate due to wind forcing events. As it will be discussed below, the mean entrainment is represented through the damping, λ, while entrainment fluctuations are represented by the weather forcing, E.

To derive a stochastic model for SST anomalies, one should take the following three steps. (i) Obtain the equation for the annual cycle (the long-term norms) by the conventional procedure of Reynolds's averaging (c.f. Monin and Yaglom 1975). (ii) Obtain the equation for departures from the norms by subtracting the equation for the norms from (9.31). (iii) Derive the equation for SST anomalies averaged in space and time over, say, 1° squares and 10-day periods, respectively; this is necessary for applications to observational data. While the first two steps are common, the focus here is on the third step which was described in more detail in section 3.1.

It is important to point out the basic differences between the present derivation and the previous ones. First, we shall not omit nonlinear terms of the form $u' \cdot \nabla T'$ which play an important role in heat anomaly transport. Second, we treat mathematically SST anomalies as (1) fluctuations from the norms and (2) mean fields with respect to subgrid processes. Though some researchers realized the importance of this separation, rigorous formulation of this principle has not been done yet.

Let each property in (9.31) be decomposed into a seasonally varying mean (norm), a departure from the norm (anomaly), as, for example, in the case of SST (9.2). A procedure done by analogy with (3.9) gives us the non-closed equation for the mean temperature field

$$\frac{\partial \overline{T}}{\partial t} + \overline{\boldsymbol{u}} \cdot \nabla \overline{T} + \overline{\boldsymbol{u}' \cdot \nabla T'} = \kappa \nabla^2 \overline{T} - \overline{\lambda} \overline{T} - \overline{\lambda' T'} + \overline{E}, \tag{9.32}$$

Subtracting (9.32) from (9.31) and keeping all terms, we obtain

$$\frac{\partial T'}{\partial t} + \boldsymbol{u} \cdot \nabla T' = \kappa \nabla^2 T' - \lambda T' + S', \tag{9.33}$$

where the new source term

$$S' = E' - \boldsymbol{u}' \cdot \nabla\overline{T} + \overline{\boldsymbol{u}' \cdot \nabla T'} - \lambda'\overline{T} + \overline{\lambda' T'} \tag{9.34}$$

includes the generation of SST anomalies due to forcing of the mean SST field by the fluctuating currents and heat fluxes. The forcing can be primarily attributed to wind stress variations leading to anomalous Ekman currents and heat flux anomalies.

The SST anomaly generation by the last two terms in (9.34) varies with season. This seasonal effect can be largely removed when we fit the statistical model to data for only one season, particularly, the cool season.

Perhaps, the most important feature is that in (9.33) the fields $\boldsymbol{u}$ and λ are original as in (9.31) and not the averaged $\overline{\boldsymbol{u}}$ and $\overline{\lambda}$. This occurs because we do not omit the nonlinear terms during the derivation.

The heat anomaly balance equation in the form (9.33) is not ready to be applied to the real data. Global SST data sets comprise arrays of spatio-temporal averages, over say 10 days and 1° geographical squares. Therefore, to make (9.33) a model for real data we should average it in time and space.

One can see that the basic equation (9.33) has the same structure as equation (3.7) from chapter 3. Thus, we can apply the averaging procedure developed in chapter 3 provided the conditions of the δ-correlation approximation are fulfilled. Let us clarify what those conditions mean in this case. First, suppose that the mentioned time space averaging can be replaced by an ensemble averaging denoted by angle brackets. In other words, we introduce a sort of the ergodic conjecture. Then, define the deviations from the ensemble means in the usual way $\boldsymbol{u}'' = \boldsymbol{u}' - \langle \boldsymbol{u}' \rangle$, $\lambda'' = \lambda' - \langle \lambda' \rangle$, and $S'' = S' - \langle S' \rangle$. The δ-correlation approximation can be used for averaging equation (9.33) if the correlation time, τ_E, of the processes $\boldsymbol{u}''$, λ'', and S'' is much less than any other time scale defined for this problem. Recall that in accordance with section 2.1 there are three such scales: the actual time (time of observation), t, the small scale diffusion time, τ_D, and the turnover time, τ_E. Notice, the Eulerian correlation time, τ_E, can be interpreted as a synoptic atmospheric scale equal to a few days. Of course $\tau_D \gg \tau_E$ and $t \gg \tau_E$ since we are interested in the observational times on the order of 10 days.

As for the turnover time given by (2.25), one could estimate it knowing the velocity space scale, l_u, and its mean square value, σ_u. While the latter can be taken as 0.1 m s^{-1}, the former can not be measured directly. Therefore it is worth making certain assumptions about the origin of the velocity fluctuations.

If the fluctuating velocity is due to atmospheric forcing of the time scale, τ_E, of 3-5 days then in accordance with chapter 1 the output space scale, l_u, is close to that of the forcing, i.e. l_u has the order of 1000 km. Under this assumption the turnover time $\tau_T = l_u/\sigma_u = 10^7$ s, a value which is much bigger than 3-5 days. If one assumes that the fluctuating velocity is due to synoptic eddies in the upper ocean then $l_u \sim 100$ km and the value of τ_T becomes 10 times less but the scale separation $\tau_E \ll \tau_T$ still holds. In brief, to date our knowledge of the time and space scales of the velocity fluctuations in the upper ocean are too poor for far reaching conclusions. Nevertheless, the basic assumptions of the δ-correlated approximation may be satisfied by the above estimates.

Thus, by applying the procedure from section 3.1 to equation (9.33) we obtain an equation for averaging anomalies similar to (3.24) but written in a slightly different form

$$\frac{\partial\langle T'\rangle}{\partial t} + \boldsymbol{u}_0 \cdot \nabla\langle T'\rangle = \nabla \cdot \boldsymbol{D}_0 \nabla\langle T'\rangle - \lambda_0\langle T'\rangle + S_0, \tag{9.35}$$

where

$$\boldsymbol{u}_0 = \boldsymbol{u}_0(t,\boldsymbol{r}) = \langle \boldsymbol{u}\rangle - \int_{-\tau_E}^{\tau_E} \langle \boldsymbol{u}''(0,\boldsymbol{r})\lambda''(s,\boldsymbol{r})\rangle ds, \tag{9.36}$$

$$\begin{aligned}\lambda_0 = \lambda_0(t,\boldsymbol{r}) = \langle\lambda\rangle - \frac{1}{2}\Big(\int_{-\tau_E}^{\tau_E} (\langle \nabla \boldsymbol{u}''(0,\boldsymbol{r})\lambda''(s,\boldsymbol{r})\rangle \\ +\langle\lambda''(0,\boldsymbol{r})\lambda''(s,\boldsymbol{r})\rangle)ds\Big),\end{aligned} \tag{9.37}$$

$$\boldsymbol{D}_0 = \boldsymbol{D}_0(\boldsymbol{r}) = \frac{1}{2}\int_{-\tau_E}^{\tau_E} \langle \boldsymbol{u}''(0,\boldsymbol{r})\boldsymbol{u}''(s,\boldsymbol{r})^T\rangle ds, \tag{9.38}$$

$$\begin{aligned}S_0 = S_0(t,\boldsymbol{r}) = \langle S'\rangle - \frac{1}{2}\Big(\int_{-\tau_E}^{\tau_E} (\langle\lambda''(0,\boldsymbol{r})S''(s,\boldsymbol{r})\rangle \\ +\langle \boldsymbol{u}''(0,\boldsymbol{r})\cdot\nabla S''(s,\boldsymbol{r})\rangle)ds\Big).\end{aligned} \tag{9.39}$$

Notice, small-scale diffusion was neglected in formula (9.38).

At first glance it seems that the integrals on the right-hand side of (9.36)-(9.39) are small because τ_E is small. In fact, this is not correct because the processes under consideration are supposed to be δ-correlated and hence the corresponding integrals can give substantial contributions.

Let us introduce a new time scale τ_0 as follows

$$\int_{-\tau_E}^{\tau_E} (\langle \boldsymbol{u}''(0,\boldsymbol{r})\lambda''(s,\boldsymbol{r})\rangle ds = \tau_0\langle \boldsymbol{u}''\lambda''\rangle. \tag{9.40}$$

Then the net advection can be given by

$$\boldsymbol{u}_0 = \langle \boldsymbol{u} \rangle - \tau_0 \langle \boldsymbol{u}'' \lambda'' \rangle. \tag{9.41}$$

Advection was discussed in this form in Ostrovskii and Piterbarg (1995). Relations (9.37)-(9.39) also can be written in such form.

In the advection term (9.41) all the motions have periods longer than 10 days and length scales larger than ~ 100 km $\times$ 100 km, respectively. The advection in (9.40) occurs not only as a result of the mean currents, but also due to the contributions of other processes, for example, Rossby waves. Furthermore, the advection is affected by the term $\tau_0 \langle u'' \lambda'' \rangle$. The effect of this term on the advection direction seems to be neither evident nor negligible. We can only note that the distribution of λ'' fluctuations is symmetrical with respect to zero. Thus an important fact is that, using (9.35), one can estimate the net advection velocity rather than the mean current velocity. A priori we do not know which process makes a major contribution to SST anomaly propagation. Conclusions about the driving processes can be made after application of the statistical model to observational data.

Therefore, the proposed model estimates the heat anomaly advection rather than the mean current velocities. While there are conventional techniques for calculating the ocean currents, this model was developed to evaluate the heat anomaly transport.

The diffusion in (9.38) is produced by subgrid processes with periods less than 10 days and length scales smaller than 1° square.

Finally, for the sake of clarity, let us omit subscripts and superscripts and replace $\langle T' \rangle$ by T and S_0' by S in (9.35) as follows

$$\frac{\partial T}{\partial t} + \boldsymbol{u} \cdot \nabla T = \nabla \cdot \boldsymbol{D} \nabla T - \lambda T + S, \tag{9.42}$$

This is the familiar form of the equation for the heat anomaly balance in the upper ocean mixed layer (see Frankignoul 1985, Herterich and Hasselmann 1987).

On the left-hand side of (9.42) the second term represents the net horizontal advection of the SST anomalies; $\boldsymbol{u}$ is the seasonally varying net advection velocity integrated over the mixed layer.

The third term represents the joint effect of SST anomaly damping due to mean entrainment at the bottom of the mixed layer and atmospheric feedback

(Frankignoul 1985). The seasonally varying Newton cooling coefficient is given by

$$\lambda = \lambda_E + \lambda_Q, \tag{9.43}$$

where $\lambda_Q = \frac{\partial}{\partial T}\langle H/\rho_w C_{pw} h\rangle$, Q is the net heat flux anomaly including turbulent and radiation fluxes, and $\lambda_E = w_E/h$, where w_E and h are the seasonally varying means of the entrainment velocity and the mixed layer depth, respectively.

The mean entrainment is zero in the seasons when the mixed layer normally does not deepens. Considering the weathership P observations, Frankignoul (1985) suggested that λ_E increases from zero in summer to a maximum of about 2×10^{-7} s^{-1} in early fall and then becomes small again by the end of December. Frankignoul (1985) followed the assumption that there is no temperature anomaly in the seasonal thermocline. Though this assumption is not strictly valid, the lack of data prevents us from explicitly describing the entrainment. However, extension of the model is straightforward if time series of the seasonal thermocline temperature become available. Both the mean entrainment and the atmospheric feedback control decay of the SST anomalies. The reciprocal, λ^{-1}, of the feedback parameter is called the local relaxation time.

On the right-hand side of (9.42) the first term describes horizontal mixing where $\boldsymbol{D}$ is the heat diffusivity varying with season. The diffusion is produced by subgrid turbulence, mostly by mesoscale eddies, which are smaller in size than 1° latitude × 2° longitude box of the numerical grid used below.

The last term of (9.42) represents the effect of atmospheric heating, anomaly entrainment, and anomaly advection of the mean temperature gradients. The source $S = S(t, \boldsymbol{r})$ is assumed to be a random process with zero mean.

9.7 AN EQUATION FOR THE MEAN HEAT BALANCE

To complete the description of the heat transport in the upper ocean mixed layer let us review the closed equation for the mean heat balance. Let us remind the reader that in the above section the non-closed equation (9.32) was used. Applying the same averaging technique to (9.31), we arrive at the following equation given by

$$\frac{\partial \overline{T}}{\partial t} + \boldsymbol{u}_1 \cdot \nabla \overline{T} = \nabla \cdot \boldsymbol{D}_1 \nabla \overline{T} - \lambda_1 \overline{T} + S_1, \tag{9.44}$$

whose parameters have a form similar to (9.40)

$$\boldsymbol{u}_1 = \overline{\boldsymbol{u}} - \tau_1 \overline{\boldsymbol{u}'\lambda'}, \tag{9.45}$$

$$\lambda_1 = \overline{\lambda} - \frac{\tau_1}{2}[\overline{\lambda'^2} + \overline{\boldsymbol{u}' \cdot \nabla \lambda'}], \tag{9.46}$$

$$\boldsymbol{D}_1 = \frac{\tau_1}{2}\overline{\boldsymbol{u}'\boldsymbol{u}'^T}, \tag{9.47}$$

$$S_1 = \overline{E} - \frac{\tau_1}{2}[\overline{\lambda' E'} + \overline{\boldsymbol{u}' \cdot \nabla E'}]. \tag{9.48}$$

but with a different time scale τ_1. The overbar denotes the seasonal average as in (9.1). We also assume that the velocity field is nondivergent.

A rigorous derivation illuminates the sources of heat which are omitted in the traditional results (see for discussion section 1.2). In addition to the term with effective heat transfer, we have a term determined by the fluctuations of both the ocean current velocity and the feedback factor. Also, now the mean values of the velocity and the feedback factor are defined more accurately. It is worth mentioning that when $\lambda' = 0$, the assumption on infinitesimally small correlation time leads to a closure equation of the form

$$\overline{\boldsymbol{u}'T'} = -\boldsymbol{D}_1 \nabla \overline{T} + \frac{\tau_1}{2}\overline{\boldsymbol{u}'E'}. \tag{9.49}$$

As it follows from (9.45)-(9.49), besides norms such as the average air-sea heat flux, various nonseasonal processes contribute to the mean SST field. Particularly, wind fluctuations contribute in many ways. The most obvious is a joint effect of the wind and heat flux fluctuations, $\overline{\lambda' E'}$. Among others one can mention the anomaly Ekman current which advects both the spatial gradient of the heat flux variability, $\overline{\boldsymbol{u}' \cdot \nabla E'}$, and the gradient of the average temperature (through the term $\overline{\boldsymbol{u}'\lambda'}$). Finally, wind fluctuations may either strengthen or weaken the damping by contributing to the term $\overline{\boldsymbol{u}' \cdot \nabla \lambda'}$. Piterbarg (1989) evaluated 'additional' terms in equation (9.44). It was shown that under certain assumptions about the wind and the heat flux the additional terms can make a significant contribution to the mean heat balance on a monthly scale.

9.8 SUMMARY

This chapter was devoted to a discussion of the stochastic model for SST anomaly transport (9.42) which is an extension of the local model by Frankignoul and Hasselmann (1977). After observing the spectra of climate variables and summarizing results of correlation estimates between SST anomalies, sea level pressure and wind stress, Frankignoul and Hasselmann (1977) hypothesized that SST anomaly (nonseasonal) variability in midlatitudes is primarily associated with the integral response of the upper ocean to continuous random excitation by short period 'weather' disturbances. Local models were extended by Frankignoul and Reynolds (1983), Piterbarg and Ostrovskii (1984), and Herterich and Hasselmann (1987) who aimed at accommodating nonlocal effects due to advection and diffusion. The present model (9.42) is not radically different from those used in the previous studies but features thorough definitions for the advection and diffusion processes. For example, from a rigorous derivation we learned that the heat advection is not merely a result of the mean currents but in fact consists of the two main parts: (a) all motions with scales exceeding the given averaging scales and (b) motion due to the joint effect of fluctuations in the feedback factor and the current velocity. Likewise, we described the diffusion, feedback, and source terms.

Obviously (9.42) is the same model as (6.1) and (7.1) which describe transport and dissipation of a passive scalar (tracer). This does not imply that the ocean temperature can always be viewed as a tracer of ocean dynamics. Furthermore, we can not apply (9.42) to the analysis of anomaly transport itself. It governs only the transport of SST anomalies on certain spatio-temporal scales. However the study of the SST variability is very important in practice, for example the SST field is often taken as a proxy for a boundary condition in atmospheric general circulation models.

In a wider sense, our approach to estimation of the advection, diffusion, and feedback can be regarded as a diagnostic model for the tracer balance governed by the stochastic partial differential equation (9.42). In physical oceanography, diagnostic models (Sarkisyan 1969) operate with the equations of motion. Those models provide a basic tool for current field simulations from temperature and salinity data but are not suitable for reconstructing the diffusivity field. At least in this respect, our approach complements conventional diagnostic models.

10

ADVECTION AND DIFFUSION INFERRED FROM SST ANOMALY TIME SERIES

In chapter 9 we introduced the background physics for the stochastic model originated by Hasselmann (1976). So far Hasselmann's theory is the only theory that describes SST nonseasonal variability in the middle latitudes. The advection-diffusion equation (9.42) governs SST anomaly transport and dissipation. Since (9.42) is identical to (6.1) and (7.1), we can use the autoregressive and maximum likelihood estimators developed in chapters 6-8 for extracting SST anomaly advection, diffusion, feedback, and the atmospheric forcing from observations. We will focus on the estimation of the velocity and diffusivity.

The statistical models (7.13) and (8.21) derived from the stochastic partial differential equation (9.42) are to be fitted to a time series of measurements. Fitting the models to the data results in estimates of variables, such as the mixing coefficient, which are difficult to measure in nature. The validated model should yield a scalar field which fits the observations with a minimum uncertainty. As with any data assimilation scheme, the estimates obtained will not be optimal under conditions different from those under which the used observations were made.

Our motivation is twofold. First, we will furnish a guideline for the empirical parameter evaluation whenever model (9.42) is suitable. For example, application of the maximum likelihood inversion technique to scalar fields over different grids may provide us with diffusivity estimates as a function of space scales. Also, using empirical diffusivity estimates one may want to examine relationships between the mixing coefficients and characteristics of the mean fields. Thus, in brief, we will consider our inversion technique as a new tool for the analysis of tracer time series data. Secondly, we have strong interest in the transport of the heat anomalies and its impact on climate. Since midlatitude SST anomalies are stochastic in nature, it is very difficult to detect their directional transport. Although the propagation of SST anomalies has been reported several times (see review by Frankignoul 1985 and more recent studies by Halliwell et al. 1991a, 1991b and Aoki et al. 1995), there is still no clear picture about the SST anomaly transport in the world ocean away from the equator. There is very limited information about heat anomaly diffusion by mesoscale and submesoscale processes in the upper ocean. We know neither the seasonal variation nor the spatial distribution of mesoscale heat diffusivity so numerical modelers of ocean circulation are forced to hold the diffusivity constant over the model domain except in the eddy-resolving models.

For two reasons, we will restrict this consideration to heat anomaly transport in the North Pacific during winter. Firstly, we do not pretend to give a complete description of the heat anomaly transport throughout the world ocean but consider processes in the area where the SST variability is known perhaps better than anywhere else so we can compare our results with those obtained earlier. The second reason is due to the model validity. On the one hand, without explicit terms for the vertical processes, the SST anomaly model (9.42) is only valid in winter. On the other hand, the data set is more or less adequate only for the northern part of the ocean. However we will discuss possible extensions of the inversion techniques for application to different areas in other seasons. Since, in practice, we can not get data with any resolution we want, it will be necessary to discuss the limitations due to the data available for this study; this we will do in the next section. Then we will turn to procedures and results of model fitting by using the autoregressive and maximum likelihood estimators.

10.1 DATA

Most empirical studies of midlatitude SST anomalies are based on monthly data averaged over 5° geographical grids (for review see Frankignoul 1985 and Dobrovolski 1992). This spatial averaging not only eliminates the boundary currents (Ostrovskii and Piterbarg 1985, Herterich and Hasselmann 1987), it can largely filter out the eddy signal (Frankignoul 1985). For example, the diffusivity estimate over such a spatio-temporal grid in the North Pacific is in the range 10^3 to 10^4 m^2s^{-1} (Herterich and Hasselmann 1987), i.e. it is much smaller than that generally accepted in numerical models. A data set with higher spatio-temporal resolution is needed to describe SST anomaly transport.

One source of higher resolution data is the Comprehensive Ocean-Atmosphere Data Set (COADS). It comprises an extensive collection of surface marine data available over the world ocean for the period 1854-1991 (Woodruff et al. 1987, Woodruff et al. 1993). The observations include weather reports by merchant ships, research vessels, fishing fleets, moored and drifting buoys, and automated platforms. These in situ data amount to 10^4 reports per week during the last decade.

Among the products offered by the COADS group, is the easily handled set of 2°latitude and 2°longitude monthly summaries. The summaries have been analyzed by more than 100 research groups worldwide. However the monthly resolution is too coarse to obtain SST anomaly transport by strong currents because in this case we can not detect advection velocities larger than 0.08 m s^{-1} according with CFL condition (7.48). This restricts the analysis to the ocean interior, far away from intensive boundary currents. Additionally, the regions of strong currents are characterized by high eddy activity which determines heat diffusion. Particularly in the North Pacific, increased mesoscale activity is associated with the Kuroshio Current and its eastward extension. Mesoscale variability is also an intrinsic feature of the coastal currents near the ocean's eastern boundary, for example the California Current system. Thus, for describing the SST anomaly transport one needs higher data resolution than that of the COADS monthly summaries.

Fortunately, COADS contains the Compressed Marine Reports (CMR) for individual observations of meteorological variables: air temperature, SST, specific humidity, wind speed, and total cloudiness. Although the amount of reports varies in space and along the annual cycle, for the North Pacific it is possible to average the individual data to compile a set with higher resolution than the monthly summaries during the past decades.

We chose to average the individual reports for SST in 1°latitude × 2°longitude boxes on a 10-day mean basis from 1965 to 1990. As was shown in chapter 8, for the autoregressive inversion the minimum grid spacings Δx and Δy are related to the maximum advection velocity by CFL-type conditions

$$\Delta x > |u|\Delta t, \; \Delta y > |v|\Delta t. \tag{10.1}$$

According to (10.1), it is better to take Δx, Δy large but Δt small to achieve the maximum possible value for the estimate of the advection velocity. However, the chosen scales are at the upper limit of resolution, which one can attain now by compiling long time series of COADS SST data over the North Pacific. The chosen scales provide a maximum value of the zonal advection velocity of about 15 – 20 km day^{-1}. From such data it is hardly possible to estimate the

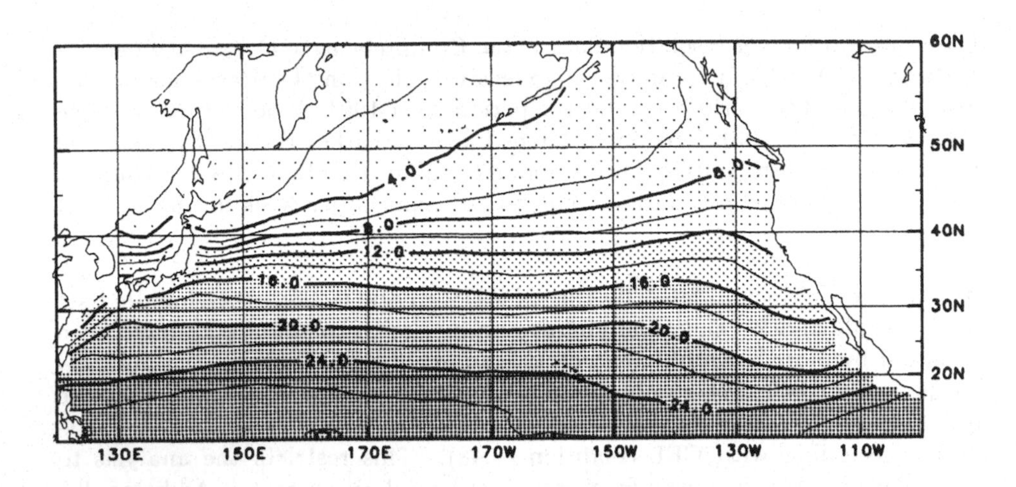

Figure 10.1 Climatological mean SST field in the North Pacific for Jan. 11-20.

meridional advection in the Kuroshio current along the ocean western boundary. Nevertheless, these data are suitable for deriving the zonal advection velocity. The advantage of this time series is the higher resolution provided over the monthly data on a 5° grid which were used earlier. First the data must pass a quality check. Then, the linear trend is removed from the time series. It indicates slight warming in the subtropical gyre of the northwest Pacific. The long-term norms are computed for each 10-day period along the annual cycle (see example for January 11-20 on Fig. 10.1). Finally, to reveal the anomalies, the 10-day norms are subtracted from the SST time series.

One typical example of the resulting SST anomaly field is drawn in Fig. 10.2. It demonstrates the large-scale warm anomaly in the central North Pacific in the late winter 1971/72. The anomaly is surrounded by relatively cold waters. The overall pattern resembled the shape of the first EOF derived by Kawamura (1994) and Tourre and White (1995) (Fig. 9.8). The pattern was dominant during the whole winter but by the end of March the warming weakened and the cold anomalies coalesced in subtropics. Analyzing the winter data, we often noticed this pattern, though with different polarities. Apparent in Fig. 10.2 are gaps in the data south of 20°N in the central Pacific.

As was discussed in section 9.4, the 10-day period seems to be the smallest scale for which the framework of Hasselmann's (1976) approach can be extended. The spectral slope of the atmospheric forcing changes at the frequency

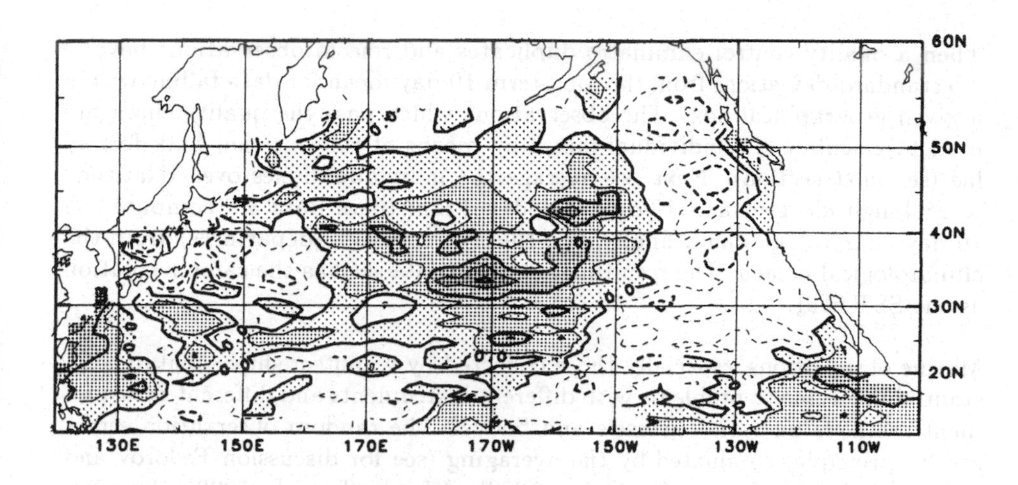

Figure 10.2 SST anomaly field in the North Pacific in Jan. 11-20, 1972. The dashed lines indicate negative anomalies.

0.3 cpd (Fig. 9.10) so that the spectrum is practically white at lower frequencies but looks like a color noise spectrum at higher frequencies. At periods shorter than 10 days, eastward traveling cyclones dominate atmospheric variability. At longer periods the atmospheric spectra are symmetric with respect to wavenumber, and there is no preferred propagation direction.

Willebrand (1978) noted that the SLP spectrum is more 'red' than the wind stress spectra; that is, it decreases faster with increasing frequency. Since some redness can appear near the 10-day period in the atmospheric spectra, we will introduce atmospheric forcing correlations into the autoregressive inversion scheme. As was shown in chapter 8, the autoregressive approach does not require the assumption of white noise forcing as in earlier studies (Frankignoul and Reynolds 1983, Ostrovskii and Piterbarg 1985, Herterich and Hasselmann 1987). Instead, it is assumed that the variability of the forcing fields can be represented as a first-order Markov process in time.

The heat flux component in the atmospheric forcing needs special treatment. To calculate 10-day heat flux, we do not use the averages but the individual observations of the marine variables (for references see Ledvina et al. 1993). More or less standard procedures for this processing can be found in (Iwasaka and Hanawa 1990 and Cayan 1992b). The calculation involves the following steps. At the beginning each meteorological variable is partitioned into geographical boxes of 1°latitude × 2° longitude and sequential 10-day periods.

Then a quality control eliminates duplicates and rejects observations beyond 3.5 standard deviations from the long-term 10-day mean for data falling within a given geographical box. The observations which pass the quality check are used to calculate the individual fluxes by means of conventional bulk formulae (see next section). Next, we average the individual fluxes over 1°latitude × 2° longitude grid and 10-day periods. From these data we compute the 10-day climatologies and, finally, the 10-day anomalies (departures) from the climatological means. The resulting heat flux data set has the same resolution as the SST field.

Marine observations, especially those collected by volunteer ships, contain substantial noise due to problems with different instruments and different measurement methods on various vessels and buoys. The random observation errors are, in principle, eliminated by the averaging (see for discussion Fedorov and Ostrovskii 1986, Weare 1989, Cayan 1992b, Woodruff et al. 1993). In practice, however, the choice of time-space averaging scales is often inadequate for this purpose (Gulev 1994). One more deficiency arises from so-called 'box-car' averaging when observations within a given spatio-temporal box equally contribute to the mean value attributed to the center of the box (Larin 1982). Such averaging disregards the fact that correlations of marine variables decay with increasing both distance and time. Notice, however, that for our analysis the 'box-car' averaging problem is lessened by explicitly averaging the heat anomaly balance equation in space (section 9.6). Other uncertainties in the heat flux estimates are associated with incomplete sampling of naturally varying weather fluctuations (see Cayan 1992b, among others).

Since the detailed study of air-sea heat fluxes is beyond this work scope, we restrict our consideration to conventional computations and suggest the above cited references as a starting point for further reading. To complete this discussion we would like to mention the pioneering efforts by Seager et al. (1988), Frankignoul et al. (1989), Blumenthal and Cane (1989), and Sennichael et al. (1994) where the SST models are tuned in order to determine poorly known parameters, particularly, air-sea heat fluxes.

As was mentioned above, winter is the best season for SST anomaly stochastic modeling. For the cold season we took the period of December-March. The spatial distribution of winter SST observations in 1°latitude × 2° longitude boxes over the North Pacific is shown in Fig. 10.3. Gaps in the time series often occur in the tropics, especially in the area between 150°E and 110°W. The data are sparse in the high latitudes, too. The most detailed observations are in the midlatitude western Pacific. Also well covered with data is the region near the eastern coast of the North America.

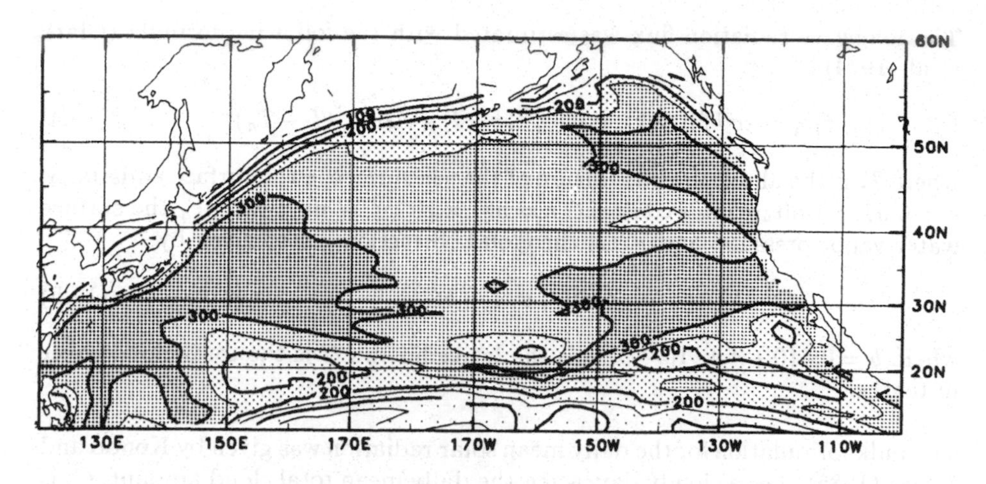

Figure 10.3 The number of 10-day mean SST observations averaged at 1°latitude × 2° longitude boxes for winters of 1965-66, 1966-67, ..., 1989-90.

10.2 OBSERVATIONS OF ATMOSPHERIC FORCING

10.2.1 Air-Sea Heat Flux

Since there is no universally accepted bulk transfer coefficient scheme (Blanc 1985), the bulk formulas for air-sea heat fluxes proposed by different authors usually differ by constants. Consequently there might be a discrepancy between the estimates given below and those obtained in other studies, especially for radiative fluxes. For these calculations we adopted the method suggested by Iwasaka and Hanawa (1990).

The latent and sensible heat fluxes were computed with the well known formula

$$H_E = \rho_a L C_E |\boldsymbol{w}_a| (q - q_a), \tag{10.2}$$

$$H_S = \rho_a C_{pa} C_S |\boldsymbol{w}_a| (T - T_a), \tag{10.3}$$

where q is the surface saturation humidity and q_a is the air humidity, L is the latent heat of water, C_E and C_S are the transfer coefficients for the latent and sensible heat, respectively. The coefficients C_E and C_S are defined as in Kondo (1975).

The longwave radiation flux was estimated with the following formula (Clark et al. 1974)

$$H_{LR} = \varepsilon\sigma T_s^4(0.39 - 0.05e_a^{0.5})F(C) + 4\varepsilon\sigma T_s^3(T - T_a), \tag{10.4}$$

where T_s is the absolute temperature of the sea surface, ε the surface emissivity, $\varepsilon = 0.97$, σ Boltzmann's constant, $s = 567 \times 10^{-10}$ W m^{-2} K^{-4}, e_a the surface water vapor pressure, and $F(C)$ the cloud correction factor given by

$$F(C) = 1 - bC^2, \tag{10.5}$$

where $b = 0.510 + 4.4 \cdot 10^{-3}\varphi$, φ is the latitude, and C is the total cloud cover in tenths.

The bulk formulation for the daily mean solar radiation was given by Kondo and Miura (1985). For a cloudy day, when the daily mean total cloud amount, C, is equal or bigger than 0.3, the daily mean solar radiation is calculated as follows. Let Cl be the daily mean value of low cloudiness, and H_0 be the incoming shortwave radiation received at the sea surface under cloudless conditions. Then, for $C < 1.0$ or $Cl \leq 0.8$

$$H_{SR} = H_0(0.737 \ln(1.22 - 1.02\alpha) + 0.521\alpha + 0.846), \tag{10.6}$$

and

$$H_{SR} = 0.2H_0 \tag{10.7}$$

when $C = 1.0$ and $Cl > 0.8$. In (10.6)-(10.7), the parameter α is given by

$$\alpha = C - 0.4e^{-3.0\ Cl} \tag{10.8}$$

Cl is the daily mean value of the low cloudiness, and H_0 the incoming shortwave radiation received at the sea surface under cloudless conditions.

The clear sky solar radiation, H_0, is adjusted for the transmissivity of the atmosphere and varies with astronomical parameters. It is defined as follows

$$H_0 = H_{a0}(C_0 + 0.7 \cdot 10^{-fm})(1.0 - c)(1.0 + d), \tag{10.9}$$

where

$$C_0 = 0.21 - 0.20C_T, \quad C_T < 0.3 \tag{10.10}$$

$$C_0 = 0.15, \quad C_T \geq 0.3 \tag{10.11}$$

$$f = 0.056 + 0.160C_T^{0.5}, \tag{10.12}$$

$$c = 0.0061(m_a + 7.00 + 0.87 \ln(w_p)) \ln(w_p), \tag{10.13}$$

$$d = 0.090(0.066 + 0.340C_T^{0.5}), \tag{10.14}$$

$$m_a = km_0, \tag{10.15}$$

$$m_0 = 1.0/\cos(\varphi - \phi), \tag{10.16}$$

and

$$k = 1.402 - 0.026\ln(C_T + 0.02) - 0.1(m_0 - 0.91)^{0.5}. \tag{10.17}$$

In (10.9)-(10.17), H_{a0} is the daily mean solar radiation at the top of the atmosphere, C_T the turbidity coefficient, taken to be 0.026 (Iwasaka and Hanawa 1990), m_a the effective air mass, m_0 the air mass at noon, ϕ the sun declination, and w_p the perceptible water, which can be evaluated from the dew point temperature T_d measured near the sea surface

$$\begin{array}{l} \log(w_p) = 0.0350T_d - 0.031, \ T_d \ < \ 18°\text{C} \\ \log(w_p) = 0.0222T_d - 0.200, \ T_d \ \geq \ 18°\text{C} \end{array} \tag{10.18}$$

Iwasaka and Hanawa (1990) computed monthly heat fluxes (10.2)-(10.4), (10.6)-(10.7) over a 5° grid in the North Pacific for the period 1950-79 and compared the results with estimates obtained in other studies. They also tested the rather complicated scheme (10.6)-(10.18) versus the solar radiation formula by Reed (1977). Although the estimates (10.6) could disagree with others by as much as 25%, it is difficult to decide which method is more reliable. This uncertainty is so acute because the solar radiation has a large absolute value. Anyway, the bulk formulae either by Kondo and Miura (1985) or Reed (1977) are the only tools available to evaluate the radiation flux from marine reports.

Finally, we calculate the net heat flux as follows

$$H_{net} = H_{LR} + H_E + H_S - H_{SR}. \tag{10.19}$$

The values of H_{net} are averaged over 1°latitude × 2° longitude boxes.

Long-term 10-day means were computed (see example for January 11-20 in Fig. 10.4) and compared with the monthly net heat fluxes of Iwasaka and Hanawa (1990). It was confirmed that, in general, there is very good agreement between these two studies, though the present analysis gives us a more detailed description of the heat flux annual cycle throughout the North Pacific. The difficulty with such a detailed data set is the higher noise level.

The North Pacific loses heat to the atmosphere in winter (Fig. 10.4). The largest losses (H_{net} can be up to 400 W m^{-2}) occur in the Kuroshio, its extension region, and the Kuroshio - Oyashio frontal zone due to increase in the

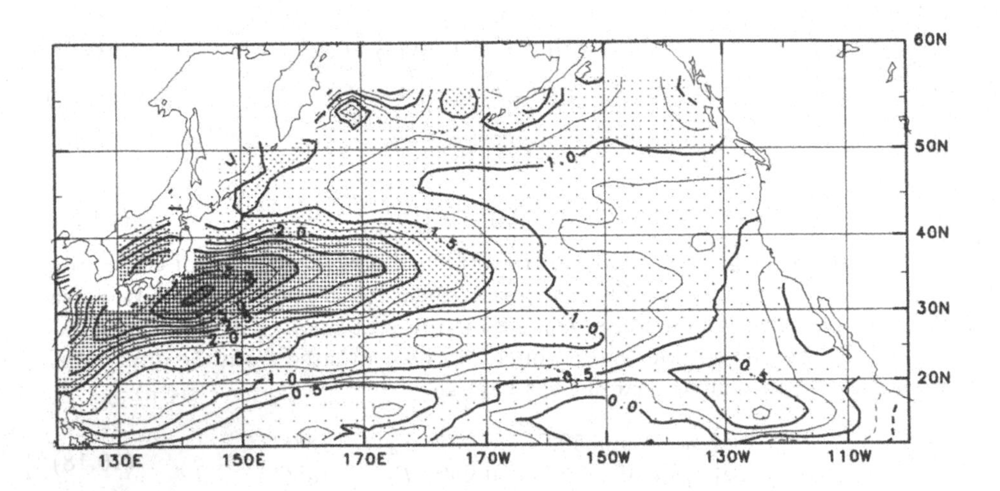

Figure 10.4 Climatically mean heat flux, H_{net} (10^2 W m^{-2}) from sea to air over the North Pacific in Jan. 11-20.

turbulent heat fluxes ($H_E + H_S$). Iwasaka and Hanawa (1990) have noticed that turbulent fluxes dominate in the western North Pacific during winter because the Asian Monsoon carries cold and relatively dry air from the continent. The long wave radiation (H_{LR}) is a minor contributor to heat flux loss. It normally is smaller than the turbulent heat flux by factor of 3-4. The solar radiation (H_{SR}) is a source of heat gain but it is 2-3 times smaller than the turbulent flux and is distributed zonally in the western North Pacific. In the eastern North Pacific, H_{net} is usually less than 150 W m^{-2} with a minimum in the southeastern part of the basin. Moreover, heat gains occurs locally offshore along the California coast in the second half of winter.

The anomalies H'_{net} were obtained by subtracting the long-term 10-day norms from the net heat flux time series. Fig. 10.5 demonstrates the geographical distribution of the standard deviation $\sigma_{H'}$ for the net heat flux anomalies in winter. The values $\sigma_{H'}$ are high in the western North Pacific. Particularly, $\sigma_{H'} \approx 140$ W m^{-2} in the Kuroshio region, though this value is only 1/3 of H_{net} itself. The variance is enhanced along the Subtropical Convergence Zone, too. The lower values $\sigma_{H'} \approx 40 - 60$ W m^{-2} are found in the subpolar regions of the western North Pacific and along the ocean eastern boundary south of 50°N.

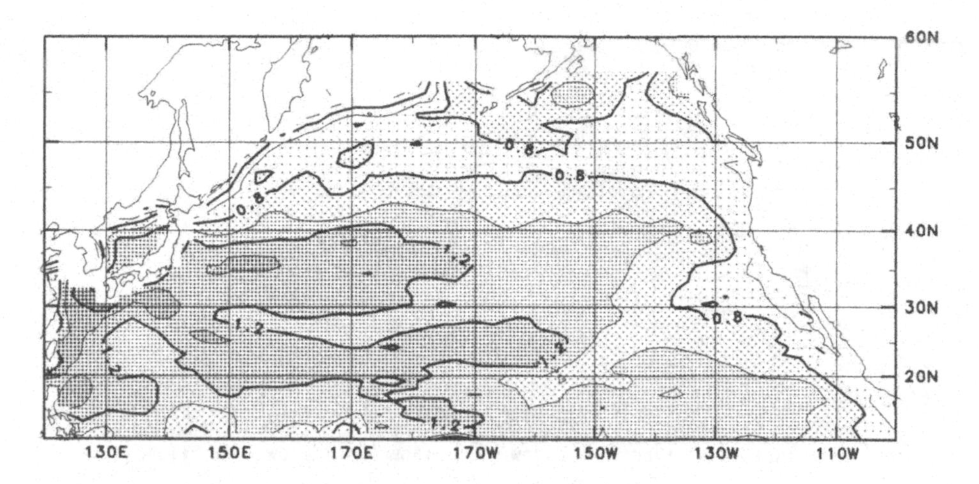

Figure 10.5 The distribution of the net heat flux standard deviation, $\sigma_{H'}$ (10^2 W m^{-2}) in winter.

10.2.2 Atmospheric Feedback

It should be mentioned that the net heat flux anomaly is not pure forcing but includes an unknown contribution from the atmospheric feedback $\lambda_{T'}$ (9.22). This impact seems to be merely in terms of the heat flux anomaly inertia at given time-space scales in the North Pacific. Indeed, the autocorrelation of H'_{net} with a 10-day lag $r_{H'H'}$ is found to be less than 0.1 almost everywhere except east of Kurils, Philippines, and Japan, northeast of the Hawaii Islands and near the eastern boundary north of 45°N. Notice that for the given time series length, the value 0.05-0.07 represents the standard deviation for the autocorrelation estimate (Davis 1973).

As was emphasized by Frankignoul (1985), the estimate for the parameter λ is obscured by difficulties in separating cause and effect in the averaged data since the air temperature adjusts to SST. Frankignoul and Hasselmann (1977) suggested a rough estimate for λ in the local stochastic SST anomaly model via the cross correlation, $r_{T'H'}$, between SST and heat flux anomalies as follows

$$\begin{aligned} r_{T'H'} &\approx (\lambda\tau)^{1/2} e^{s/\tau}, \; s \leq 0, \\ r_{T'H'} &\approx (\lambda\tau)^{1/2} (2e^{-s\lambda} - e^{-s/\tau}), \; s \geq 0, \end{aligned} \tag{10.20}$$

where τ is the correlation time of the anomalous heat flux and s is the time lag. From (10.20) it follows that the cross correlation is negligible when SST leads atmospheric forcing and largest when atmospheric forcing leads SST at small

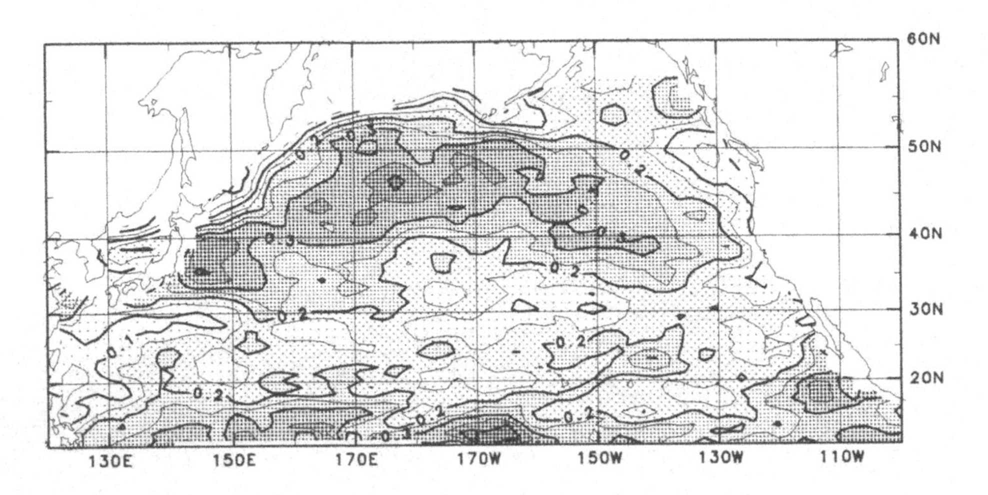

Figure 10.6 The cross correlation between the SST and net heat flux anomalies in the North Pacific during winter: *a*) zero lag correlation.

lags. The correlation decreases slowly on SST anomaly time scales. The values predicted by (10.20) were found to be markedly similar to empirical estimates (Davis 1976) for SLP and SST anomalies.

Unfortunately, we do not know the time scale τ which is needed to derive the empirical estimates for the parameter λ in (10.20) (see also Frankignoul 1985). The available data, for example NOAA Marine Environmental Buoy Database, are not enough to obtain this statistic reliably. So, one can evaluate λ by fitting local (Reynolds 1978) or non-local models (Frankignoul and Reynolds 1983, Ostrovskii and Piterbarg 1985, 1995, and Hasselmann and Herterich 1987) to the SST data alone. These estimates also contain an oceanic contribution λ_E which seems to be smallest in winter (see section 9.6). For this reason it is better to use only the winter data in order to derive the feedback. Below, in section 10.4, to derive λ, we apply an autoregressive estimator to SST time series for the cool season alone.

When the anomalous heat flux lags the SST anomaly by 10 days the correlation between them is usually small $|r_{T'H'}| < 0.1$ with irregular variation over the North Pacific. If there is no lag between two fields the cross correlation is positive everywhere (Fig. 10.6a), especially in tropics and between 40°N and 50°N where the largest values $r_{T'H'} > 0.4$ are found. In the subtropics, the correlation is less than 0.2. Positive correlation simply reflects simultaneous changes in both fields: the warmer (cooler) is the upper mixed layer, the larger

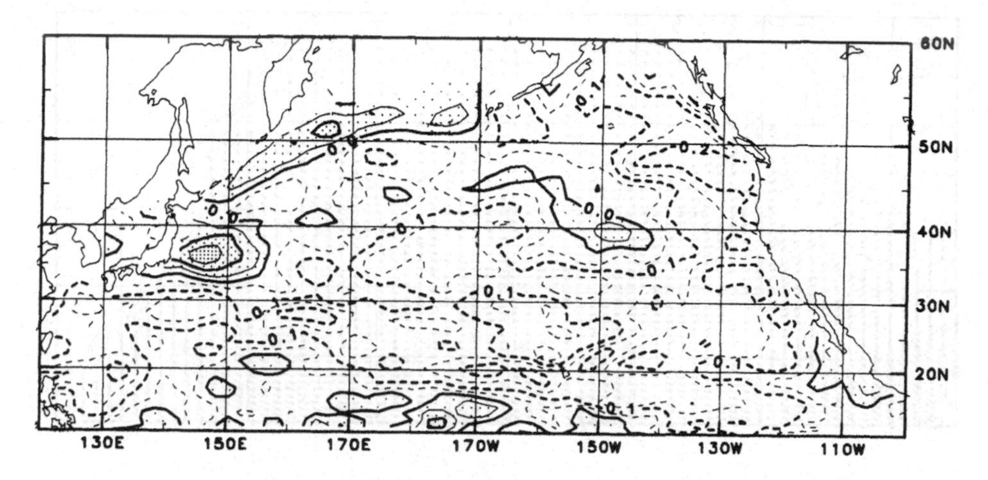

Figure 10.6 *(continued)* *b*) **The anomalous heat flux leads by 10 days.**

(smaller) the heat loss to the air above. However, since the correlation is relatively small in absolute value, the oceanic contribution should be minor compared with that of the atmosphere.

When the anomalous heat flux leads the anomalous SST by 10 days, the cross correlation abruptly becomes negative throughout the North Pacific (Fig. 10.6b). The values of $r_{T'H'}$ mostly are between 0.0 and -0.2. This change of sign in the cross correlation can be associated with the prolonged oceanic memory of anomalous heat flux forcing. The increase (decrease) in heat loss leads to cooling (warming) of the upper ocean mixed layer on SST anomaly time scales. Negative correlation also dominates when the lag, s, is 20 days, but the connection becomes weaker. These observations are in accord with the result by Cayan (1992a) which states that positive (negative) anomalous surface heat fluxes are correlated with anomalous cooling (warming) on the time scale of about 1 month over the North Pacific in winter.

10.2.3 Wind-driven Currents

Besides the anomalous heat flux, the forcing term S' (9.34) contains a contribution due to anomalous near surface currents which are usually associated with wind-drift (Frankignoul 1985). To estimate this contribution we used the COADS CMR surface wind data. The wind field was obtained over 1° latitude

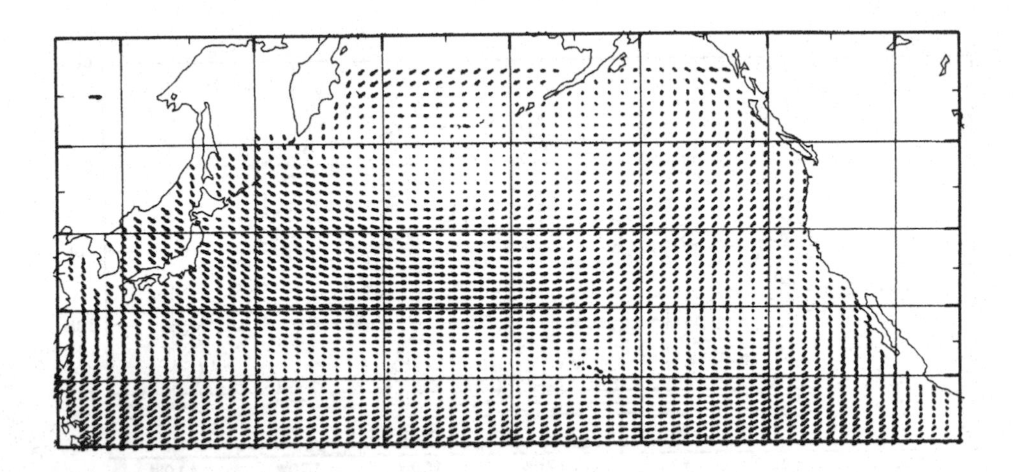

Figure 10.7 Climatological mean vector wind field over the North Pacific in Jan. 11-20. The arrow in the upper left corner indicates a wind speed of 10 m s^{-1}.

$\times$ 2° longitude grid on a 10-day mean basis from 1965 to 1990. The long-term norms were computed for each 10-day interval during the cold season.

Although the surface wind field itself is beyond the scope of this study, its main features are worth mentioning. The long term norms of the vector wind field over the North Pacific are shown for January 11-20 in Fig. 10.7. It agrees broadly with the wind fields derived in other studies (Hellerman and Rosenstein 1983, Kutsuwada and Teramoto 1987, Adamec et al. 1993). The relatively high resolution data set correctly resolves the wind's direction and amplitude. The Asian Winter Monsoon is evident in the western part of the observational domain. The transition from the midlatitude westerlies to the lowlatitude easterlies occurs between 20°N and 25°N. In the high latitudes the easterly flow has a strong southern component as predominantly cyclonic motion is stretched along the Aleutian Islands. Anticyclonic motion is centered at 30°N, 130°W matching the location of the subtropical high.

The autocorrelation of the anomalous wind is stronger than that of the anomalous heat flux. Fig. 10.8 demonstrates the autocorrelations $r_{w'_a w'_a}$ of the anomalous winds with 10-day lags over the North Pacific in winter. The autocorrelation is usually low, below 0.2, and its spatial coherence is of relevance to the mean wind speed (Fig. 10.7). The autocorrelation increases in the broad zonal belts of strong zonal wind, more specifically in the areas of the tropical easterlies, the midlatitude westerlies, and north of 50°N. The autocorrelation

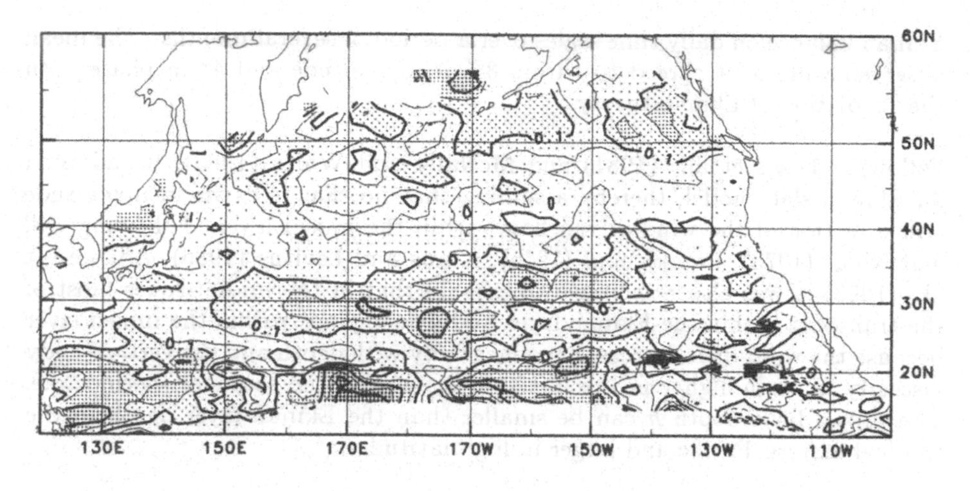

Figure 10.8 The 10-day lag autocorrelations of the anomalous wind in winter.

weakens in the midlatitude regions where the meridional wind component is enhanced. The decrease in the mean wind speed, naturally, leads to a decline in the autocorrelation. The overall relationship between the autocorrelation distribution and the mean wind field implies that the zonal wind is more stable than the meridional wind for time scales on the order of 10 days.

The surface wind is needed to compute the wind-drift current. Pure wind driven current in the whole frictional layer can be defined as in Adem (1970):

$$
\begin{aligned}
u &= c_1 \frac{0.0126}{\sqrt{\sin\varphi}} w_y, \\
v &= -c_1 \frac{0.0126}{\sqrt{\sin\varphi}} w_x,
\end{aligned}
\tag{10.21}
$$

where u and v are, respectively, the zonal and meridional components of the wind driven current, w_x and w_y are the zonal and meridional components of the wind speed, and the constant $c_1 = 0.225$. Equations (10.21) are derived from the Ekman's archetypal solution (see Neumann and Pierson 1966) provided that a steady stress acting together with the Coriolis force generates a water transport to the right of the wind. Experimental evidence for the classical Ekman current has been found by Price et al. (1987) in the western Sargasso Sea (34°N, 70°W). More recently, Chereskin (1995) found Ekman type wind-driven flow at a site in the California Current system. That flow was in the

Ekman balance on daily time scales over a period of several months. The mean observed wind-drift agreed to within 3% in magnitude and 4° in phase with the prediction of Ekman theory.

Following Davis et al. (1981) and McNally and White (1985), we adopted an Ekman slab model, thereby assuming that no change exists in magnitude and direction of the wind-driven flow within the mixed layer. Following Gill and Niiler (1973), Frankignoul (1985), Kraus and Levitus (1986), Adamec et al. (1993) it was also reasonable to assume that in the midlatitudes most of the transport within the Ekman layer would take place within the mixed layer because the wind-drift decreases exponentially with depth and the vertical eddy viscosity is much higher in the mixed layer than in the seasonal thermocline. The mixed layer depth h can be smaller than the Ekman layer depth in the tropical eastern Pacific and larger in high latitudes.

For the depth of the upper-ocean mixed layer, h, we took rather moderate estimates suggested by Bathen (1972). According with that study, the mixed layer is generally about 100 m deep in the midlatitudes whereas it is shallower, $h \approx 60 - 90$ m, in the eastern tropical Pacific and deeper, $h \approx 110 - 150$ m, north of 40°N. Empirical formulas (see Neumann and Pierson 1966) indicate that the Ekman layer is about 90 m deep when the wind speed is 10 m s^{-1} at 45°N. In lower latitudes, for the same wind, the Ekman layer becomes deeper as the Coriolis parameter decreases; it could be 105 m at 30°N. In the high latitudes the effect of the growth in the Coriolis parameter is, obviously, less important so the Ekman depth could be 80 m at 60°N.

These estimates, as well as formulas (10.21), contain two uncertainties due to the unknown variation of the vertical eddy viscosity with depth (see Krauss 1994) and the parameterization for the surface drag coefficient. However, we believe, the effects of these uncertainties are much less than the effects of errors and noise in the COADS wind reports. Anyway, concerning the wind-drift currents within the upper ocean mixed layer, we should bear in mind that the Ekman transport can be somewhat overestimated for latitudes north of 40°N.

Climatological norms were computed for the wind-drift currents. Fig. 10.9 demonstrates such a mean field for January 11-20. Its main features are the Subtropical Convergence Zone, divergence near the Aleutian Islands, and intensification of the monsoon-driven transport. The latter is in a direction opposite to those of the Kuroshio current and its extension but is in accord with recirculation in the subtropical gyre. The data also indicate a wind-driven flow from the North American coast south at about 35°N. The wind-drift velocity is

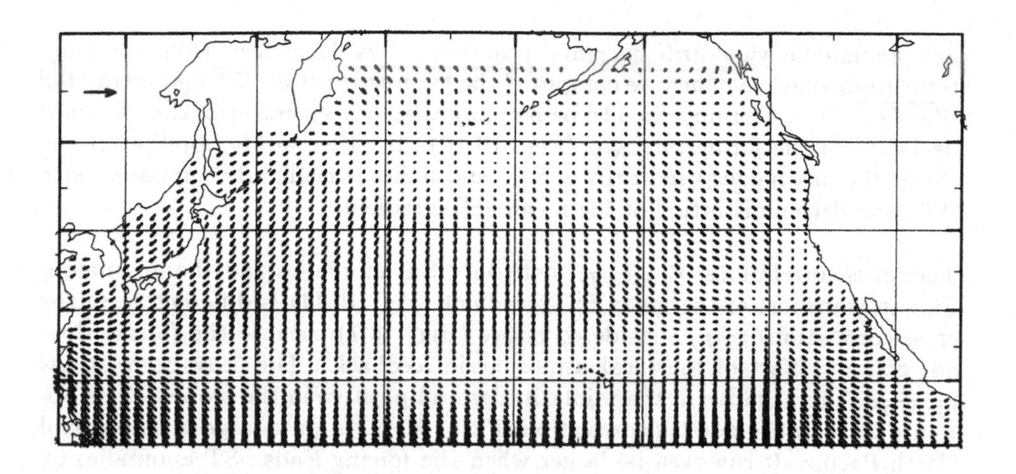

Figure 10.9 Climatological mean wind-driven currents in the North Pacific ocean upper mixed layer in Jan. 11-20. The arrow in the upper left corner indicates a current speed of 0.1 m s^{-1}.

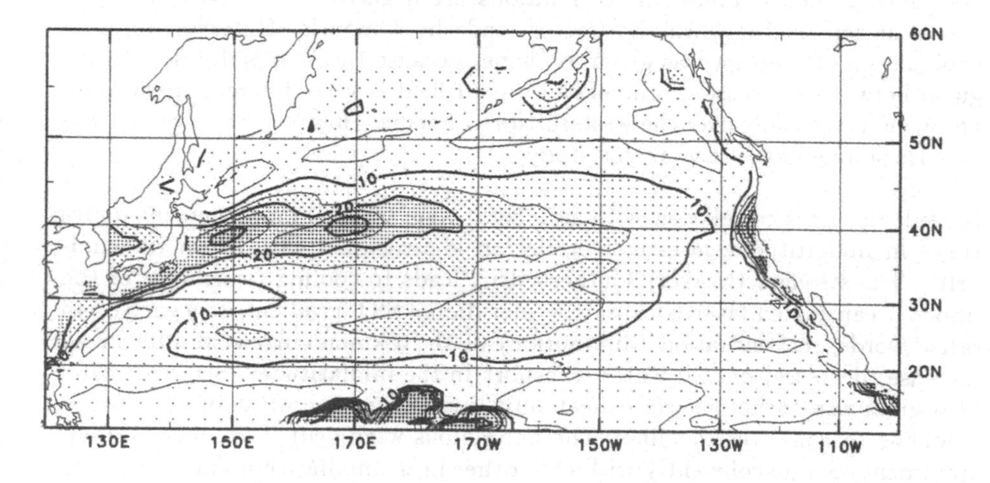

Figure 10.10 Standard deviation of the anomalous wind-driven advection of the mean SST gradient (10^8 °C s^{-1}).

usually less than 0.03 m s^{-1}. In general it increases towards the low latitudes where maximum values can be as high as 0.04 m s^{-1}.

The anomalous wind-drift currents were derived as departures from the long-term 10-day norms. Then we computed the time series of $\boldsymbol{u}' \cdot \nabla \overline{T}$ and estimated $\overline{\boldsymbol{u}' \cdot \nabla T'}$. The latter appeared to be no more than a few percent of the standard deviation of the former, $\sigma_{\boldsymbol{u}' \cdot \nabla \overline{T}}$ (Fig. 10.10). In other words, for the forcing (9.34) the anomalous wind-drift of SST anomalies is negligible compared with the wind-driven distortion of the mean SST gradient.

The cross-correlations, $r_{T'\overline{\boldsymbol{u}' \cdot \nabla T'}}$, between SST anomalies and the anomalous Ekman transport of the mean SST gradient (Fig. 10.11) is different from the cross-correlations $r_{T'H'}$ considered above (Fig. 10.7) because the former does not contain effect of the local atmospheric feedback. The cross-correlations $r_{T'\overline{\boldsymbol{u}' \cdot \nabla T'}}$ are negative in the midlatitudes and positive in the tropics. The absolute value of the zero lag correlation (Fig. 10.11a) is quite large in the central North Pacific. It can even be larger when the forcing leads SST anomalies by 10 days (Fig. 10.11b) and then decreases slowly. In other words, the oceanic response to anomalous wind reaches a peak at about 10 days. This is akin to the results of Frankignoul and Reynolds (1983) who related EOF's of monthly SST anomalies and anomalous Ekman heat transport. According to Frankignoul and Reynolds (1983), the correlations are negative with largest absolute values at zero lag or when the forcing leads by 1 month. Unfortunately, no explanation of this fact was given, perhaps, because the EOF's did not distinguish between the seasons whereas the wind field is very different in summer and winter especially over the western North Pacific. However, the explanation is quite simple in the case of this study.

Indeed, the cross-correlation with zero lag is akin to the Ekman current direction. In midlatitudes, negative correlation is associated with southern wind-drift. The stronger (weaker) southern drift leads to cooling (warming) of the upper ocean mixed layer because the meridional SST gradient is negative. In other words, intensification (diminution) of the Ekman current results in an increase (decrease) of cold water transport in the midlatitudes. Stronger wind also gives rise to both surface heat flux loss and entrainment of cold water from the seasonal thermocline. The anomalous wind-drift, heat flux and entrainment operate coherently with each other in the midlatitudes in winter. By contrast, in the tropics, stronger wind-drift leads to warming since the Ekman current is usually northward. Interestingly, in terms of the cross-correlations, the strongest oceanic response to the wind is found in area bounded approximately by 25° and 40°N, 160°E and 140°W. This strong response area coincides with region of negative loading in the first EOF SST mode by Tourre and White (1995) (Fig. 9.8). It seems that the global wind field, which itself may depend on the tropical SST (Philander 1990), largely drives SST variability in the interior of the North Pacific.

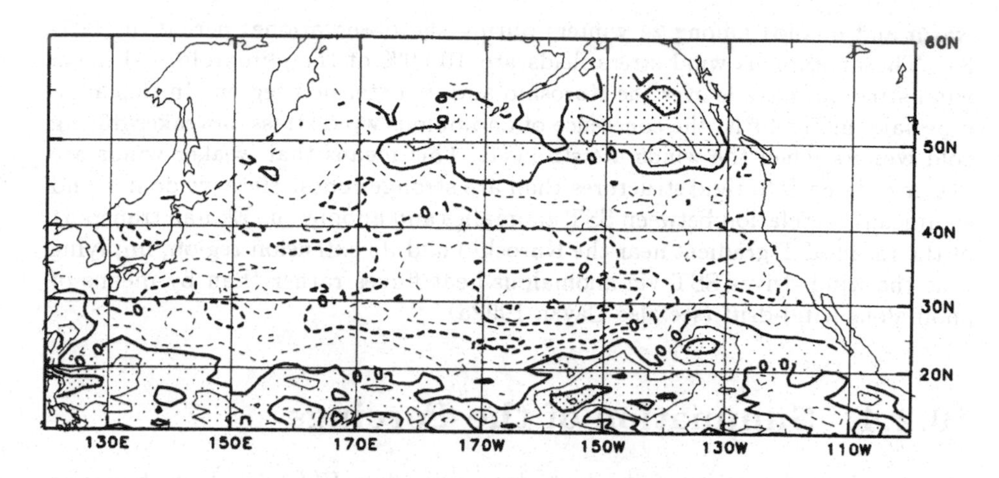

Figure 10.11 Distribution of the lagged cross-correlations between SST anomalies and anomalous Ekman transport of the mean SST gradient: *a*) zero lag correlation. The dashed lines indicate negative correlations.

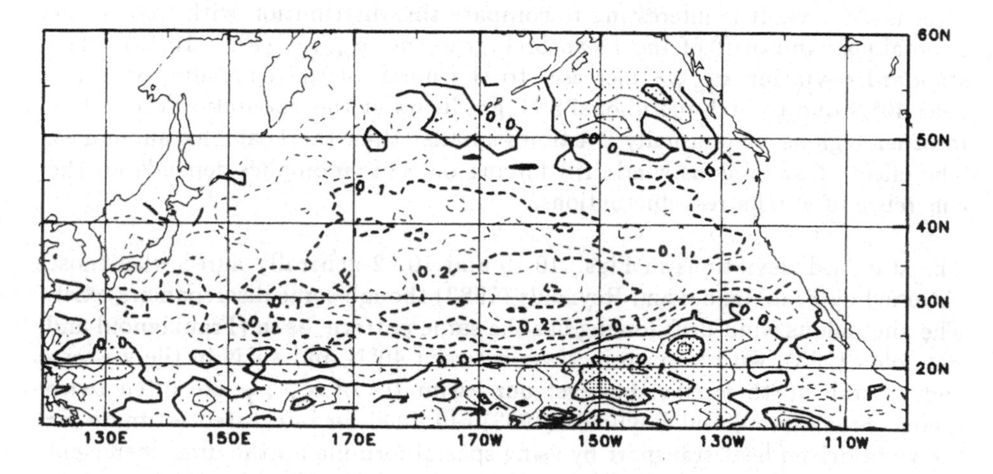

Figure 10.11 *(continued)* *b*) SST anomaly lags by 10 days.

The relationships between the wintertime wind stress and SST anomalies in the western North Pacific were considered by Hanawa et al. (1989a). They concluded that when SST is abnormally warm the westerlies shift northward and weaken but when SST is abnormally cold the westerlies shift southward and strengthen. Such abnormal conditions were observed in 12 winters (6 warm and 6 cold) among 24 winters during the observational period of 1961-84. The anomalous wind stress fields are 10-50% of the climatological mean wind stress amplitude over the Kuroshio and its extension region. In this area, especially along 40°N, the coherence of anomalous wind stress is weaker during cold winters when the winds are stronger. This means that weaker winds are better ordered into rigid structures than are stronger ones. Here we do not find significant correlation between SST anomalies and anomalous Ekman transport of the mean SST gradient near the Kuroshio and its extension region, implying that the wind drives SST via anomalous heat fluxes rather than by means of anomalous wind-drift (see also Cayan 1992a).

10.2.4 Summation of the Forcings

The standard deviation of the heat flux anomalies $H'/\rho_w C_{pw}\overline{h}$ is shown in Fig. 10.12. The values are of the order 10^{-7} °C s^{-1}. The maximum stretches along the Kuroshio current and then extends southeastward of Japan. Another maximum is located in the eastern tropical Pacific where the upper mixed layer is shallow. It is interesting to compare this distribution with that of the anomalous wind-drift of the mean SST gradient, $\sigma_{\boldsymbol{u}'\cdot\nabla\overline{T}}$ (Fig. 10.10). The standard deviation $\sigma_{\boldsymbol{u}'\cdot\nabla\overline{T}}$ increases to the north of the Kuroshio extension near 40°N due to high meridional SST gradients at the Subarctic Front while in other regions, it is usually 2-4 times smaller than the heat flux anomalies. The effect of anomalous wind-drift forcing on SST anomalies depends on the coherence of wind stress fluctuations.

The standard deviations in Figs. 10.10 and 10.12 generally agree with those obtained by Frankignoul and Reynolds (1983) though some discrepancies exist. The anomalous wind-drift forcing has the same pattern, as in (Frankignoul and Reynolds 1983), with a sharp increase between 40°N and 45°N in the western and central North Pacific, however our estimates of $\sigma_{\boldsymbol{u}'\cdot\nabla\overline{T}}$ are often twice theirs. Frankignoul and Reynolds (1983) obtained an integrated estimate for the wind-driven heat transport by using special formula for the drag coefficient (Kondo 1975). Their estimates were monthly over a $5° \times 5°$ grid making the data variance smaller and the larger averaging scales resulted in more smoothness. Additionally, in data poor regions, climatology was used when no measurement were available. Although the upper mixed layer depth was usually smaller than in the present analysis, north of 40°N it was 120 m, i.e. as in winter. All these factors contributed to the smaller values of $\sigma_{\boldsymbol{u}'\cdot\nabla\overline{T}}$ used in (Frankignoul and Reynolds 1983)

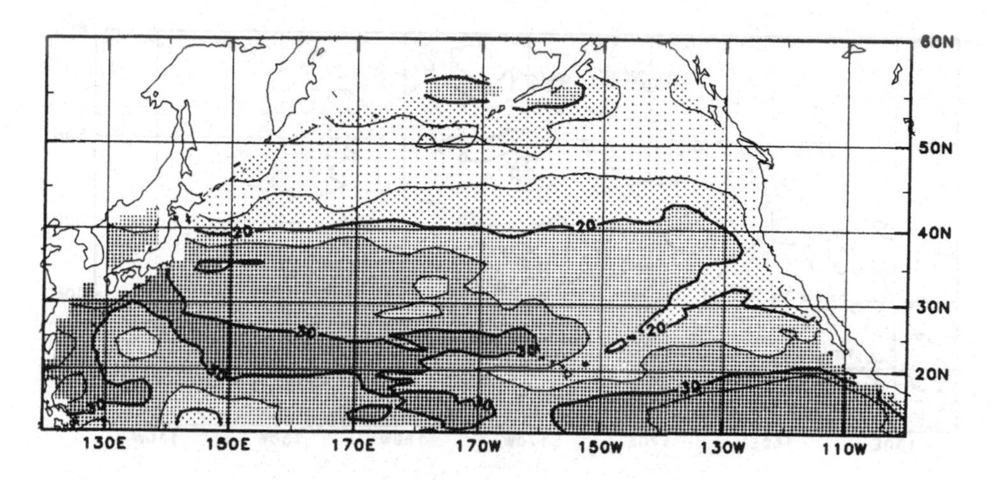

Figure 10.12 Distribution of the standard deviation for the heat flux anomalies $\frac{H'}{\rho_w C_{pw} \overline{h}}$ (10^8 °C s^{-1}) over the North Pacific in winter.

As for the heat flux variance, Frankignoul and Reynolds (1983) reported larger values in the southwestern North Pacific. This difference can be attributed to poor data coverage for this region. However in general, despite the differences in the data sets and data processing, the sum of the two forcing terms, the anomalous heat flux and the anomalous advection of the mean SST gradient, computed by Frankignoul and Reynolds (1983) agrees broadly with that adopted in the present study.

Fig. 10.13 demonstrates the autocorrelation $r_{Q'Q'}$ with a 10-day lag for the sum of the two forcing terms, $Q' = H'/\rho_w C_{pw}\overline{h} + \boldsymbol{u}' \cdot \nabla \overline{T}$. The autocorrelation $r_{Q'Q'}$ is generally below 0.1 but can be approximately 0.15 near the western boundary and off the North American coast near 50°N. In the central North Pacific, it is higher than the autocorrelation of the anomalous heat flux but lower than the autocorrelation of the anomalous wind speed, $r_{w'_a w'_a}$ (Fig. 10.8), implying that the anomalous heat flux whitened the color-noise effect of the anomalous wind-drift forcing. Near the western boundary, the autocorrelation in the anomalous heat flux is affected by SST anomaly inertia. Therefore, pure atmospheric forcing should be less correlated in this region, especially east of Japan.

The forcing term S' (9.34) also contains a contribution due to feedback fluctuations $\lambda'\overline{T}$. Unfortunately, it is difficult to estimate this contribution directly.

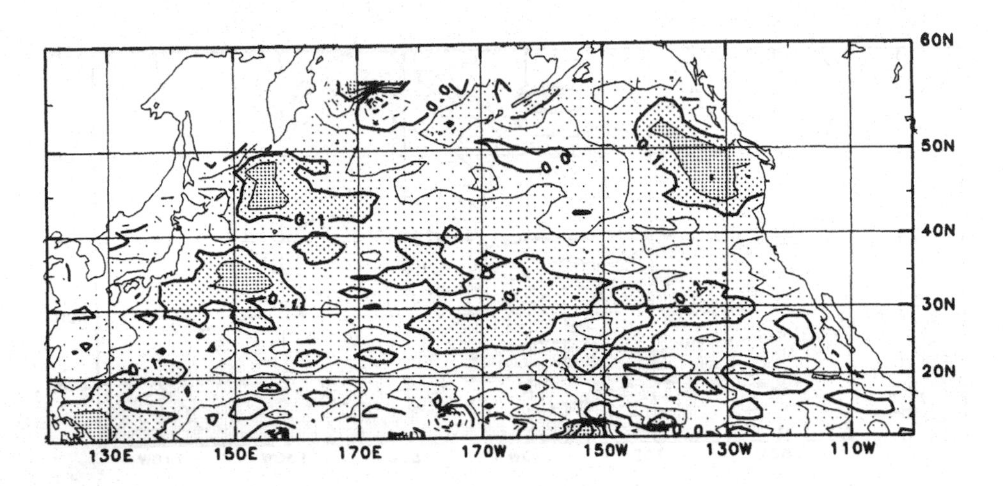

Figure 10.13 The 10-day lag autocorrelation for the sum of anomalous heat flux and anomalous wind-drift of the mean SST gradient.

In order to do this, one can subtract the contribution of $\overline{\lambda_Q T'}$ from the above forcing terms and calculate the variance of the residuals, refined forcing terms. Then, if the mean product $\overline{\lambda' T'}$ is negligible, as well as the term $\overline{\boldsymbol{u}' \cdot \nabla T'}$, and the correlations between atmospheric forcing terms is small then the variance of $\lambda' \overline{T}$ could be found by subtracting the variances of the refined forcing terms from the noise variance prescribed by the autoregressive model.

Overall, this section provides us with empirical estimates of the relationships between the atmospheric forcing and SST needed for the inversions. The obtained cross-correlations are in a good agreement with theoretical predictions and previous estimates. It is important that the cross-correlations have large-scale patterns thereby confirming earlier results by Cayan (1992a), which stated that although the heat flux forcing of SST anomalies operates locally, the connection between the two fields are organized over spatial scales. Here, in addition to the heat flux forcing, we also examine the wind-drift of SST. The basically white-noise behavior of the atmospheric forcing on a 10-day scale is associated with the anomalous heat flux rather than the anomalous wind-drift of the mean SST gradient.

But why should SST anomalies be organized over spatial scales in certain areas whereas the EOF of anomalous SLP modes span scales of basin size? Or, in other words, why should large-area SST anomalies be formed often in particular locations such as the region centered at 30°N, 145°W, as has been suggested by several other investigators? Now it seems that we know the answer only in part by exposing the connections between atmospheric forcing and SST anomalies. We still do not know which processes in the upper-ocean provide background conditions favorable for the generation of the large-scale SST anomalies by spatially coherent atmospheric forcing. However we anticipate that advection in the upper-ocean should act coherently or be sufficiently small on large scales to provide such conditions. Both horizontal and vertical heat exchange in the upper ocean must be uniform enough to allow large scale SST anomalies to develop. The horizontal transport of anomalous heat will be considered in the next sections.

10.3 AUTOREGRESSIVE MODEL FITTING

Here the objectively consistent estimation (see Fig. 9.13) consists in applying the autoregressive model to derive the net advection velocity, diffusivity, feedback and atmospheric forcing parameters from the observational data. As was shown in chapter 8, the similarity between the stochastic equation (8.1) and equation (9.34) implies that discrete observations $T_n(\boldsymbol{r})$ of the SST anomaly field satisfy the autoregressive model (8.21), which we rewrite here for the sake of convenience

$$T_n(\boldsymbol{r}_0) = \sum_{m=0}^{8} \alpha_m[T_{n-1}(\boldsymbol{r}_m) - \beta T_{n-2}(\boldsymbol{r}_m)] + \beta T_{n-1}(\boldsymbol{r}_0) + b_n(\boldsymbol{r}_0), \tag{10.22}$$

where $\boldsymbol{r}_0$ is arbitrary, $\boldsymbol{r}_1, ..., \boldsymbol{r}_8$ are its neighbors (see Fig. 8.2), b_n is white noise, α_m are the autoregressive coefficients related with the physical parameters via formulas (8.20) and β is the autoregressive coefficient of the forcing process. Therefore, we have to estimate the unknown coefficients α_m and β from SST time series and then compute the physical parameters of interest by using (8.20).

The estimation of α_m and β is based on the mean square method. Namely, after multiplying (10.22) by $T_n(\boldsymbol{r}_\mu)$ and averaging through time, we arrive at the following relation

$$\gamma_1^{0\mu} = \sum_{m=0}^{8} \alpha_m[\gamma_0^{\mu m} - \beta\gamma_1^{\mu m}] + \beta\gamma_0^{0\mu}, \quad \mu = 0, ..., 8, \tag{10.23}$$

where, as in (8.23),

$$\gamma_s^{ij} = \langle T_n(\boldsymbol{r}_i)T_{n-s}(\boldsymbol{r}_j)\rangle. \tag{10.24}$$

This autoregressive inversion differs in essential details from the inversion proposed by Herterich and Hasselmann (1987) even though the earlier method was derived from the same heat anomaly balance equation as (9.42). Herterich and Hasselmann (1987) spatially discretized (9.42) with central differences. Then, they replaced the temperature anomaly and the source with the corresponding Fourier expansions in the frequency domain. We preferred another approach: the heat anomaly balance equation is solved analytically and the solution is replaced by its discrete form (see chapter 8) which is to be fitted to the time series over the space grid.

Discretizing the solution in time, we assume that the forcing field is a first-order Markov process. This is important because the averaging period in the time series is 10 days, i.e. close to the period at which the atmospheric forcing spectrum becomes like color noise. Thus, a priori, we can not consider the atmospheric forcing as a white noise process as Herterich and Hasselmann (1987) did. For spatial discretization, we introduce a polynomial approximation of the gridded observations $T_n(\boldsymbol{r})$ whose coefficients represent advection and diffusion.

10.3.1 Seasonal Tuning

The autoregressive model (10.22) has an advantage over the model of Herterich and Hasselmann (1987). Whereas the latter yields the long-term annual means of the parameters, the former is suitable for seasonal means. Indeed, it can be fitted to seasonally sampled data. The model (10.22) is used for the winter season alone, so the seasonal features of the atmospheric forcing are retained. According to Cayan (1992a), SST anomalies in winter are governed, to the utmost, by the response of the upper ocean to heat flux forcing.

While the SST anomalies $T_n(\boldsymbol{r})$ are departures from the 10-day climatological norms, we can evaluate their covariances at matched seasonal period (c.f. Davis 1973), for example in winter

$$\begin{aligned}\gamma_s^{ij} = {} & \frac{1}{(M-1)N}\sum_{d=2}^{M}\sum_{k=1}^{N} T_{d+36k}(\boldsymbol{r}_i)T_{d-s+36k}(\boldsymbol{r}_j) \\ & -\frac{1}{(M-1)^2N^2}\sum_{d=2}^{M}\sum_{k=1}^{N} T_{d+36k}(\boldsymbol{r}_i)\sum_{d=2}^{M}\sum_{k=1}^{N} T_{d+36k}(\boldsymbol{r}_j)\end{aligned} \tag{10.25}$$

where M is the number of 10-day intervals in the matched season, N is the total number of observational years, and $s = 0, 1$ is the lag in 10-day intervals.

The value γ_s^{ij} represents the covariance of the available overlapped portions of the time series.

The covariance (10.25) describes the relationship between time series shifted relative to one another with lag s in the same seasonal period. The sense of the seasonal covariance is similar to that of a regime-dependent covariance function (see Zwiers and von Storch 1989). Formula (10.25) defines either the cross-covariances between the time series at different geographical locations when $i \neq j$, or the autocovariance, when $i = j$.

As was shown in section 9.2, seasonal nonstationarity is an intrinsic feature of SST variability. For example, in the Kuroshio and its extension regions, the standard deviation σ_T is about the same in autumn, winter, and spring but, in summer, it can be twice as larger (Ostrovskii and Piterbarg 1995). In the tropics, the autocovariances (10.25) drastically decrease in late spring and early summer, apparently, quite in agreement with a reduction in skill in predicting central Pacific SST anomalies in numerical models (Cane et al. 1986, Latif and Graham 1992).

The need to distinguish between the seasons becomes evident if one takes into account the annual modulation of the atmospheric forcing and the seasonal variation of the mean entrainment (see formula 9.43). As long as we have no data on the temperature anomalies in the seasonal thermocline, it is better to consider only the cool season when, in the extratropics, the anomaly temperature gradient below the mixed layer is small or the seasonal thermocline actually disappears. So, for this study we selected December, January, February, and March when, according to Bathen (1972), the mixed layer depth h is greater than 100 m throughout the North Pacific except in the eastern tropics. These estimates actually seem to be on the low side for the Kuroshio extension region (Hanawa and Hoshino 1988) and the area north of the Subarctic Front (c.f. GarciaOrtiz and RuizdeElvira 1995).

10.3.2 Numerical Schemes

In order to obtain the coefficients α_m and β, one must estimate the correlations (10.23) from the data and solve the system (10.23) with respect to α_m and β. In other words, the system (10.14) should be fitted to the covariances of SST anomalies at the points shown in Fig. 8.2.

Then, to estimate the advection velocities $\boldsymbol{u}$, the mixing coefficient D, and the feedback parameter λ, one substitutes (8.20) into (10.23) and solves the new system numerically. Notice, the system is over determined with respect to the unknowns $\boldsymbol{u}, D, \lambda$, and β, but not with respect to the coefficients α_m and β. If the value of β is known then the model (8.22) becomes exactly determined for the coefficients α_m.

Recently Ostrovskii and Piterbarg (1995) considered the problem with unknown coefficient β. In order to avoid uncertainty in the solution, the problem was constrained by fixing the unknowns D and λ over spatial scales larger than those of the numerical frame (Fig. 8.2). Although this assumption is not strictly valid, it seems reasonable for many ocean regions. The obtained advection velocities basically yield the overall pattern of the near surface ocean currents in the western North Pacific.

Interestingly, the coefficient β was found to vary from -0.05 to 0.25 in the western North Pacific (Ostrovskii and Piterbarg 1995). It exceeded 0.1 near the western boundary, where the SST anomaly field demonstrated prolonged memory of earlier winds thereby implying that the atmospheric forcing might be correlated at 10-day lags during the Asian Winter Monsoon. Noticeably, there was an increase in the values of β near the southern boundary of the analysis domain. Over this area, strong wind stress anomalies occur when the westerlies weaken (Hanawa et al. 1989b).

Since the parameter β was found to be small, Ostrovskii and Piterbarg (1995) suggested that the 10-day memory of the forcing field has a small impact on the SST anomaly behavior over the northern part of the subtropical gyre far away from the western boundary. The small coefficient β was anticipated because the spectral analysis by Willebrand (1978) showed that the longitudinally averaged wind stress spectrum is practically 'white' over periods lasting more than 3 days. Nevertheless, it was emphasized, that the correlations in the forcing process on a 10-day scale could not be always neglected.

In fact, both the autocorrelation $r_{Q'Q'}$ for the anomalous heat flux forcing along with the anomalous wind-drift of the mean SST gradient and the autocorrelation of the anomalous wind itself $r_{w'_a w'_a}$ determine the distribution of the parameter β. As will be discussed below, the role of the atmospheric forcing memory seems to be more complex than it was thought because of the effect of the forcing on the diffusivity and feedback. The response of the SST field to variations in β is noticeable in the diffusivity and feedback estimates rather than in those of the advection.

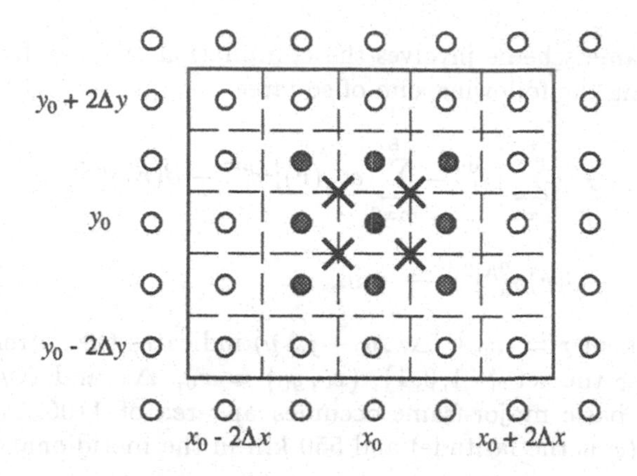

Figure 10.14 The major-frame for autoregressive model fitting. Individual models are defined on sub-frames (of 3 × 3 grid boxes, as in Fig. 8.2) with centers at the shaded circles. The parameters of the heat anomaly balance equation are found at the central box, (x_0, y_0). The divergence norms can be defined for locations between the centers of the sub-frames. These locations are marked by crosses, ×.

In the calculations below, we will take β to be a known constant either as the autocorrelation of the sum of the anomalous heat flux forcing and the anomalous wind-drift of the mean SST gradient, $r_{Q'Q'}$ (Fig. 10.13), or the autocorrelation of the anomalous wind, $r_{w'_a w'_a}$ (Fig. 10.8) since the wind is a major player in the atmospheric forcing. Therefore we have an exactly determined system for the coefficients α_m. The results relevant to this case will be considered at the beginning of the next section.

We also use a slightly more complicated algorithm that involves the inversion for a set of the frames instead of one frame. The frames or, for the sake of clarity, 'sub-frames' can be compiled into a set called a 'major-frame' as shown in Fig. 10.14. Notice, sub-frames are overlapped and embedded into the major-frame. One motivation for this 'frame scheme' was a concern about divergence norm minimization.

The simplest frame scheme involves the compilation of 9 sub-frames into one set by minimizing the following sum of squares:

$$\sum_{r}(\sum_{\mu=0}^{8}|\gamma_1^{0\mu}-\sum_{m=0}^{8}\alpha_m(r)[\gamma_0^{\mu m}-\beta(r)\gamma_1^{\mu m}] \\ -\beta(r)\gamma_0^{0\mu}|^2) \rightarrow \min, \tag{10.26}$$

where the argument $r = (x_0 + i\Delta x, y_0 + j\Delta y)$ indicates the current grid point, i and j run over the set $\{-1, 0, 1\}$, $(x_0, y_0) = r_0$, Δx and Δy are the grid spacings. This basic major-frame occupies an area of $1100 \cos\varphi$ km in the zonal direction (φ is the latitude) and 550 km in the meridional direction.

The continuity equation can be introduced for the horizontal divergence ∇u between the centers of the sub-frames (Fig. 10.14). In this case, to find the unknowns u, D, λ, and β, it will be necessary to minimize the following sum of squares in addition to (10.26):

$$\sum_{r}|(u(x+\Delta x, y+\Delta y)\Delta x + v(x+\Delta x, y+\Delta y)\Delta y \\ -u(x,y)\Delta x - v(x,y)\Delta y - v(x, y+\Delta y)\Delta x + u(x, y+\Delta y)\Delta y \\ +v(x+\Delta x, y)\Delta x - u(x+\Delta x, y)\Delta y)/(\Delta x^2+\Delta y^2)^{1/2}|^2 \rightarrow \min. \tag{10.27}$$

While introducing the horizontal divergence norm (10.27), we assume that the seasonal mean depth of the upper mixed layer does not vary on scales smaller than the frame size. When using this norm it is reasonable, perhaps, to consider larger sets of sub-frames, say 5×5 sub-frames.

The test experiments demonstrated the usefulness of minimizing (10.27) in the inversion for the advection velocities. This minimization is commonly used for treating uncertainties in tracer inversions. Kelly (1989) and Kelly and Strub (1992) combined minimization of the heat equation and the horizontal divergence norm to get more realistic near-surface ocean current velocity fields from satellite infrared images.

Generally speaking, the optimization (10.26)-(10.27) can be solved by, for example, the Levenberg-Marquardt-Morrison method so that no equation in (10.23) would be satisfied exactly. Notice, only the estimates for the central cells r_0 have to be kept for further analysis.

While estimating unknowns in practice, we must locate the center of a major-frame at each grid point, so the major-frame is translated over the data grid to cover the whole observational domain. A numerical solution is sought at each location and the estimates for the central cells $\boldsymbol{r}_0$ of the overlapped major-frames are compiled to form the fields for the advection velocity, diffusivity, and feedback parameter. Notice, in order to enhance the presentation near the boundaries of the observational domain, one uses solutions for the cells on the sides of the major-frame.

It is worth noting that in addition to (10.27), other norms can be minimized during the inversion. For example, following Kelly (1989), it is easy to minimize the kinetic energy and the relative vorticity. We do not discuss this or other norm minimization here because in equation (9.42) we use the net advection velocity rather than the current velocity.

As in Ostrovskii and Piterbarg (1995), one can additionally constrain mesoscale mixing and Newtonian cooling by assuming that the diffusivity and the feedback parameter are constant over major-frames. Although this assumption is not strictly valid, it helps to improve the solution. In this case, one obtains an overdetermined system of nonlinear equations with respect to the unknowns $\boldsymbol{u}, D$, and λ, among which D and λ are fixed over major-frames.

To close this discussion, it is worth making one more remark on the autoregressive model fit. If one attempts to solve the underdetermined system for the coefficients α_m and β

$$\gamma_1^{0\mu} = \sum_{m=0}^{8} \alpha_m [\gamma_0^{\mu m} - \beta \gamma_1^{\mu m}] + \beta \gamma_0^{0\mu}, \quad \mu = 0, 1, ..., 4 \tag{10.28}$$

which is exactly determined with respect to the unknowns $\boldsymbol{u}, D, \lambda$, and β then certain information about the northwest - southeast and northeast - southwest components in (10.23) is neglected and the obtained solution may not be satisfactory.

On the other hand, it is possible to overdetermine the basic system. For example, one may suggest to add the equations defined for the second, third and so on covariances in order to get a new system like

$$\begin{aligned}
\gamma_1^{0\mu} &= \sum_{m=0}^{8} \alpha_m [\gamma_0^{\mu m} - \beta \gamma_1^{\mu m}] + \beta \gamma_0^{0\mu}, \\
\gamma_2^{0\mu} &= \sum_{m=0}^{8} \alpha_m [\gamma_1^{m\mu} - \beta \gamma_0^{m\mu}] + \beta \gamma_1^{0\mu}, \\
\gamma_3^{0\mu} &= \sum_{m=0}^{8} \alpha_m [\gamma_2^{m\mu} - \beta \gamma_1^{m\mu}] + \beta \gamma_2^{0\mu}, \\
&\dots\dots\dots\dots,
\end{aligned} \tag{10.29}$$

where $\mu = 0, 1, ..., 8$. However, according to model (10.22) all the information about the past is contained in two prior SST anomaly fields, i.e. the value T_n is completely determined by the knowledge of T_{n-1} and T_{n-2}. Therefore the excess in the system such as (10.29) does not bring in new knowledge. Instead, the solution would be complicated due to additional degrees of freedom in the minimization and the growth of noise in the covariance estimates at longer lags.

Perhaps, the worst scheme that combines the disadvantages of the two previous ones (10.28) and (10.29) is the system

$$\begin{aligned}
\gamma_1^{00} &= \sum_{m=0}^{4} \alpha_m [\gamma_0^{0m} - \beta \gamma_1^{0m}] + \beta \gamma_0^{00}, \\
\gamma_2^{00} &= \sum_{m=0}^{4} \alpha_m [\gamma_1^{m0} - \beta \gamma_0^{m0}] + \beta \gamma_1^{00}, \\
\gamma_3^{00} &= \sum_{m=0}^{4} \alpha_m [\gamma_2^{m0} - \beta \gamma_1^{m0}] + \beta \gamma_2^{00}, \\
&\dots\dots\dots\dots
\end{aligned} \tag{10.30}$$

It should be emphasized that the only purpose of showing the systems (10.28)-(10.30) is to warn of their deficiencies. On the contrary, it appears that the scheme (10.26)-(10.27) maximizes the chance of achieving a correct solution by fitting an autoregressive model to the time series of observations.

Finally, one more warning is necessary. By solving the system (10.26)-(10.27) we obtain estimates for the unknowns $\boldsymbol{u}, D$, and λ, but the system (10.23) is not satisfied exactly. The coefficients α_m can be found by substituting the values of $\boldsymbol{u}, D$, and λ into (8.20). Since such estimates for the autoregressive parameters would contain errors, it bothered us whether the evaluation of the model (10.23) alone provides better estimates or not. It was thought that with a fixed value of the parameter β, the coefficients α_m could be readily obtained as a solution of the linear system (10.23). Numerically, however, conventional procedures often returned sets of coefficients that were wrong, as was uncovered by direct substitution back into the original equations (10.23). We faced these troubles while trying to solve the system (10.23) on the real SST anomaly data. The misfits of the 'exact' solutions were often larger by one order of magnitude than those obtained via the optimization procedure (10.22) with the statistical parameters substituted into (8.20).

10.4 AUTOREGRESSIVE ESTIMATES OF THE VELOCITY AND DIFFUSIVITY

The study of heat anomaly advection should answer the question, from where do the SST anomaly propagate? We want to know the redistribution of heat anomalies within the upper-ocean mixed layer and the relative importance of the processes responsible for SST anomaly behavior. This knowledge is needed for better an understanding of interactions in climate system.

Earlier studies indicate that SST anomalies propagate eastward in the middle latitudes. More specifically, Namias (1959) and Favorite and McLain (1973) suggested that SST anomalies are advected along the main currents in the North Pacific. Michaelsen (1982) used EOF techniques to demonstrate that the main anomaly pattern expands both in size and in intensity as it moves across the North Pacific. Ostrovskii and Piterbarg (1985) and Herterich and Hasselmann (1987) estimated the heat anomaly advection in the North Pacific. In both cases, data were monthly mean SST fields in 5° squares. In the northeast Pacific the analysis revealed circulations which resemble the overall structure of the surface current field inferred from ship-drift observations. In contrast, for the northwest Pacific those studies did not exhibit advection patterns consistent with the circulation in the subtropical gyre and the Kuroshio extension region. The authors speculated that the spatio-temporal averaging of the SST data considerably smoothed the effects of strong currents. The present analysis and the previous one (Ostrovskii and Piterbarg 1995) provide us with much more realistic advection fields than earlier studies though, admittedly, our understanding of heat anomaly advection benefited from the ideas of Namias (1959) and the other researchers mentioned.

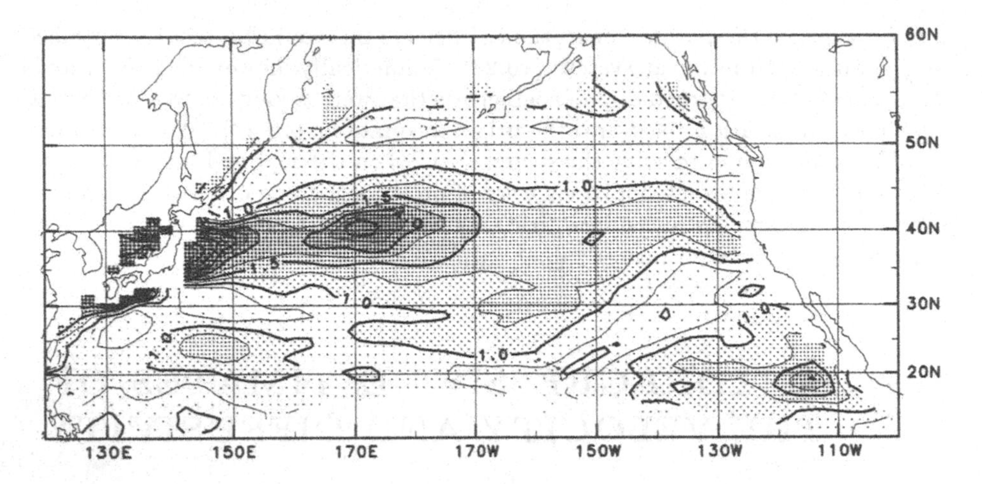

Figure 10.15 Lagged SST anomaly time series autocovariances over the North Pacific in winter: *a*) lag is zero.

10.4.1 Covariance in SST Anomaly Time Series

To estimate the SST anomaly transport one should fit the autoregressive model (10.22) to the covariances in the time series. The result of the model fitting or, in other words, the solution of the inverse problem sought, depends on the differences between the covariances as in (8.20). Roughly speaking, in (8.20) we approximated the analytical solution by taking differences of the SST anomaly field over a discrete grid. Multiplying (8.20) by the previous SST anomaly and taking sums over the time series, we obtained the discrete form (8.24) or (10.23) in terms of the covariances. In a sense, we obtained a sort of difference approximation for the stochastic SST anomaly model. Since the covariances and their differences determine the solution, it is worth describing their empirical estimates briefly now.

Fig. 10.15 shows the autocovariances of the SST anomaly time series in winter. The autocovariance with zero lag (Fig. 10.15a) is simply the variance of SST anomaly. Notice, the SST anomaly variance as obtained from the COADS data is somewhat high compared with that of the NMC's analysis (Fig. 9.5). It means, that either the NMC's analysis largely smoothes the data or the COADS marine reports are noisy. However the overall patterns on both Fig. 10.15a and Fig. 9.5 are similar. In terms of the COADS data, there are several

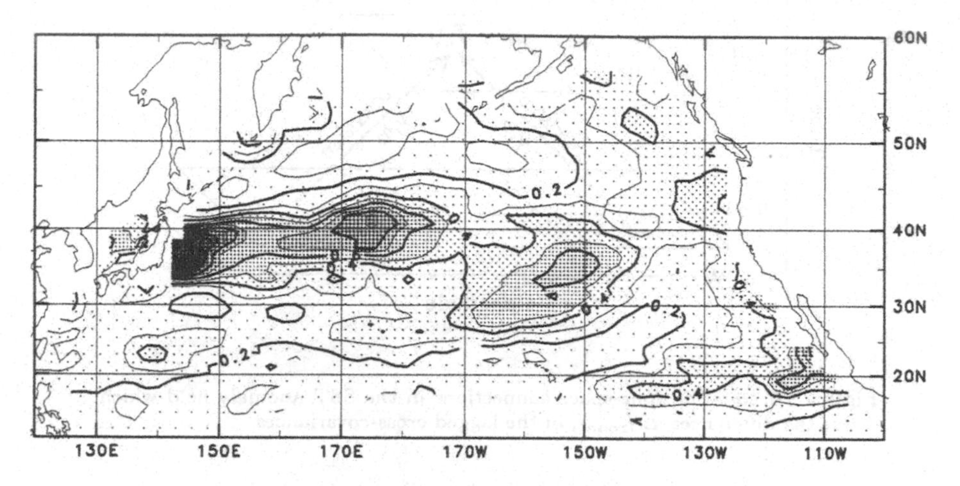

Figure 10.15 *(continued)* *b*) 10-day lag.

regions where SST anomalies often have an amplitude bigger than 1°C. Among them is the most active region around Japan where the variance $\sigma_T^2 > 3°C$. The SST anomaly amplitude is also high near California and in the central North Pacific between 35°N and 45°N.

The autocovariance slumps when the lag increases to 10 days (Fig. 10.15b). This decrease varies throughout the North Pacific, though the pattern of the autocovariance changes little. The decrease is only marginal in the regions where the variance is small, for example northeast of Hawaii, but up to 60% are possible if the variance is high. In general, Fig. 10.15b shows that the SST anomaly inertia is much stronger than that of the atmospheric forcing (see section 10.2). If one extends the lag to 20, 30 days and so on then the autocovariance decreases slowly. This slow decrease is a characteristic feature on SST anomaly time scales (see Frankignoul 1985).

For heat anomaly transport, the differences in the lagged cross-covariances are important. Since the system (10.26) depends on the 10-day lag cross-covariances, it is instructive to compute the differences such as

$$\Delta\gamma_{zonal} = \gamma_1^{04} - \gamma_1^{02}. \tag{10.31}$$

The value $\Delta\gamma_{zonal}$ indicates whether the current SST anomaly at the central cell, $T_n(\boldsymbol{r}_0)$, has a stronger connection with the previous SST anomaly to the west of the central cell, $T_{n-1}(\boldsymbol{r}_4)$, or with the previous SST anomaly located

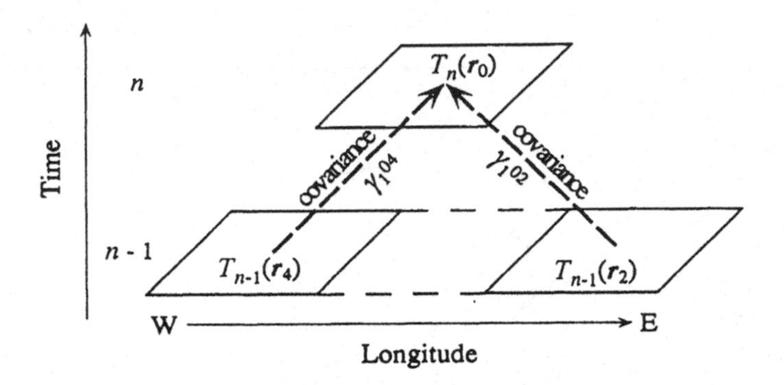

Figure 10.16 The time-space connections in the SST anomaly field which define the differences, $\Delta\gamma_{zonal}$, in the lagged cross-covariances.

east of the central cell (Fig. 10.16), $T_{n-1}(\boldsymbol{r}_2)$. In other words, the value $\Delta\gamma_{zonal}$ shows which propagation direction, eastward or westward, is preferable for SST anomalies. The autoregressive model (10.22) weights the propagation tendencies in each direction and evaluates a resultant vector.

Fig. 10.17 shows the distribution of $\Delta\gamma_{zonal}$ over the North Pacific. Considering Fig. 10.17 suggests that SST anomaly propagation is generally a function of latitude: SST anomalies tend to propagate westward north of 45°N and in the latitudinal belt of 20 – 35°N whereas the drift is basically opposite, eastward, between 35° and 45°N. This pattern is overlaid with local propagation tendencies such as those near the North American coast. More definite conclusions about the direction of SST anomaly translation can be drawn by using model (10.22) which takes into account transport in different directions such as the southwest-northeast represented by terms such as

$$\Delta\gamma_{SW-NE} = \gamma_1^{08} - \gamma_1^{06}. \tag{10.32}$$

It should be noted that studies of SST anomaly propagation often rely on the longitude-time and latitude-time correlation diagrams (c.f. Aoki et al. 1995 among others). Propagation is then determined by correlation averaged over long sections, for example along every latitude over 10° longitude or more. As a result, the effects of surface current variations at high wavenumbers are somewhat filtered out, so effects of planetary waves may emerge in the correlation diagrams. Here, instead, no a priori propagation is assumed and the analysis is aimed at inferring the net advection due to different processes including the long waves.

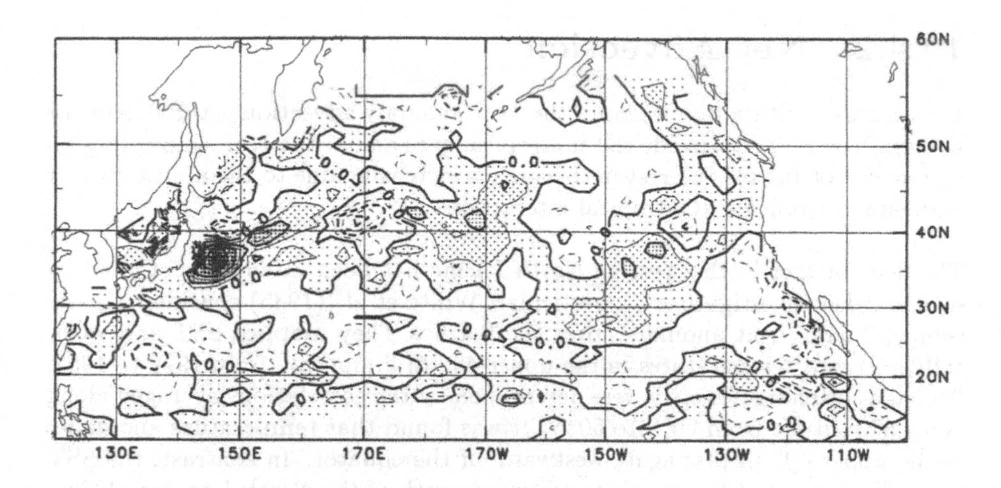

Figure 10.17 Distribution of the differences in the lagged cross-covariances $\Delta\gamma_{zonal}$. Contour interval is 0.05. The dashed lines indicate negative covariances.

It is worth smoothing the cross-covariances by using some kernel. This smoothing results in growth of coherence in the SST anomaly propagation detected with the model (10.23) whereas the smoothing of the original data could eliminate information concerning the directional transport. Notice, however, that in the earlier study, Ostrovskii and Piterbarg (1995) did not smooth the covariances at all and arrived at basically the same results as those demonstrated below.

Finally, let us notice two important features of the autoregressive estimates. First, this inversion procedure is robust. Almost everywhere in the North Pacific, misfits are obtained for only a few percent of the values of the observed covariances $\gamma_1^{0\mu}$. The inversion usually converges to the same solution though he minimization was tried with different directions of the initial advection velocity $\boldsymbol{u}_0$. Otherwise, the solution is chosen in accordance with the smallest misfit. When choosing the solution one can apply constraints to the parameters, such as the current speed should be less than 1 m s^{-1} or the mixing coefficient should be always positive. The second feature, is that the basic distributions of the advection, diffusion, and feedback can be inferred from the simplest inversion defined for a set of 3×3 cells, i.e. fit a system of 9 equations (10.23). Consideration of the results for this simple inversion begins below.

10.4.2 Net Advection

Let us begin, with a discussion of the heat anomaly advection estimates in the western North Pacific inside the subtropical gyre and in the Kuroshio extension region east of Japan. Interest in this area is increasing due to the need to better understand tropical-extratropical interactions.

The heat budget in the western North Pacific has been intensively studied by several other investigators. In particular, White et al. (1985) considered zonal propagation of heat anomalies over the Pacific. They mapped SST and vertically averaged temperatures in the upper 400 m bimonthly from 1979 to 1982. Propagation characteristics were determined by the correlations averaged along every 5° latitude from 20°S to 50°N. It was found that temperature anomalies in the upper 400 m propagate westward off the equator. In contrast, the SST anomalies appeared to propagate eastward north of the North Equatorial Current. As was mentioned above, the satellite data analysis by Aoki et al. (1995) indicated that west of date line south of 36°N, the SST anomalies propagate westward with the Rossby waves. It is noteworthy that a similar phenomenon was observed earlier in the northwest Atlantic near the Subtropical Convergence Zone (Halliwell et al. 1991a, b). Also, Qiu and Kelly (1993) analyzed the upper-ocean heat balance in the Kuroshio extension region during 1986-89. They presented estimates of the monthly heat balances averaged either zonally or over the Kuroshio extension region. That study concerned the seasonal heat balance. Here, instead, we shall investigate the nonseasonal (anomaly) heat balance of the upper-ocean in the subtropical gyre and Kuroshio extension.

Fig. 10.18 shows the net advection velocities as derived for the region of the western North Pacific bounded by 115° and 160°E, and 10° and 50°N. The net advection velocities were obtained from the model with the atmospheric parameter β defined as the correlation coefficient for the wind velocity time series with 10-day lag. In another experiment, we define β as the autocorrelation coefficient for the sum of the anomalous heat flux forcing and the anomalous wind-drift of the mean SST gradient. Since in both cases the autocorrelation coefficients are small, the results obtained are practically identical.

As is evident from Fig. 10.18, the SST anomaly drift basically conforms to the well known features of the ocean circulation field. Apparent is the westward propagation in the Philippine and Parece Vela Basins, between 122° and 142°E. The highest velocities are found in the northern part of the tropics where the North Equatorial Current is known to be dominant. The North Equatorial Current appears on the sea surface between the latitudes of 7° and 18°N (Nitani

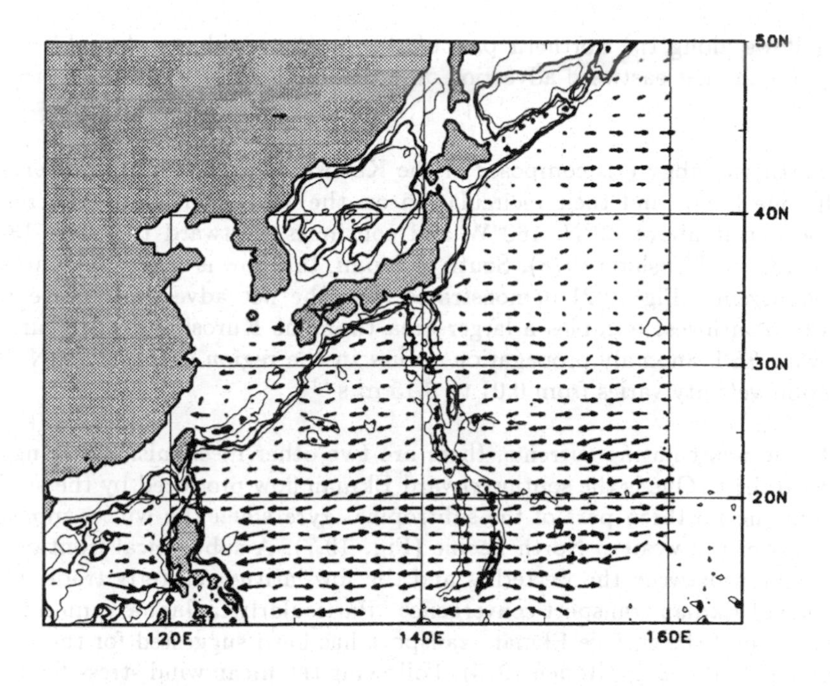

Figure 10.18 The distribution of the heat anomaly advection velocity $\boldsymbol{u}$ (m s^{-1}) estimates in the northwest Pacific. The arrow in the upper-left corner indicates a velocity of 0.1 m s^{-1}.

1972, Qiu and Joyce 1992). It is directed westward and has a maximum speed of about 0.25 m s^{-1} at 10−12°N. The volume transport of the North Equatorial Current is strongly affected by interannual fluctuations of the wind-stress curl (Qiu and Joyce 1992). According to our estimates, the net advection velocity varies between 0.10 and 0.20 m s^{-1} at along 15 − 16°N in the North Equatorial Curren region.

Inside the subtropical gyre, eastward SST anomaly propagation emerges in the region 132 – 136°E by 16 – 18°N. This transport can be associated with the semipermanent eastward flow called the Subtropical Countercurrent (Yoshida and Kidokoro 1967) driven by the dynamic topography of the subtropical gyre which splits into two subgyres, northern and southern, connected together in the west (Hasunuma and Yoshida 1978, Tsuchiya 1982). In winter the southern ridge is weak, but the northern ridge is well developed. The southern ridge originates from the western boundary near 20°N and first extends in the direction of 18°N, 135°E where it turns to the northeast. The Subtropical Countercurrent follows along the northern part of the southern ridge and could be the reason for the net eastward advection observed in the center of the subtropical gyre.

The northern subgyre is composed of the Kuroshio current and its extension, on the one hand, and their recirculation on the other. In winter the return flow begins at about 32°N, 160°W and continues westward to 22°N, 130°E (Hasunuma and Yoshida 1978). South of Japan, this flow is called the Kuroshio countercurrent. Fig 10.21 demonstrates that the net advection to the west and west-southwest occupies a larger area than the Kuroshio countercurrent. Westward SST anomaly propagation occurs in the region south of 29°N. The advection velocity varies from 0.01 to 0.15 m s^{-1}.

Beside the recirculating currents, there are two other mechanisms driving the westward drift. One is the southwestward Ekman flow produced by the surface winds in the northern part of the subtropical gyre when the winter monsoon prevails over the western North Pacific (Fig. 10.7). A subtropical front occurs in the region between the westerlies and the low-latitude easterly trade winds in a zone of Ekman transport convergence. Particularly in January, meridional convergence of the surface Ekman transport has been suggested for the region between 20° and 28°N (Roden 1975). Following the mean wind stress field, the Ekman transport should be northwestward (Fig. 10.9) between 10° and 22°N and southwestward at higher latitudes in the western North Pacific.

In addition to the wind-driven current, westward advection can be produced by the mean effect of wave-like eddies. As was discussed above (section 9.6), all motions on scales longer than 10 days and larger than 1° latitude × 2° longitude can contribute to the net advection estimated by this inversion from the SST anomaly covariances. In the Subtropical Convergence Zone the SST anomaly covariances contain information about transport by wave-like eddies.

As was mentioned above, this effect was uncovered by Halliwell et al. (1991a, b) near the Subtropical Convergence Zone in the Sargasso Sea of the North Atlantic where SST anomalies propagated westward at 0.03 - 0.05 m s^{-1}. The propagation was persistent from fall to spring and less evident during the summer. The anomalies propagated in a manner consistent with the theoretical zonal dispersion properties of first-mode baroclinic Rossby waves. The wave-like features had wavelengths of about 800 km and periods of about 200 days.

Westward propagation of SST anomalies was suggested recently on the basis of satellite observations in the latitudinal belt of 20 – 30°N of the northwest Pacific, too (Aoki et al. 1995). Longitude-time lag correlation diagrams indicate westward propagation patterns for anomalies in both SST and sea surface dynamic topography. The propagation speed estimate was in the range of 0.06 to 0.08 m s^{-1}, i.e., close to the phase speed of first-mode baroclinic Rossby waves.

The importance of planetary waves in SST anomaly transport can be supported by estimates of the net advection velocities over the sea mountain ridges at 140 – 144°E. These ridges are very high, often rise above the depth of 1000 m, and even culminate in islands. One may anticipate an effect of these tall ridges on planetary waves by analogy with reflection and blockage of Rossby waves by the Hawaiian island chain (Magaard 1983, Jacobs et al. 1993). What follows from Fig. 10.18 is that the westward transport is canceled at distances of up to several hundred kilometers east of the ridges. Above and in front of the ridges, from the eastern side, SST anomaly propagation is basically towards the east. In front of the ridges, westward transport is not just blocked but eastward transport dominates thereby implying that vertical heat flux is important near the ridges.

Overall, the westward drift of SST anomalies in the subtropical gyre of the northwest Pacific can be attributed to the joint effect of the surface currents and the dispersion properties of the eddy field. The westward propagation of long Rossby waves should decrease near the southern boundary of the North Equatorial Current, where the thermocline becomes shallower in the vicinity of the North Equatorial Countercurrent (see Kessler 1990). North of the subtropical convergence zone, wave-like eddy propagation must diminish, too (see Halliwell et al., 1991a, b).

South of Japan in the Kuroshio current region the estimated net advection velocity is directed eastward with speed of about 0.10-0.15 m s^{-1} (Fig. 10.18). These values are smaller than the Kuroshio current velocity, perhaps, because of the spatio temporal averaging of the SST data. Near 33°N and 141°E, a local westward transport seems to be induced by a large meander path of the Kuroshio current (Taft 1972, Kawai 1972). Downstream of the large meander, the eastward drift widens and becomes two branches in the net advection velocity field. The first separates from the coast and follows along the Kuroshio extension axis. The second extends east of Hokkaido, and is partially fed by the Oyashio extension.

It is well known that the Kuroshio extension is characterized by distorted meanders and numerous eddies of various size and duration, especially on its north side (Kawai 1972). Typical horizontal sizes of the quasi-stationary meanders and warm-core eddies are larger than the averaging scales of the SST data adopted for this study. Therefore, the large meanders and eddies contribute to the estimates of the net advection velocity.

The anticyclonic eddies on the north side of the Kuroshio extension should have a profound effect on the surface temperature field with sharp horizontal gradients. Kuroshio warm-core rings often detach from quasi-stationary meanders located near 144°E and 150°E (see for references Yasuda et al. 1992). Most of the warm-core rings move northward towards the Oyashio current (Mizuno and White 1983, Kawamura et al. 1986, see also for references, Yasuda et al. 1992). The northeastward transport by the anticyclonic rings is revealed in the net advection velocity field north of the Kuroshio extension axis east of Japan (Fig. 10.18). The northward transport implies translations of large warm-core eddies. Notice, also, that near 144°E and 152°E local recirculations in the net advection velocity field coincide with quasi-stationary meanders.

Strong eastward SST anomaly transport appears south of 34°N on the southern side of the Kuroshio extension (Fig. 10.18). South of the Kuroshio axis, the cyclonic eddies are less evident on the surface because these cold eddies are beneath the warm upper ocean mixed layer. The effects of the eddies are more evident at larger depths. Mizuno and White (1983) found that at 300 m depth, the seasonal temperature anomalies propagate eastward at ≈ 0.01 m s^{-1} in the region $140 - 155$°E, and westward at $\approx 0.01 - 0.02$ m s^{-1} in the region 155°E - 175°W. The zonal propagation of mesoscale perturbations (200-600 km) was found to be westward in the region $140 - 180$°E (Bernstein and White 1981, Tai and White 1990, Qiu et al. 1991). It seems that the eastward SST anomaly drift south of the Kuroshio extension axis is associated mainly with the geostrophic current itself. Downstream of the second quasi-stationary meander, the SST anomaly propagation exhibits a southward deflection. Before the Shatsky Rise, near 160°E, the SST anomaly drift changes to the opposite direction.

As to the north of the Kuroshio current, the SST anomaly transport is largely affected by the Oyashio current, which separates from the coast near Hokkaido and then couples with the northward branch and eddies of the Kuroshio current. The resulting flow produces strong eastward transport of SST anomalies with velocities of up to 0.1 m s^{-1} (Fig. 10.18).

There is remarkable agreement between the overall net advection pattern and that of the simulated surface velocity field by the eddy-resolving model of Semtner and Chervin (1992). Our inversion demonstrates eastward flows at the same latitudes near the western boundaries as in the numerical model although the absolute values of the simulated velocities are higher. We found the bifurcation of the eastward flow near the large meander path of the Kuroshio current and the secondary separation from the coast near Hokkaido. In the numerical model, the subtropical recirculating gyre separates from the coast south of Honshu and the second separation occurs south of Hokkaido, where it is partially fed by the Oyashio current.

It should be stressed, however, that this analysis is not capable of detecting the northward Kuroshio flow between Taiwan and Japan because its velocity is much higher than the upper limit defined from CFL-condition for the data resolution of 10 days and 1° latitude × 2° longitude adopted in this study.

Another intriguing region for heat anomaly transport research is that of the eastern North Pacific near the California Current system, which geographically lies between the polar region and the tropics. The large-scale structure of the California Current system is largely determined by combined effect of temperature and salinity on density (Batteen et al. 1995); the geostrophic circulation has a significant equatorward component along the coast of California and a strong offshore component adjacent to Baja California.

The California Current system is known for its high mesoscale activity. Numerous observations (see Bernstein et al. 1977, Hickey 1979, Rienecker et al. 1985 and Huyer and Kosro 1987 among others) clearly show that the California Current is unstable; its meanders often develop into cyclonic and anticyclonic eddies with dimensions of 100-200 km.

Winds off the coast of California exhibit strong seasonal changes. In particular, near northern California, the winds are predominantly poleward in winter and equatorward during summer, resulting in the coastal upwelling events. It is noteworthy that Chereskin (1995) provided Ekman theory observational evidence at a site in the California Current. While the mean flow within upper 150 m is dominated by the cyclonic geostrophic eddies of long temporal scales (about 150 days), the wind-driven current, calculated as the velocity relative to about 50 m, was shown to be in Ekman balance on a daily time scale for at least 4 months.

Heat budget over the northern California shelf was estimated by Dever and Lentz (1994) from early December 1988 through early May 1989. It was found that fluctuations in the heat budget at periods of days to weeks are largely associated with the response of cross-shelf heat transport due to local alongshelf wind forcing. This wind-driven cross-shelf heat flux is the main cause for changes in heat content whereas the along-shelf temperature gradient flux is only occasionally significant in February, March, and late April.

Norton and McLain (1994) collected observational evidence to check earlier observational and model results on the links between northern California Coastal Current and both the remote forcing from the eastern equatorial Pacific and local forcing from the North Pacific. Earlier, it had been noticed that the anomalously warm years in the California Current correspond to El Niño conditions in the eastern tropical Pacific (McCreary 1976, Enfield and Allen 1980, Chelton et al. 1982). However, it had been also suggested that the Aleutian Low atmospheric cyclogenesis, which is often augmented during El Niño years, can be directly responsible for warming episodes in the California Current system (Emery and Hamilton 1985, Huyer and Smith 1985, Simpson 1983, 1992).

Norton and McLain (1994) used EOF techniques to distinguish between locally and remotely forced effects. The emphasis was placed on events of spatial scale greater than 10^3 km. The monthly mean SLP at 45°N, 165°W in the northeastern Pacific was selected as an indicator of local forcing because this location is near the center of maximum seasonal Aleutian Low variation. The analysis showed that the California Coastal Current warming episodes linked with equatorial Pacific events are associated with interannual temperature changes at 100 m and below, whereas warming associated with North Pacific influences can be detected in the surface layers (0-50 m), rather than at depth. Interestingly, the lag between the North Pacific local forcing and ocean temperature change in the California Current system is less than 2 months.

Our analysis showed that the net advection pattern is complex off the coast of North America (Fig. 10.19). However, there are two distinct large-scale tendencies. The first is eastward advection with speeds up to 0.10-0.15 m s^{-1} from the midlatitude ocean interior. This advection seems to be associated with the broad North Pacific Current (Sverdrup et al. 1946). It takes several months to translate warming and cooling events from the interior Pacific to the North American coast. Since the response in the California Current system to SLP changes of the Aleutian Low occurs on a shorter time interval, the atmospheric variability in the vicinity of the California Current system should be primarily responsible for the SST anomaly behavior along the coast. This is possible via the local forcing of the cross-shelf heat transport as suggested by Dever and Lentz (1994).

The second tendency in the net advection field (Fig. 10.19) is the westward propagation originating from the coast south of 35°N. This westward transport intensifies towards the tropics reaching speed of 0.20 m s^{-1} near the North Equatorial Current. Although the westward propagation can be locally disturbed, it dominates throughout the tropics. It is known that one branch of the California Current departs from the coast and extends in a west-southwest

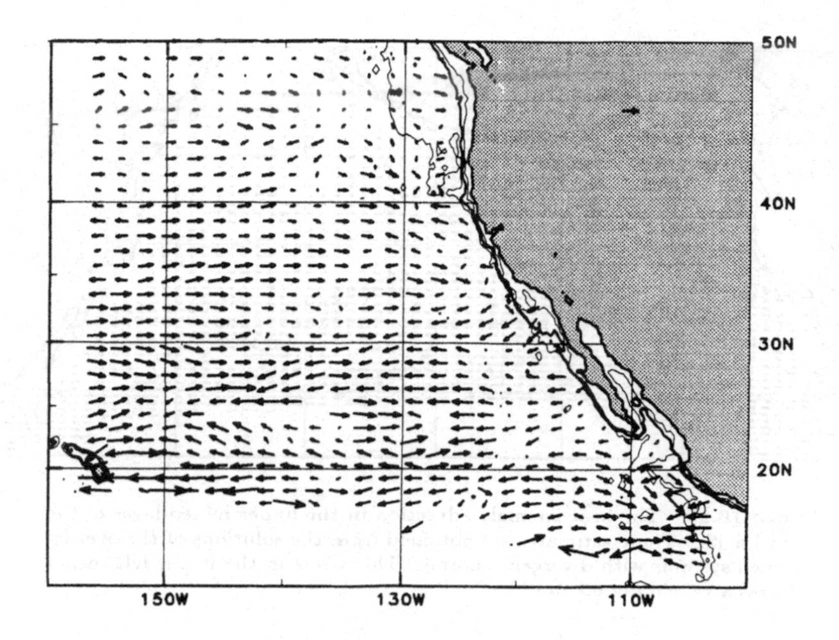

Figure 10.19 The distribution of the heat anomaly advection velocity $\boldsymbol{u}$ (m s^{-1}) estimates in the northeast Pacific. The arrow in the upper-left corner indicates a velocity of 0.1 m s^{-1}.

direction (Batteen et al. 1995). However, its velocity is somewhat smaller than that of the SST anomaly transport. Noteworthy, is the offshore wind drift in the upper mixed layer south of 30°N which has a speed of only 0.03-0.04 m s^{-1} (Fig. 10.9). Therefore the propagation towards the southwest should be in part associated with the Rossby waves generated near the coast (see Jacobs et al. 1993).

The westward transport sharply converges with the North Pacific Current pinpointing the Subtropical Convergence Zone (Fig. 10.19). The existence of heat anomaly convergence has important implications. It explains why the rotated first mode EOF for SST anomalies (Tourre and White 1995) (see Fig. 9.8) shows a distinct boundary between the positive and negative loadings that stretches from California to Hawaii. As follows from Fig. 9.8 and Fig. 10.19, usually SST anomalies of opposite signs are translated to the convergence zone where they cancel each other out, thereby resulting in the lowest SST anomaly variance (Fig. 10.15a).

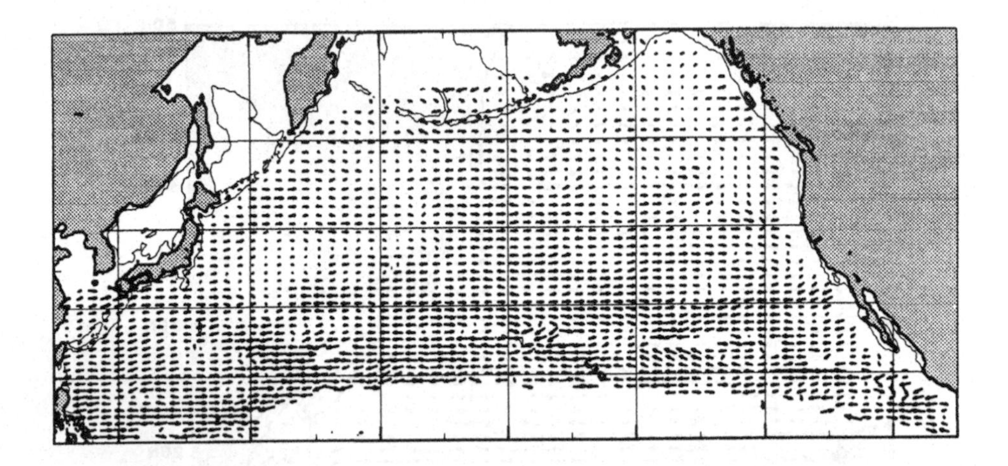

Figure 10.20 The heat anomaly advection in the upper mixed layer of the North Pacific. These estimates are obtained from the solutions of the overdetermined systems with divergence norms. The arrow in the upper-left corner indicates a velocity of 0.1 m s^{-1}.

Fig. 10.20 shows the net advection velocity estimates obtained from solutions of the overdetermined systems with the divergence norms (10.26)-(10.27) for the North Pacific. Application of the divergence norm results in a certain smoothness of the advection velocity field though the main features are retained as in the simplest case considered above (Figs. 10.18 and 10.19).

Notice, in order to elucidate the fact that small variations of β have no considerable effect on the advection velocities when the parameter β itself is small, we show in Fig. 10.20 the estimates $\boldsymbol{u}$ for the case when β is taken to be the autocorrelation of the sum of the anomalous heat flux forcing and the anomalous wind-drift of the mean SST gradient rather than the autocorrelation of the anomalous wind alone as in the simplest case (Figs. 10.18 and 10.19). In the regions with more or less uniform advection, for example at $146-150°$W and $30-35°$N, the estimates $\boldsymbol{u}$ obtained in these two cases are approximately the same.

In the interior of the North Pacific, the heat anomaly advection is basically eastward in the latitudinal belt $37-47°$N (Fig. 10.20). At these latitudes westward transport appears locally, particularly near 170°E between 42° and 46°N. These variations look strange but the differences in the cross-covariances $\Delta\gamma_{zonal}$ also change sign at the corresponding locations (Fig. 10.17). The advection is unstable and weak in the northeastern part of the North Pacific.

Overall, it is remarkable that the advection velocity field is in a perfect agreement with the patterns of the differences in the cross-covariances of SST anomalies. This consistency serves to convince us of the method's robustness. It seems that further progress in the inversion hinges on improving the raw data. Hopefully, the new COADS Release 1a will provide a better data set due to a number of important corrections and substantial augmentation in the number of data reports (Woodruff et al. 1993).

10.4.3 Diffusivity

Now let us discuss the diffusivity estimates. Fig. 10.21 shows the distribution of the heat diffusion coefficient, D, over the North Pacific. These estimates were obtained via the optimization procedure (10.26) constrained with the divergence norm (10.27). The estimates D are in the range of 1×10^3 to 8×10^3 m^2s^{-1}, i.e. are of the same order of magnitude as those commonly used in ocean general circulation models with approximately the same spatial resolution. These estimates are about three times smaller than those derived by Herterich and Hasselmann (1987) from the monthly SST anomaly data on a 5° grid. Notice, the uncertainty in the mixing coefficient in the model by Herterich and Hasselmann (1987) is of the same order of magnitude as the estimate itself.

In theory (chapter 8), the autoregressive method evaluates diffusivity reasonably well. Practically, however, the diffusivity estimates are sensitive to the forcing which contains uncertainty due to the unknown contribution of the feedback. Taking the parameter β to be either the autocorrelation of the anomalous heat flux forcing along with the anomalous wind-drift of the mean SST gradient, $r_{Q'Q'}$, or the autocorrelation of the anomalous wind itself, $r_{w'_a w'_a}$, we found interesting responses of the diffusivity to atmospheric forcing (the diffusivity estimates for these two cases are shown in Fig. 10.21 and Fig. 10.22, respectively).

As was discussed above, the distributions of the autocorrelations $r_{Q'Q'}$ and $r_{w'_a w'_a}$ are different in some details. In particular, the autocorrelation $r_{w'_a w'_a}$ (Fig. 10.8) is higher by a factor 1.5 than the autocorrelation $r_{Q'Q'}$ (Fig. 10.13) in the central North Pacific near 30°N and in the tropics. The inversion shows that at the same locations, the estimate of D is higher if the value of β is larger. For example, in the central North Pacific near 30°N and in the tropics, the estimates D are about 10-20% larger for the inversions with the autocorrelations $r_{w'_a w'_a}$ compared with those obtained with the autocorrelations $r_{Q'Q'}$.

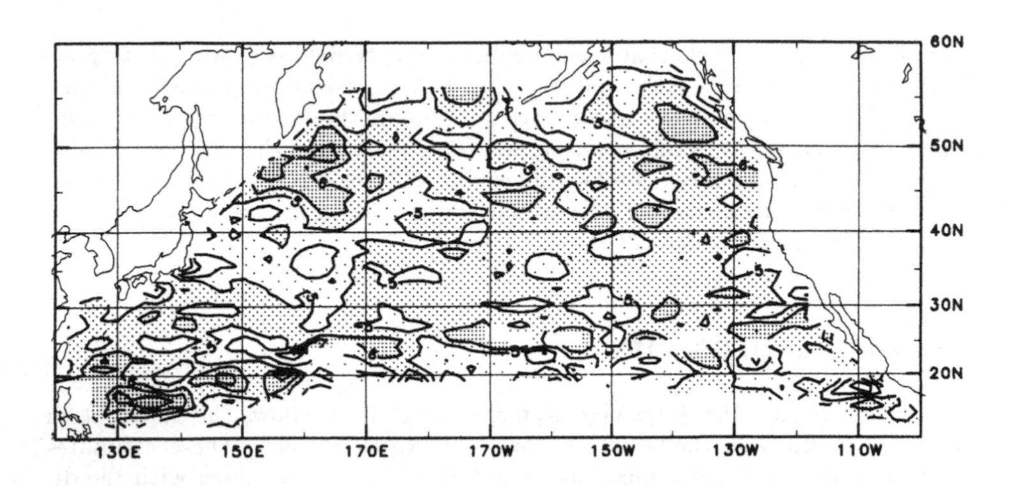

Figure 10.21 The distribution of heat diffusivity D (m^2 s^{-1}) over the North Pacific. These estimates were obtained in the case when the parameter β was defined as the autocorrelation of the anomalous heat flux forcing along with the anomalous wind-drift of the mean SST gradient, $r_{Q'Q'}$.

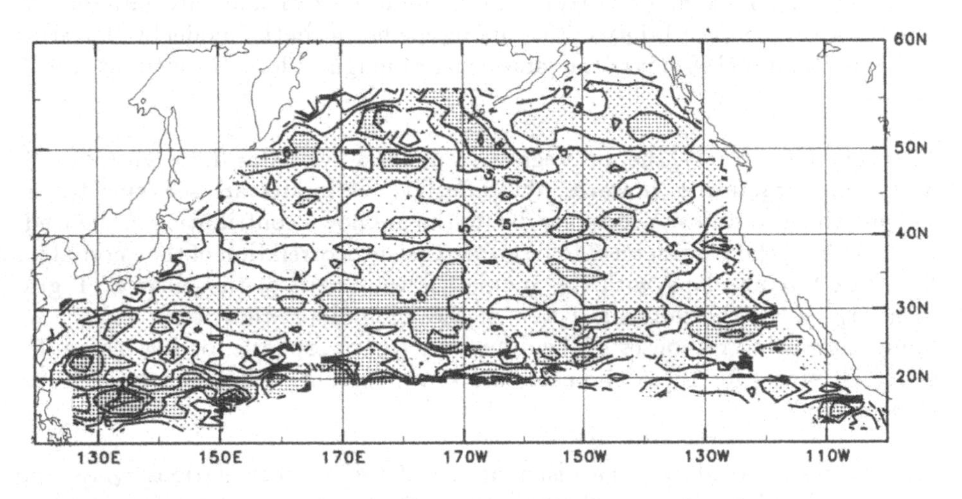

Figure 10.22 The same as in Fig. 10.21 but for the case when the parameter β was defined as the autocorrelation of the anomalous wind, $r_{w'_a w'_a}$.

This response can be explained as follows. The higher β reveals prolonged forcing events of the same sign. Under forcing events of different durations and intensities the upper-ocean responses, i.e. the SST anomaly fields, are identical if strong dissipation counteracts the effects of intense forcing. In other words, under different external forcing, the model output is essentially the same if the model dissipation is enhanced when the forcing is stronger. Since the SST anomaly field is identical in the two cases concerned but the forcing is different, the inversion provides two versions for the dissipation field. As will be discussed later, the Newton cooling coefficient estimates are also higher when the forcing memory is longer.

Returning to the diffusivity distribution, one notices that the high values of the diffusion coefficient, $6 - 8 \times 10^3$ m^{2}s^{-1}, occur near 20°N in the subtropical gyre. This enhancement is consistent with an increase in the mesoscale kinetic energy near the Subtropical Convergence Zone (Shum et al. 1990, Aoki et al. 1995). According to the autoregressive inversion, the diffusivity is enhanced also south of the Kuroshio current and near the Aleutian Islands.

The diffusivity estimates along the Kuroshio current and its extension are of special interest. The values of D are higher south of the Kuroshio axis compared with the northern side. As was emphasized above, the model diffusion is produced by subgrid motions with time and space scales less than 10 days and 1° latitude × 2° longitude. In the Kuroshio extension region, the horizontal scale of the quasi-stationary meanders and eddies is often bigger than the grid size of the SST anomaly data set used. Bernstein and White (1981) estimated the temporal and longitudinal decorrelation scales in the Kuroshio extension region from 300-m temperature data. The scales were found to be 2 months and 2 degrees. Also, Qiu et al. (1991) analyzed Geosat altimeter data and identified the decorrelation scales of sea surface height fluctuations in the along-track and zonal directions to be 140 km and 225 km, respectively, with a decorrelation time scale of about 90 days. Therefore, this inversion is able to resolve the large meanders and eddies in the Kuroshio extension region by including the corresponding transport into the net advection, as was discussed above. The estimates of D are low on the north side of the Kuroshio extension because the model diffusivities represent motions with scales smaller than the dominant scales of mesoscale fluctuations. In contrast, the diffusivity is enhanced south of the Kuroshio extension axis, where cyclonic eddies are largely hidden below the upper-ocean mixed layer.

The characteristic feature of these diffusivity estimates is an increased variability at scales of a few hundred kilometers. It is particularly evident in the feedback estimates (see below). Since the used time series are rather short, the inversion estimates exhibit mesoscale diffusion spatial variations. In order to obtain the background conditions for the mesoscale diffusivity one should smooth this variability of the high wavenumbers by fixing the mixing coefficient

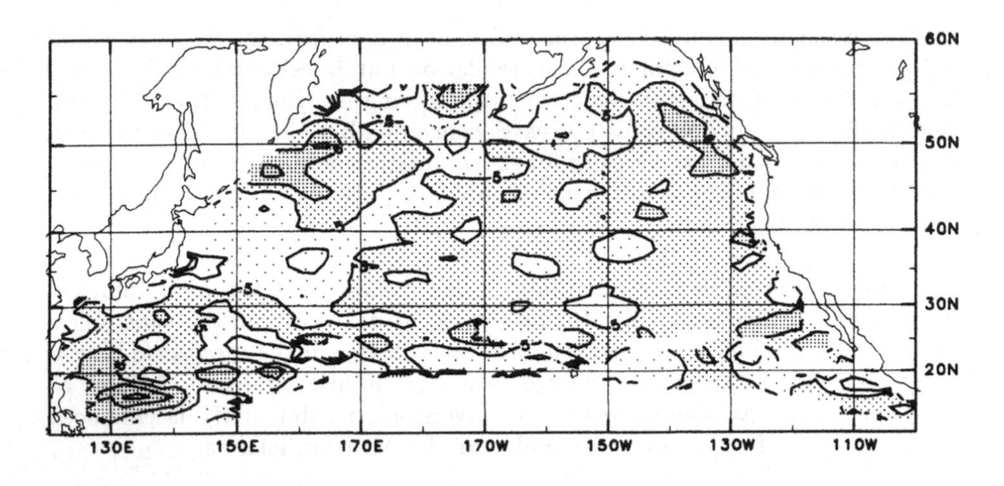

Figure 10.23 The same as in Fig. 10.21 but for the case when the coefficient D was fixed over the major-frame.

D over the major-frame as in (Ostrovskii and Piterbarg 1995). Fig 10.23 shows such estimates in the case of the inversion of the overdetermined system (10.26) with the divergence norm (10.27) forced by the sum of the anomalous heat flux and the anomalous wind-drift of the mean SST gradient. The diffusivity field in Fig. 10.23 has the overall diffusivity pattern of Fig. 10.22 but contains fewer mesoscale disturbances. One can further smooth the estimates by expanding the major-frame, but we would better postpone this discussion and resume it in the next section devoted to the maximum likelihood estimates for diffusivity.

10.4.4 Newton Cooling

One more parameter still remains to be discussed in this section. This is Newton's cooling coefficient, λ, whose reciprocal, λ^{-1}, is called the local relaxation time or the decay time. Fig. 10.24 displays the geographical distribution of the local relaxation time for SST anomalies. These estimates were obtained form the solution of the inverse problem for the simplest case for individual sets of 3×3 cells. In calculating this field, we defined the forcing as the sum of the anomalous heat flux and the anomalous wind-drift of the mean SST gradient.

The outstanding feature of the relaxation time field in Fig. 10.24 is its smoothness. It is particularly clear when compared with the distribution of diffusivity estimates obtained in the same inversion (Fig. 10.21). This smoothness implies

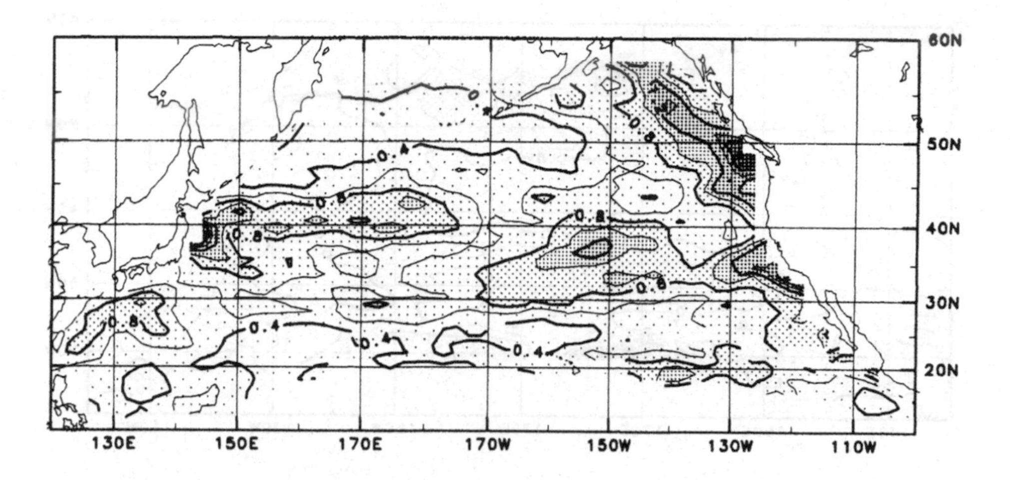

Figure 10.24 Distribution of the SST anomaly decay time λ^{-1} (months) over the North Pacific.

that the cooling coefficient estimates are very stable and distributed throughout the large scales.

The decay time λ^{-1} usually lies in the range of 15 days to 1 month. These values are two or three times smaller than the estimates by Herterich and Hasselmann (1987), who analyzed SST anomalies on larger scales (monthly, in 5° grid). The relaxation is longest near the North American coast, in the region bounded by 30° and 40°N, 140° and 170°W, and on the northern side of the Kuroshio extension region. Particularly, it is longer than 1 month near Vancouver Island, off the California cost and east of Honshu, Japan.

Noteworthy are the prolonged relaxation regions coinciding with the locations of loadings with large absolute values provided by the first mode rotated EOF for SST anomalies (Fig. 9.8). Extreme loadings reveal large-area SST anomalies, while prolonged relaxations indicate long-term SST anomalies. Since the large-scale SST anomalies exist longer, the patterns in Fig. 9.8 and Fig. 10.24 match.

Fig. 10.25 shows the logarithm of the ratio of the estimates λ'_w obtained for the case of wind forcing to the estimates λ'_Q for the case of the anomalous heat flux forcing along with the anomalous wind-drift of the mean SST gradient (the latter were shown in Fig. 10.24). In Fig. 10.25, the positive values stand for $\lambda'_w > \lambda'_Q$, and the negative ones for $\lambda'_w < \lambda'_Q$. While larger values of

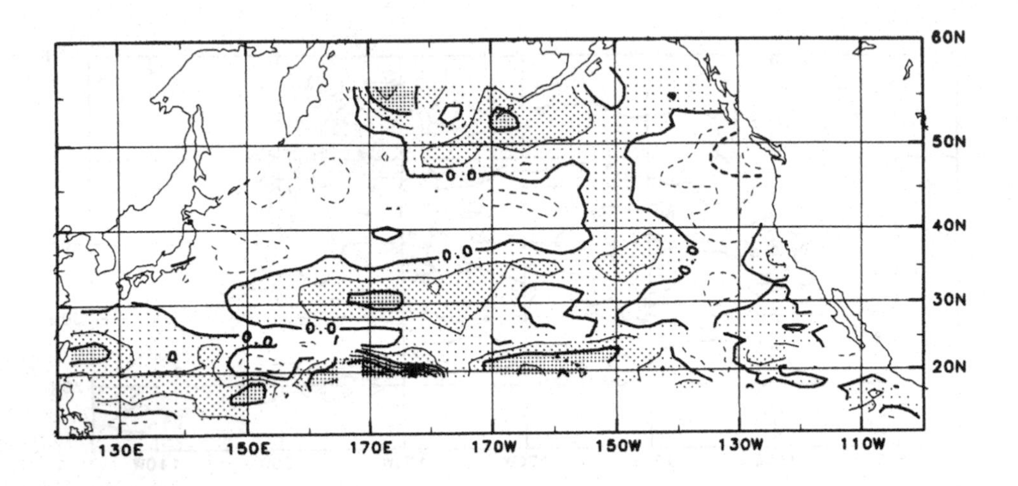

Figure 10.25 Distribution of log (λ'_w/λ'_Q). See text for explanation. Contou interval is 0.05.

λ indicate faster dissipation of SST anomalies via sea-air feedback, one can distinguish changes in the dissipation for two different cases of atmospheric forcing. As follows from Fig. 10.25, the dissipation λ'_w is bigger than the dissipation λ'_Q in the regions that almost exactly coincide with the areas where the autocorrelations of 10-day lag in the wind speed are relatively high (Fig. 10.8). This means that enhanced forcing is partly compensated for by the growth of the Newtonian cooling in the model (10.22) thereby providing an output which is adequate in terms of SST anomalies.

10.4.5 Control Experiment

Finally, a control experiment with unknown β was carried out for the inversion. This experiment confirmed the feasibility of making advection estimates. The net advection velocities were not significantly changed being basically constrained by the cross-correlations between SST anomalies at neighboring locations. The diffusivity estimates appeared to be more vulnerable, though certain main features retain, for example the enhancement of diffusion towards the tropics. In general, the diffusivity estimates slightly decreased compared with those obtained in the experiment with fixed β. The Newton cooling coefficient estimates, on the contrary, increased by 10-20% though their distribution was essentially the same as in the basic experiment.

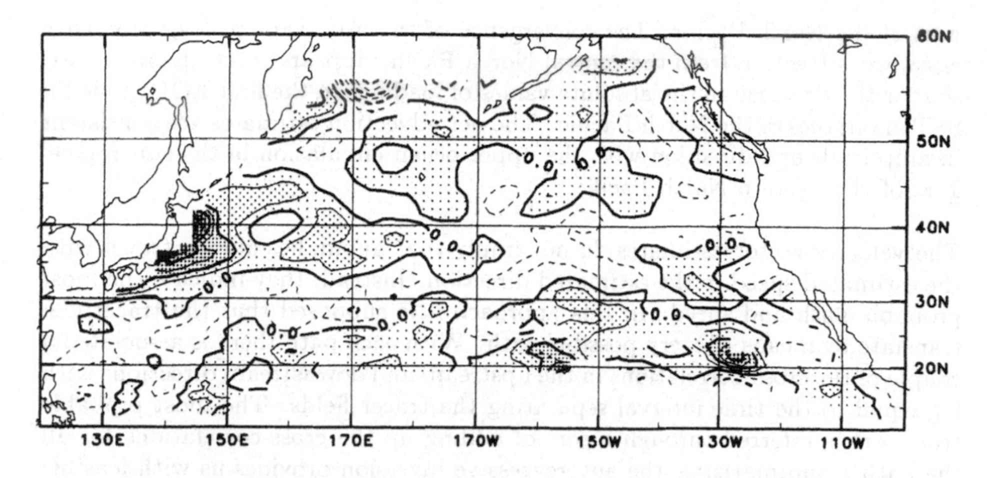

Figure 10.26 Distribution of the model-forcing correlation coefficient β. Contour interval is 0.05. The dashed lines indicate negative correlations.

As for the coefficient β itself, the estimates are shown in Fig. 10.26. These were found to vary in the range from -0.15 to 0.15 over the North Pacific. An increase in the β estimates occurs east of Japan, similar to the increase in the autocorrelations $r_{Q'Q'}$ (Fig. 10.13). In general, the values of β were closer to the autocorrelations $r_{Q'Q'}$ rather than to the autocorrelations $r_{w'_a w'_a}$. Since the coefficient β was found to be about zero in the central North Pacific we can conclude that the red noise forcing term yields no improvement compared with white noise forcing in this inversion model for the midlatitudes of the North Pacific far away from the western boundary. Thus our analysis supports Hasselmann's (1976) theory in which the generation of midlatitude SST anomalies lasting the dominant time scale of atmospheric processes was primarily attributed to short period stochastic weather forcing. However, the analysis indicates that the inertia of SST anomalies, their 'memory' of earlier winds, can not be neglected in the vicinity of the western boundary and in the tropics.

Overall, we successfully derived advection, diffusion, feedback, and the parameter β via the autoregressive inversion from the time dependent distributions of SST anomalies in the North Pacific in winter. Perhaps, the most interesting were the advection estimates. It was found that heat anomaly advection is coherent in the central North Pacific, thereby providing conditions favorable for the generation of large-scale SST anomalies in the region centered at 30°N, 145°W. Strong offshore transport of anomalous heat occurs in the southeastern part of the North Pacific. The convergence of this westward propagation with eastward advection from the central North Pacific appears to be the main reason for the decrease in the absolute values of loadings in the first EOF mode for SST anomalies of the North Pacific. The net advection estimates are consistent in amplitude and direction with the upper-ocean circulation in the subtropical gyre of the western North Pacific.

The velocity vector estimates do not imply that anomalies always travel with the estimated speed in the estimated direction. Instead, they indicate the most probable path and speed. In this approach it is supposed that the tracer can translate by means of every possible path. With each path there is associated a couple of the cross-correlations in the upstream and downstream directions with lag equal to the time interval separating the tracer fields. The most probable transport is inferred through a sort of adding up the cross-correlations for all the paths. Summarizing, the autoregressive inversion provides us with feasible estimates of $\boldsymbol{u}, D$, and λ. Hence we conclude that the model (9.42) contains the essential physics of the processes contributing to the heat anomaly balance of the upper ocean mixed layer in the North Pacific midlatitudes in winter.

10.5 MAXIMUM LIKELIHOOD ESTIMATES OF DIFFUSIVITY

It is known that in some parts of the ocean, the mesoscale heat flux is an important factor of regional thermodynamics. By reviewing the results of Vonder Haar and Oort (1973), Trenberth (1979), Bryden and Hall (1980), Fu et al. (1982) and others, Gill (1983) concluded that the mesoscale heat transport can be substantial in energetic areas like the Gulf Stream recirculation and the Drake Passage, but is not significant in the less active regions. The main reason for this geographic variation is attributed to eddy variability. Indeed, from analysis of historical ship drift files (Wirtky et al. 1976) and Seasat altimeter measurements (Cheney et al. 1983) it follows that the ocean eddy kinetic energy exhibits marked differences between various regions (see, also, reviews by Emery 1983 and Koshlyakov 1986).

A need to describe eddy motions and their spatial variations has led to the development of eddy-resolving numerical models of ocean general circulation. Among the questions posed by Holland (1983) for such models, is the problem of how to parameterize the effects of mesoscale circulations. Satisfactory parameterization of subgrid turbulence in terms of mean variables would improve the relatively coarse spatial resolution of ocean calculations which do not resolve the mesoscale explicitly. The classical 'constant eddy viscosity' hypothesis suggests that mesoscale eddy terms are represented as constant times the Laplacian of the mean field ('Laplacian closure'). This hypothesis for subgrid mixing has been routinely used in coarse resolution ocean general circulation models. The advanced eddy-resolving model by Semtner and Chervin (1992) also uses this kind of parameterization during the first two phases of the numerical calculations to obtain quasi-steady, near-equilibrium ocean circulation which is consistent with the known density structure of the ocean. The eddy-resolving state is attained by substituting scale-selective biharmonic mixing in the horizontal for Laplacian mixing. Thus, knowledge of the Laplacian closure coefficients is vital at begining of calculations even in eddy-resolving models. Lack of information usually forces modelers to use constant coefficients throughout the computational domain.

As was emphasized by Holland (1983), there is a problem in judging the quality of subgrid parameterizations. One reason for this difficulty seems to be associated with the complex structure of eddy terms compared with Laplacians computed from the observed mean fields. Also, it is an unfortunate fact that up-to-date information about eddy heat fluxes in the ocean is rather sparse. This section is aimed at adding empirical insights which will hopefully lead to a better description of ocean mesoscale heat diffusivity in terms of the Laplacian closure.

This study is focused on maximum likelihood inversion for mesoscale heat diffusivity in the upper ocean mixed layer which overlies the deep ocean where the main bodies of the mesoscale eddies and vortices are situated. For this reason one can not relate these estimates directly to the isopycnal diffusivity that is often regarded as a good approximation for the oceanic mesoscale transport. Another important issue is that the estimates of diffusivity are always scale dependent. Here we consider the mesoscale only. The obtained estimates hinge on the spatial resolution of the data at a particular latitude because the zonal resolution of 2° longitude increases with latitude. But when reliable data of different resolution become available it would be plausible to consider a set of scales. One candidate for such a study would be the NOAA/NASA Pathfinder Ocean and Atmosphere data set that is compiled now (The NOAA/NASA Pathfinder Program 1994).

Additionally, we have to stress the following. There is a notion that it is a notoriously difficult matter to infer the velocity and mixing coefficient from the advection-diffusion equation, so the approaches to this problem are a priori unreliable. In particular, there is a belief that the quality of the inversion is crucial to the task of finding the diffusivity because the advective terms are much bigger than the diffusive ones. One cannot, however, really argue with the theorem proven in chapter 6, which provides that it is, in principle, possible to derive the diffusivity correctly via ML estimators. Furthermore, the numerical experiments in chapter 7 confirmed the theory by demonstrating that ML estimates for the diffusivity in transient 2D flows are feasible when the forcing is known, even if the data are limited to two sequential fields over a small spatial grid. It is important that the ML inversion does not seek the diffusive terms as residuals in the tracer balance! In fact, expansion of the signal into a spectral-type basis is a tremendous asset, allowing us to operate with the amplitudes of the modes attributed to dissipation while knowledge of the velocity is localized in the mode phases. The amplitudes are related to each other by means of the time-space correlations. The likelihood function describes how the correlations are defined by the advection-diffusion equation.

In practice, observational data contain random observational noise. So, in order to get the correct correlations for the tracer modes, one has three options: take a long time series of observations, increase the observational domain or find a compromise solution for a limited time series length and grid size. Since the spatial resolution of the climatic data, COADS for example, is rather coarse, one can not enlarge the grid size much and has to extend the time series length. After numerical experiments with the time series of different lengths on various spatial grids we choose to consider long time series (up to 325 sequential fields) over a relatively small spatial grid.

It should be emphasized also that the ML estimator for velocity was proven to be inconsistent in chapter 6. For this reason we do not discuss the velocity estimates in this section. Perhaps, other estimators give better estimates for the velocity. The dispute about the feasibility of the inversion approach for the velocity was raised a decade ago. Wunsch (1985) considered the questions of how to use tracer data and the degree to which an underdetermined system is tractable. It was explained that although a complete solution is not always possible to obtain, an inversion can deliver a solution which has a straightforward and useful interpretation. In the above chapter 9 we saw that this is indeed so.

Finally, the results obtained below would further serve as a reference in the study Lagrangian measurements. Also, for the purposes of heat transport modeling, the estimates derived from SST data themselves may provide an appropriate heat diffusion field.

10.5.1 A Practical Solution

Now let us turn to the maximum likelihood estimates of SST anomaly transport in the North Pacific. We focus on the zonal, D_x, and meridional, D_y, diffusion coefficients because both the theory and numerical experiments have proved consistency of the ML estimator for diffusivity (see above chapters 6 and 7). The conservation equation (9.42) is fitted by means of the ML estimator to the COADS data described in section 10.1.

The basic underlying idea of the ML inversion technique is to expand the forcing and tracer fields in terms of superpositions of the time-dependent mode amplitudes, $C^{(j)}_{m,n}(t)$, and a Fourier-type basis, $\varphi^{(j)}_{m,n}(x,y)$, in space

$$T(t,x,y) = \sum_{m,n=0}^{\infty} \sum_{j=1}^{4} C^{(j)}_{m,n}(t)\varphi^{(j)}_{m,n}(x,y), \tag{10.33}$$

where the basis functions are given by (7.5).

This treatment is similar to that often adopted in numerical spectral models. Richards et al. (1995), for example, treated the advection-diffusion equation in terms of Chebyshev polynomials. When the expansion is completed, the advection-diffusion equation (9.42) can be fitted by means of ML estimators to a time series of amplitudes for the higher space modes of the data fields. The likelihood function (7.27) describes how the time-space correlations between the amplitudes are incorporated in the advection-diffusion equation (9.42).

It is assumed that advection, diffusion, and feedback do not vary over the region for which the expansion (10.33) is set up. Thus, the ML method provides us with a sort of integral estimate. If one defines the expansion for part of the observational domain then the derived estimate is an integral characteristic for this region. From numerical experiments with ML estimators (chapter 7), it follows that the proper size of the observational grid is 7×7 ($m = 0, 1, ..., 7, n = 0, 1, ..., 7$) or more grid boxes. For this reason, we divide the data fields into the overlapping sets of 7×7 geographical boxes of 1° latitude $\times$ 2° longitude (Fig. 10.27). In each of these sub-domains, the forcing and the tracer distributions are expanded by superposition of time-dependent modes with a Fourier-type bases in space. Then the advection-diffusion equation (9.42) is fitted by means of ML estimators to a time series of amplitudes for the higher space modes at each sub-domain. Notice that the subset size of 7×7 limits the maximal wavenumber to $M_1 = M_2 = 3$ while the number of the eigenfunctions is restricted to $(2M_1 + 1) \times (2M_2 + 1) = 49$.

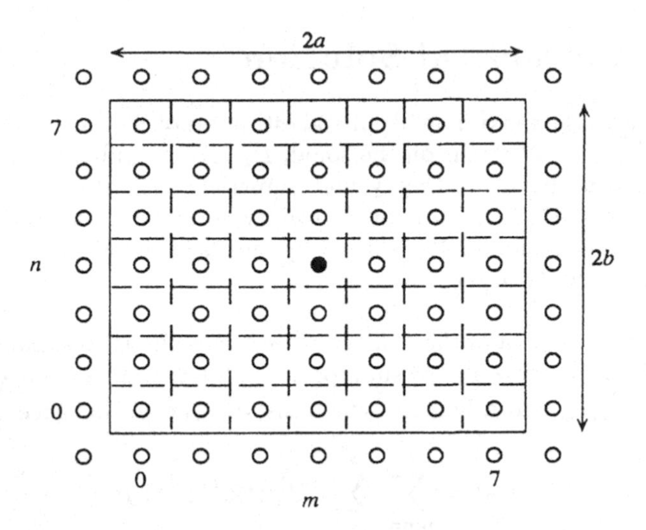

Figure 10.27 A set of grid boxes for ML inversion.

Since the subsets overlap, their centers are located at each grid point, i.e. the frame runs throughout the data grid covering the whole observational domain. The results of individual inversions are further referenced to the centers of the overlapping frames so as to cover the whole observational grid in the North Pacific. Obviously, in the vicinity of the ocean boundaries, the subsets for the boxes can not be compiled, so the unknown parameters are evaluated in the ocean interior at distances of about 300 - 600 km offshore.

What are the advantages and disadvantages of ML inversion compared with the autoregressive inversion? The main advantage is basically the convenient mathematical form. For example, the likelihood function was obtained explicitly (7.27) and the ML estimator for the diffusivity was proven to be consistent ever in infinitely large observational domain (chapter 6, Theorem 6.2).

The drawbacks emerge in practical applications. First, the spatial resolution is limited by the subset size chosen for the expansion (10.33). Although the moving frame technique can be used as in the autoregressive inversion case, the derived estimate is characteristic of the entire subset which is believed to have a minimum size of 5×5 grid boxes, as the numerical experiments demonstrated in chapter 7. This estimate would correspond to that obtained via the autoregressive estimator (10.26) for a 3×3 sub-frame over which the unknown parameters are fixed. Second, unlike the case of the autoregressive inversion, the forcing field is necessary for the ML fitting. Fortunately, certain information about the atmospheric forcing is available for this study. Third, the likelihood function (7.27) has been derived for the case of white noise forcing and not for color noise forcing. Thus, unlike the autoregressive method, the ML method, so far, is limited to the case of flow forced by a source that is uncorrelated in time. The latter condition is satisfied for SST anomaly transport on a 10-day scale almost everywhere in the North Pacific, as it was concluded in section 10.4.

Anyway, ML inversion for diffusivity is the only estimator for which consistency has been proven in theory. ML inversion is also advantageous in that it provides us with the components of the mixing coefficient, D_x and D_y, thereby allowing a study of the diffusion anisotropy. Since there are very few estimates for the heat diffusivity on the ocean mesoscale, every new estimate is interesting. A further aim of this section is to demonstrate an application of the ML technique to experimental data such as SST.

10.5.2 Atmospheric Forcing for the Maximum Likelihood Model

In section 10.2 we estimated the contribution of the anomalous heat flux and the anomalous wind-drift of the mean SST gradient into the atmospheric forcing. However, we did not evaluate the total forcing (9.34) because the term S in the heat anomaly balance equation (9.42) contains an unknown contribution associated with the product $\lambda'\overline{T}$ of the feedback fluctuations and the mean SST. Unfortunately, at present we also do not know the distribution of the parameter λ_E which represents the SST anomaly damping due to entrainment (see section 9.6). If this parameter is known then it is easy to remove an effect of the atmospheric feedback due to the term $\overline{\lambda_Q T'}$ from the anomalous heat flux, thereby providing an estimate for the pure anomalous heat flux forcing.

But what could be the distribution of the variance of the total forcing? How big is the difference between this variance and the above evaluated variance of the sum of the anomalous heat flux forcing and the anomalous wind-drift of the mean SST gradient? The answers, though preliminary, can be drawn from the results of the autoregressive model fitting.

The noise ε in the autoregressive model represents atmospheric forcing. The noise time series can be calculated from the SST anomaly time series as follows

$$\varepsilon_n(\boldsymbol{r}_0) = T_n(\boldsymbol{r}_0) - \sum_{m=0}^{8} \alpha_m T_{n-1}(\boldsymbol{r}_m), \tag{10.34}$$

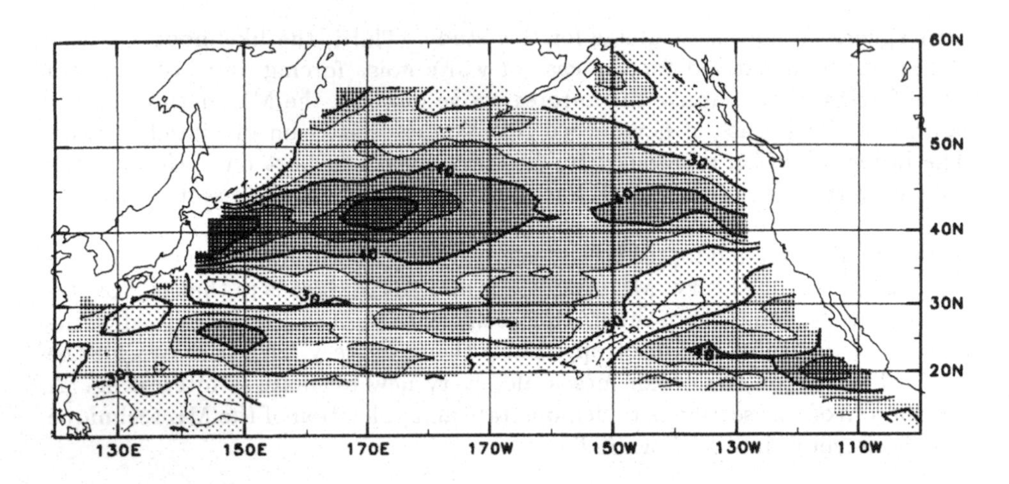

Figure 10.28 Distribution of the model-forcing standard deviation σ_ε (10^8 °C s^{-}1) obtained from the autoregressive inversion.

where the coefficients α_m can be found be substituting the estimates $\boldsymbol{u}$, D, and λ into (8.22).

Fig. 10.28 shows the distribution of the standard deviation σ_ε for the autoregressive model noise. Fig. 10.29, for comparison, displays the standard deviation $\sigma_{Q'}$ for the anomalous heat flux forcing along with the anomalous wind-drift of the mean SST gradient deduced from the observational data. The agreement in pattern of Figs. 10.28 and 10.29 is obvious, thereby supporting the stochastic forcing hypothesis for SST anomaly generation and evolution. However the differences in the absolute values of σ_ε and $\sigma_{Q'}$ should be noticed. As it was anticipated, the hypothetical forcing is generally larger. South of 45°N, σ_ε is on the average about 60% larger than $\sigma_{Q'}$. North of 45°N, σ_ε is larger by a factor 2 and more. Similar result on the increase of the hypothetical forcing variance for high latitudes has been obtained by Frankignoul and Reynolds (1983) from a stochastic SST anomaly model. The increase at high latitudes was attributed to the omission of mixed layer variability.

Therefore, we assume that the pattern of total anomalous forcing in the middle latitudes, south of 45°N, is the same as that of the anomalous heat flux forcing along with the anomalous wind-drift of the mean SST gradient, though the standard deviation of the total forcing can be about 60% larger. We also assume that the correlation structure of the total anomalous forcing field is identical

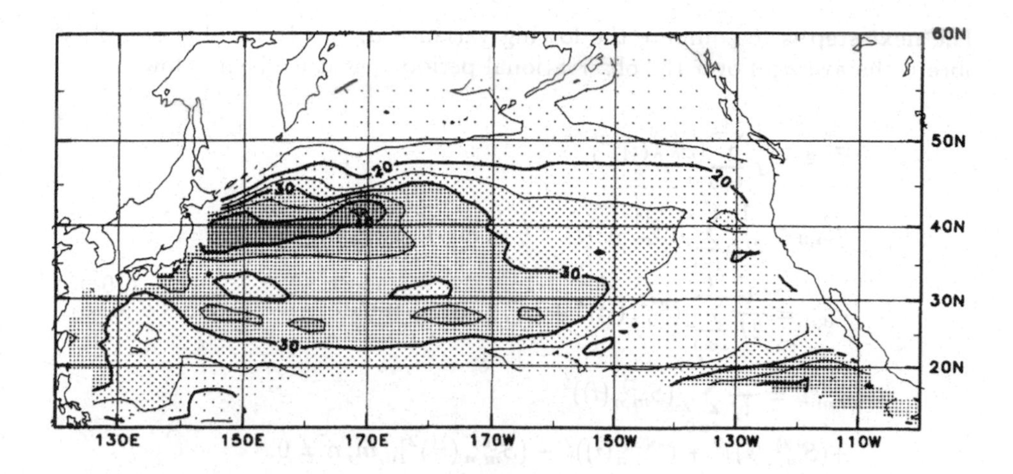

Figure 10.29 Distribution of the standard deviation $\sigma_{Q'}$ (10^8 °C s^{-1}) for the anomalous heat flux forcing along with the anomalous wind-drift of the mean SST gradient deduced from the observational data.

to that of the anomalous heat flux forcing plus the anomalous wind-drift of the mean SST gradient.

10.5.3 Numerical Procedure

A procedure of fitting ML estimators to the data was presented in section 7.2. Here we focus on certain details relevant to application for the time series of observations. For ML inversion, one has firstly to compute the time series of the amplitudes $C_{m,n}^{(j)}(i)$ and $S_{m,n}^{(j)}(i)$ from the SST anomaly and heat flux anomaly time series, respectively, by applying formula (7.39). In the latter case, obviously, we should use observations $S_{k,l}$ of the forcing on the right-hand side of (7.39). Here i is the time, (k,l) is the grid point, (m,n) is the wave number, and $j = 1, ..., 4$ is the number of basis function in the group related to (m,n). Since the time series contain gaps it is necessary to interpolate the data. In such cases we filled the gaps by using spatial linear interpolation with weights proportional to the cross-covariances in the time series.

The next step is to compute the forcing variance $\sigma^2_{m,n}$. To do this one must obtain the averages over the observational period I as it is given below

$$\begin{aligned}
\sigma^2_{0,0} &= \frac{1}{I}\sum_{i=0}^{I}(S^{(1)}_{0,0}(i))^2, \\
\sigma^2_{m,0} &= \frac{1}{2I}\sum_{i=0}^{I}[(S^{(1)}_{m,0}(i))^2 + (S^{(3)}_{m,0}(i))^2],\ m \neq 0 \\
\sigma^2_{0,n} &= \frac{1}{2I}\sum_{i=0}^{I}[(S^{(2)}_{0,n}(i))^2 + (S^{(4)}_{0,n}(i))^2],\ n \neq 0 \\
\sigma^2_{m,n} &= \frac{1}{4I}\sum_{i=0}^{I}[(S^{(1)}_{m,n}(i))^2 \\
&+ (S^{(2)}_{m,n}(i))^2 + (S^{(3)}_{m,n}(i))^2 + (S^{(4)}_{m,n}(i))^2],\ m,n \neq 0.
\end{aligned} \tag{10.35}$$

So one gets, for example,

$$\sigma_{0,0}[^{\circ}Cs^{-1/2}] = \begin{bmatrix} 0.103E-06 & 0.385E-07 & 0.232E-07 & 0.190E-07 \\ 0.433E-07 & 0.407E-07 & 0.379E-07 & 0.213E-07 \\ 0.233E-07 & 0.285E-07 & 0.219E-07 & 0.209E-07 \\ 0.268E-07 & 0.280E-07 & 0.248E-07 & 0.239E-07 \end{bmatrix}.$$

Notice, the values of $\sigma^2_{m,n}$ are not the forcing variances yet. The forcing variances are products of the $\sigma^2_{m,n}$ and τ_c, which is the atmospheric forcing correlation time. In order to assure the validity of ML model, we took the correlation time scale larger than the averaging interval of our data set (10 days), namely 15 days.

The third step is to compute the sample correlations $\hat{\rho}_s(k,l)$ of the series $C^{(j)}_{m,n}(i)$ in accordance with (7.30) for $s = 0, 1, k, l = 1, ..., 4$ and the functions $A_{m,n}$, $B_{m,n}$, and $E_{m,n}$ by using the formulas (7.28).

Further one must initialize the vector of unknowns

$$Q_0 = (u, v, D_x, D_y, \lambda)_0$$

and compute the quantities

$$
\begin{aligned}
\varphi^{(1)}_{m,n,0} &= \cos\frac{m\pi u_0 \Delta t}{a}\cos\frac{n\pi v_0 \Delta t}{b},\\
\varphi^{(2)}_{m,n,0} &= \cos\frac{m\pi u_0 \Delta t}{a}\sin\frac{n\pi v_0 \Delta t}{b},\\
\varphi^{(3)}_{m,n,0} &= \sin\frac{m\pi u_0 \Delta t}{a}\cos\frac{n\pi v_0 \Delta t}{b},\\
\varphi^{(4)}_{m,n,0} &= \sin\frac{m\pi u_0 \Delta t}{a}\sin\frac{n\pi v_0 \Delta t}{b},\\
F_{m,n,0} &= \pi^2\left(\frac{m^2}{a^2}D_{x,0} + \frac{m^2}{b^2}D_{y,0}\right)\Delta t + \lambda_0 \Delta t,
\end{aligned} \tag{10.36}
$$

included in the ML function (7.27).

Now everything is ready to proceed with an iterative minimization of the likelihood function. During the iterations the vector of the parameters $\boldsymbol{u}, \boldsymbol{D}$, and λ is renewed and the iterations stop when specified convergence criteria are satisfied.

10.5.4 The Anisotropy of Mesoscale Heat Diffusivity

The solutions of the inverse problem via ML estimators were found successfully for about 93% of the grid points. We should admit that the minimization often failed to converge in the north most and south most locations. Unfortunately also, the convergence criteria were not satisfied in the grid points next to the western boundary at 37° − 41°N. Although the reason for this failure is not absolutely clear, we suspect that the atmospheric forcing in these regions was too low to produce adequate responses in terms of the inverse model. Despite these misfortunes, in general, ML inversion quickly converged, so the computations for the whole observational domain took about a few hours at an ordinary workstation. Of coarse, the total computational time depends upon the number of minimizations for each individual grid point: the minimizations start from different initial vectors and the solution was chosen to be that of the smallest misfit.

Since we already got the diffusivity via the autoregressive technique and can compare those estimates with the ML ones, let us remind the reader that the autoregressive inversion did not require data about the amplitude of the atmospheric forcing. Furthermore the autoregressive inversion provided us with its own version of the forcing field variance somewhat different from that deduced from the available data. It is remarkable that north of 45°N the standard deviation of the autoregressive model forcing (Fig. 10.28) can be a few times larger than the standard deviation for the observed anomalous heat flux forcing along with the observed anomalous wind-drift of the mean SST gradient (Fig. 10.29). Thus, the large values of the diffusion coefficient at high latitudes were obtained under the atmospheric forcing, which was much stronger than that derived from the available observations. In this respect it is worth noticing also that the model by Frankignoul and Reynolds (1983) also experienced much larger forcing than observed north of 45°N in the Pacific.

In chapter 7 and section 10.4 we showed that the variance of the weather forcing drives diffusivity. Under different forcing conditions the SST anomaly model response is the same if the dissipation is enhanced (reduced) when the forcing is increased (decreased). Thus, underestimation of the atmospheric forcing leads here to the controversially low diffusivity

ML inversion evaluates both the zonal and meridional diffusion coefficients, D_x and D_y. These estimates are found to vary from 0.1×10^3 m^2 s^{-1} in the high latitudes to 5.6×10^3 m^2 s^{-1} south of the Kuroshio extension axis. There is good agreement between the patterns of D_x and D_y with those of a sea level mesoscale variability detected by means of the TOPEX/POSEIDON satellite altimeter for the winter seasons of 1993-95 (Minster and Brossier 1996). The higher heat diffusivity estimates match the instability of the strong Kuroshio current and its extension. As could be anticipated, the mesoscale heat diffusion is much smaller at the same latitudes in the eastern North Pacific. There are also some indications of relatively high mesoscale diffusivity near the tropics.

North of 45°N the ML diffusivity estimates decrease drastically thereby causing a major discrepancy with the pattern of the heat diffusion coefficient obtained via the autoregressive inversion. Indeed, the autoregressive estimates for the high latitudes are of the same order of magnitude, i.e. $D \approx 10^3$ m^2 s^{-1}, as those in the midlatitude North Pacific.

The reason for this contradiction is due to uncertainty in the atmospheric forcing adopted for ML inversion. As was stressed above, the atmospheric forcing is underestimated here because we do not have information about the contribution associated with the product of the feedback fluctuations and the mean SST. This effect is of the utmost importance north of 45°N where the mixed layer variability is largest. Additionally, the mesoscale variability could be af-

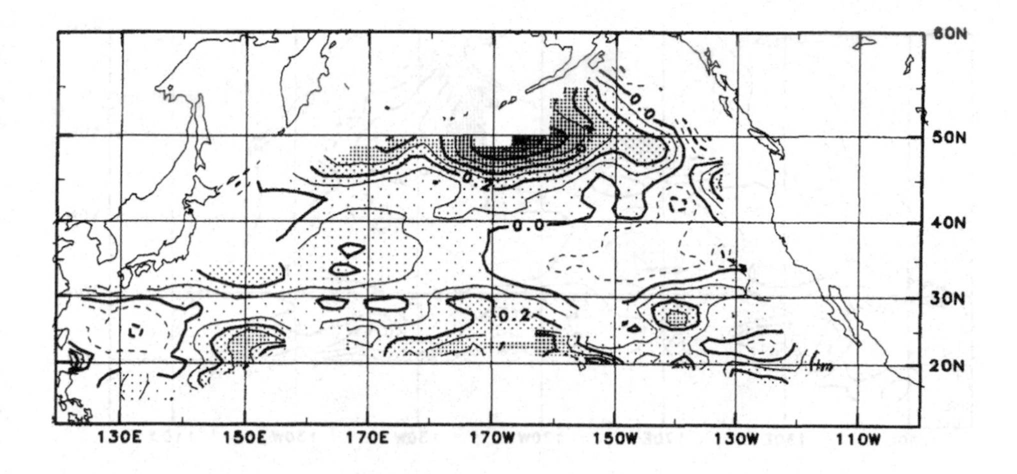

Figure 10.30 The distribution of the logarithm of the ratio the zonal diffusivity to the meridional diffusivity $\log D_x/D_y$: *a*) under the decorrelation time $\tau_c = 15$ days. The dashed lines indicate negative values.

fected by wind fluctuations at high latitudes. According to recent results of satellite remote sensing (Minster and Brossier 1996), the stochastic wind fluctuations are significant in winter at latitudes over 40°N, where they exceed 3.5 r.m.s. Such wind fluctuations can excite oceanic eddies, basically barotropic (Frankignoul and Müller 1979, Willebrand et al. 1980).

But if there is such an uncertainty about these estimates for the heat diffusivity why are they valuable? The answer is that the ratio of the zonal diffusivity to the meridional diffusivity estimates, D_x/D_y, does not change significantly under variations in the hypothetical atmospheric forcing! Or, in other words, the relative value of the diffusivity can be confidently derived via ML estimators even in the case when the absolute value of the forcing is known with some error. Obviously, if the forcing is known exactly than one can get an accurate estimate for the absolute value of the diffusivity too.

Fig. 10.30 displays the distribution of the ratio of the estimates D_x/D_y. A very similar distribution can be obtained for forcing with a somewhat underestimated or overestimated variance. One can introduce such variations by changing voluntary the decorrelation time of the atmospheric forcing. Indeed, following the discussion in section 10.4, one can conclude that the longer the forcing the larger the mesoscale mixing needed to produce an ocean response which is adequate in terms of the observed SST variability. The estimates

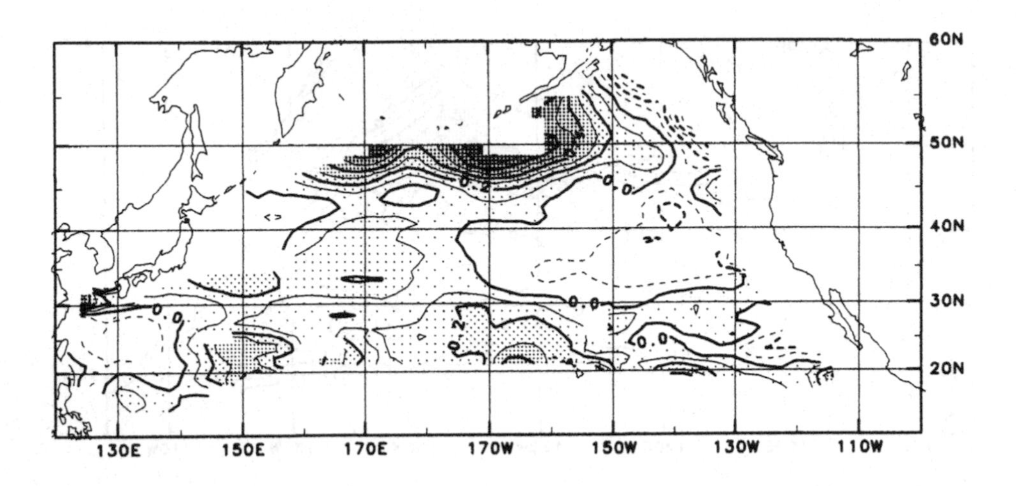

Figure 10.30 *(continued)* b) under the decorrelation time $\tau_c = 20$ days.

given in Fig. 10.30a were obtained with an atmospheric forcing decorrelation time of 15 days. By testing the ML inversion with reasonably underestimated (10 days) and overestimated (20 days) values of the correlation time scale, we found that the absolute values of the diffusion coefficients can differ by a factor of 1.5-2, but the ratio D_x/D_y does not change significantly. In proof of this, Fig. 10.30b shows the distribution of the ratio D_x/D_y computed for the case when the decorrelation time was fixed at 20 days.

An increase in the diffusivity component reflects growth of the mesoscale transport in the corresponding direction. The zonal component outweighs the meridional one by a factor of 1.5-2 south of the Kuroshio extension axis at the northern edge of the subtropical gyre. The zonal diffusivity decreases but the meridional component increases north of the Kuroshio current axis and inside the subtropical gyre, namely in the recirculation region of the Kuroshio current. However, the zonal diffusivity is relatively high again east of Philippines. In the central North Pacific, the zonal diffusivity usually dominates near the Subtropical Convergence Zone. By contrast, the meridional diffusivity is higher, $0.5 < D_x/D_y < 1.0$, in the eastern part of the midlatitude North Pacific.

The distribution of the zonal and meridional diffusivities basically correspond to the pattern of mean currents at 15 m depth obtained by Niiler (1995) from Lagrangian drifters; thus, the diffusivity is generally enhanced in an along current direction. For example, the region south of the Aleutian Islands between 175°E and 165°W where the mean currents at 15 m depth are directed eastward is characterized by local enhancement of the zonal diffusivity, $1 < D_x/D_y < 4$. Since the isotherms are stretched along the currents, these results agree with the common notion that oceanic eddy mixing tends to be along isopycnals. This assumption is used in numerical modeling of eddy induced tracer transports (e.g. Gent et al. 1995).

The ratio D_x/D_y demonstrates the anisotropy of mesoscale heat diffusivity in the upper ocean. Sub-mesoscale diffusion anisotropies have been observed by Davis (1985) locally off Northern California. The Coastal Dynamics Experiment made with current-following drifters indicated that the mean transport of a passive scalar can be modeled by eddy diffusion with an anisotropic and inhomogeneous eddy diffusivity. Also encouraging are observations by Ollitrault (1995) in the North Atlantic. An analysis of SOFAR float data collected in the depth range of 600 to 800 dbar between 30°N and 45°N allowed Ollitrault (1995) to show that the zonal and meridional diffusion coefficient behavior is different in the western and eastern parts of the observational domain. The diffusivity absolute values were on the order of 10^3 m^2s^{-1} at averaging time scales of 10 days. A larger diffusivity was observed in the western basin. It was shown that at longer time scales the zonal diffusivity dominates in the western basin, but the difference between the zonal and meridional diffusivities is only marginal in the eastern basin. In this study, spatio-temporal averaging was used and the analysis concerns the upper layer heat diffusivity, so our estimates are not directly relevant to the results of Ollitrault (1995) but, in general, the correspondence is striking.

As far as we know there are no direct estimates of the zonal and meridional heat diffusivity on scales of 100 km over the North Pacific. Hopefully the anisotropy of mesoscale heat diffusivity can be extracted from Lagrangian data in future. We also plan to use the ML method to derive the eddy diffusivity for the horizontal exchange of heat on other spatial scales such as the sub-mesoscale.

10.6 LIFETIME AND AREA OF LARGE-SCALE SST ANOMALIES

Let $T'(t,x,y)$ be the deviation from the long-term norm of the temperature observed at time t in the geographical point (x,y) and let T_0 be some prescribed level. In section 9.1, we defined large-area SST anomalies as the connected components of the set

$$\{(x,y)\ :\ T'(t,x,y) > T_0\}. \tag{10.37}$$

In other words we agreed to associate such an anomaly with an area where the SST deviation substantially exceeds the norm at a given time. The term 'connected' was used to stress that only coherent structures are considered. Such structures appear as big 'islands' on SST maps. While there are also many small unconnected 'reefs' of high deviations we do not consider them as part of the anomaly.

Our analysis is based on (i) the high level excursion theory given in chapter 1, (ii) the stochastic partial differential equation (9.42) for the field $T'(t,x,y)$, and (iii) the estimates of diffusivity and the feedback parameter obtained in this chapter. The third issue explains why only now, at the end of the manuscript, we return to the subject discussed at the beginning.

To apply results of high level excursion theory we need additional assumptions for the model (9.42). These are as follows: 1) advection is negligible; 2) the large-scale diffusion and forcing are isotropic; 3) both the diffusivity and the feedback parameter vary much slower than the anomalies themselves. Obviously, these assumptions are very restrictive but, it seems, do not greatly effect the estimates for the lifetime and the area of large-area SST anomalies.

Under these assumptions the basic stochastic equation (9.42) for anomalies is reduced to the following

$$\frac{\partial T'}{\partial t} + \lambda T' = D\nabla^2 T' + S, \tag{10.38}$$

where $S(t,x,y)$ is a Gaussian homogeneous isotropic random field with zero mean, and λ, D are constant.

While performing the estimation, one can choose the forcing spectrum in the form (1.9) which is determined by the variance σ_S^2, the space scale l_S, and the time scale τ_S. Using formulas (1.10) which connect the input and output scales of the linear stochastic system given by (10.38), one can get for the corresponding scales of the temperature anomaly field

$$l_{T'} = l_S \varphi_1(\alpha,\beta), \;\; \tau_{T'} = \lambda^{-1}\varphi_2(\alpha,\beta), \tag{10.39}$$

where

$$\alpha = \frac{D}{\lambda l_S^2}, \;\; \beta = \frac{\lambda \tau_S}{1+\lambda\tau_S}, \tag{10.40}$$

and $\varphi_1(\alpha,\beta)$ and $\varphi_2(\alpha,\beta)$ are given in (1.11).

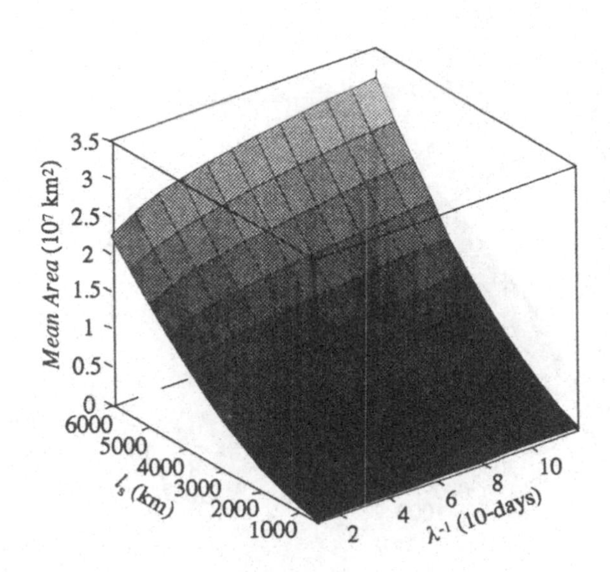

Figure 10.31 To the estimates of area of large-scale SST anomalies. See text for explanation.

Finally substituting (10.39) into (1.18) and (1.20) we get the following formulas for the parameters of interest

$$\begin{aligned} Mean\ Area &= 2\pi l_S^2 \varphi_1^2(\alpha,\beta)(\sigma_{T'}/T_0)^2, \\ Lifetime &= \sqrt{2\pi}\lambda^{-1}\varphi_2(\alpha,\beta)(\sigma_{T'}/T_0), \end{aligned} \tag{10.41}$$

where $\sigma_{T'}^2$ is the SST anomaly variance.

Let us assume $\sigma_{T'}/T_0 = 1$, $\tau_S = 3$ days, and $D = 10^3$ m^2 s^{-1}. The latter is taken accordingly with the estimates in sections 10.4 and 10.6. The computations via formulas (10.41) give us the results shown in Figs. 10.31 and 10.32.

From formulas (10.41) it follows that: 1) the area and the lifetime do not depend on the forcing variance, 2) the areas of the forcing and response are of the same order of magnitude, and 3) the lifetime is much longer than the forcing time scale. The overall conclusion is that a relatively weak atmospheric input can produce a long-term SST anomaly which occupies a large area. It should be noted that originally the relationship between the time scales of the forcing and response were found by Frankignoul and Hasselmann (1977) within the local model framework (see chapter 9). In theory, the conclusion of approxi-

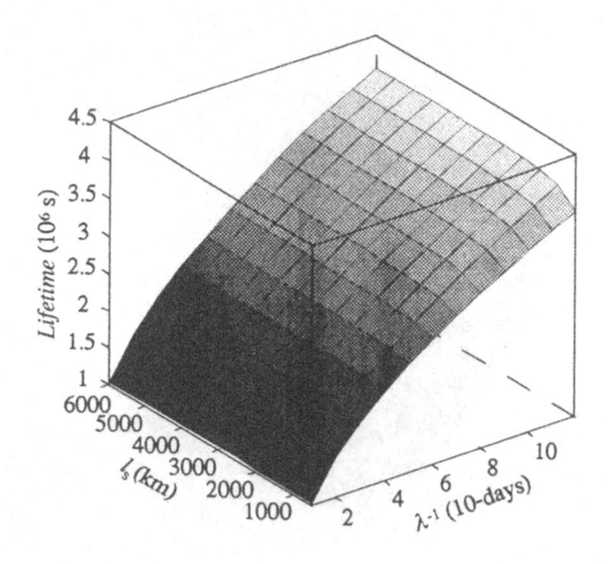

Figure 10.32 To the estimates of lifetime of large-scale SST anomalies. See text for explanation.

mate equality of the space scales of the input and the output was obtained by Piterbarg (1987). In practice, Nesterov (1991) deduced the same conclusion from an analysis of the wind velocity and SST time series for the North Atlantic. Finally, notice that one can get estimates of the time of occurrence of the large-area anomalies by applying the formula (1.23).

10.7 SUMMARY

This chapter described the second phase of our objectively consistent estimation approach, namely application of the statistical models, derived from the advection-diffusion equation, to the temperature anomaly phenomena in the upper ocean mixed layer. Using the Comprehensive Ocean - Atmosphere Data Set, we derived time series of SST, marine winds, and net heat flux across the air-sea boundary on a 10-day mean basis for the winter seasons of 1965-90 over 1°latitude × 2°longitude grid throughout the North Pacific midlatitudes. These long (up to 325 observations) time series were compiled for the autoregressive and maximum likelihood inversions. The statistical approximations of the heat anomaly balance equation were fitted to the time series. The models fits involved minimizations of a nonlinear systems of equations. The chosen minimizing algorithms and the numerical schemes were shown to be robust.

Analysis of the obtained results revealed that the statistical models capture the essential physics of the heat anomaly balance. In particular, the statistical models demonstrate a reasonable dissipation adjustment to variations in atmospheric forcing. The forcing drives diffusivity in such a way that an increase (decrease) of the former results in growth (decline) of the later. With different forcing magnitudes, the model SST anomaly field is the same if the model dissipation is enhanced (reduced) when the forcing is increased (decreased). The autoregressive inversion provided us with reasonable estimates of the heat anomaly advection in the North Pacific. Perhaps, the most remarkable result is the concise pattern of the net advection field east of Japan that was particularly difficult to obtain by earlier statistical models. Although accurate knowledge of atmospheric forcing is necessary to get absolute values of diffusivity in the maximum likelihood inversion, the ratio of the zonal and meridional heat diffusion coefficients is easily estimated. It was found that the ratio does not change significantly under hypothetical variations in the atmospheric forcing. As could be anticipated, the ratio varies over the analysis domain thereby demonstrating the anisotropy of heat diffusion in the upper layer of the North Pacific. The distribution of the zonal and meridional diffusivities is basically in accord with the pattern of near-surface ocean currents so that the diffusivity is generally enhanced in an along current direction.

Finally, by using heat diffusivity estimates we evaluated the area and lifetime of the large-scale SST anomalies in the North Pacific.

REFERENCES

[1] Adamec, D., M. M. Reinecker, and J. M. Vukovich, "The time-varying characteristics of the meridional Ekman heat transport for the world ocean", J. Phys. Oceanogr., 23, 1993, pp. 2704-2716.

[2] Adem, J., "On the prediction of mean monthly ocean temperatures", Tellus, 22, 1970, pp. 410-430.

[3] Adem, J., "Numerical-thermodynamical prediction of mean monthly ocean temperatures", Tellus, 27, 1975, pp. 541-551.

[4] Alexander, M. A., "Midlatitude atmosphere-ocean interaction during El Niño. Part I. The North Pacific Ocean", J. Climate, 5, 1992, pp. 949-958.

[5] Anderson, P. W., "Absence of diffusion in certain random lattices", Phys. Review, 109, 1958, pp. 1492-1499.

[6] Aoki, S., S. Imawaki, and K. Ichikawa, "Baroclinic disturbances propagating westward in the Kuroshio Extension region as seen by a satellite altimeter and radiometers", J. Geophys. Res., 100, 1995, pp. 839-855.

[7] Avellaneda, M., and A. J. Majda, "Mathematical models with exact renormalization for turbulent transport", Commun. Math. Phys., 131, 1990, pp. 181-429.

[8] Avellaneda, M. and A. J. Majda, "Renormalization theory for eddy diffusivity in turbulent transport", Physical Review Letters, 68, N20, 1992, pp. 3028-3031.

[9] Barnett, T. P., "On the nature and causes of large-scale thermal variability in the central North Pacific Ocean", J. Phys. Oceanogr., 11, 1981, pp. 887-904.

[10] Batchelor, G. K., "Diffusion field of homogeneous turbulence II. The relative motion of particles", Proc. Camb. Phil. Soc., 48, 1952, pp. 345-363.

[11] Batchelor, G. K., "The Theory of homogeneous turbulence", Cambridge University, Cambridge, 1982.

[12] Bathen, K. H., "On the seasonal changes in the depth of the mixed layer in the North Pacific", J. Geophys. Res., 77, 1972, pp. 7138-7150.

[13] Batteen, M. L., C. A. Collins, C. R. Gunderson, and C. S. Nelson, "The effect of salinity on density in the California Current system", J. Geophys. Res., 100, 1995, pp. 8733-8749.

[14] Belyaev, Yu. K., V. P. Nosko, and S. S. Filimonova, "On limits of applicabilty asymptotic formulas for features of the homogeneous Gaussian random fields", In: Excursions of Random Fields, Moscow State U., Moscow, 1972, pp. 46-53 (in Russian).

[15] Bennet, A. F., "A Lagrangian analysis of turbulent diffusion", Reviews of Geophysics, 25, 1987, pp. 799-822.

[16] Bennet, A. F., "Particle displacements in inhomogeneous turbulence", In: Stochastic modeling in physical oceanography, Eds. R.Adler, P.Muller, B.L.Rozovskii, Birkhauser, Boston, 1996, pp. 1-46.

[17] Bernstein, R. L., and W. B. White, "Stationary and traveling mesoscale perturbations in the Kuroshio Extension Current", J. Phys. Oceanogr., 11, 1981, pp. 692-704.

[18] Bernstein, R. L., L. Breaker, and R. Whritner, "California current eddy formation: ship, air, and satellite results", Science, 195, 1977, pp. 353-359.

[19] Bjerknes, J., "Synoptic survey of interaction of sea and atmosphere in the North Atlantic", Geoph. Norvegica, 24, 1962, pp. 115-145.

[20] Bjerknes, J., "Atlantic air-sea interaction", Adv. Geoph., 10, 1964, pp.1-82.

[21] Bjerknes, J., "A possible response of the atmospheric Hadley circulation to equatorial anomalies of ocean temperatures", Tellus, 18, 1966,pp. 820-829.

[22] Bjerknes, J., "Atmospheric teleconnections from the equatorial Pacific", Mon. Weather Rev., 97, 1969, pp. 163-172.

[23] Blanc, T. V., "Variation of bulk-derived surface flux, stability, and roughness results due to the use of different transfer coefficient schemes", J. Phys. Oceanogr., 15, 1985, pp. 650-669.

[24] Blumenthal, M. B., and M. A. Cane, "Accounting for parameter uncertainties in model verification: An illustration with tropical sea surface temperature", J. Phys. Oceanogr., 19, 1989, pp. 815-830.

[25] Bond, N. A., J. E. Overland, and P. Turet, "Spatial and temporal characteristics of the wind forcing of the Bering Sea", J. Climate, 7, 1994, pp. 1119-1130.

[26] Box, G. E. P. and G. M. Jenkins, "Time Series Analysis Forecasting and Control", 2nd ed., Holden-Day, San Francisco, 1976.

[27] Bryden, H. L. and M. M. Hall, "Heat transport by ocean currents across 25 N latitude in the Atlantic Ocean", Science, 207, 1980, pp. 884-886.

[28] Bunimovich, L. A., A. G. Ostrovskii, and S. Umatani, "Observations of the fractal properties of the Japan Sea surface temperature patterns", Int. J. Remote Sensing, 14, 1993, pp. 2185-2201.

[29] Bunker, A. F., "Computations of surface energy flux and annual air-sea interaction cycles of the North Atlantic Ocean", Mon. Weather Rev., 104, 1976, pp. 1122-1140.

[30] Bunker, A. F., "Trends of variables and energy fluxes over the Atlantic Ocean from 1948 to 1972", Mon. Weather Rev., 108, 1980, pp.720-732.

[31] Businger, J. A., "Interactions of sea and atmosphere", Rev. Geophys. Space Phys., 13, 720-726, 1975, pp. 817-822.

[32] Cane, M. A., S. E. Zebiak, and S. C. Dolan, "Experimental forecasts of El Nino", Nature, 321, 1986, pp. 827-832.

[33] Careta, A., F. Sagues, L. Ramires-Piscina and J. M. Sancho, "Effective diffusion in a stochastic velocity field", J. Statist. Physics, 71, 1993, pp. 235-242.

[34] Cayan, D. R., "Latent and sensible heat flux anomalies over the northern oceans - Driving the sea surface temperature", J. Phys. Oceanogr., 22, 1992a, pp. 859-881.

[35] Cayan, D. R., "Variability of latent and sensible heat fluxes estimated using bulk formulae", Atmosphere-Ocean, 30, 1992b, pp. 1-42.

[36] Chave, A., D. S. Luther, and J. H. Filloux, "Variability of the wind stress curl over the North Pacific, Implications for the oceanic response", J. Geophys. Res., 96, 1991, pp. 18361-18379.

[37] Chechetkin, V. R., V. S. Lutovinov, and A. A. Samokhin, "On the diffusion of passive impurities in random flows", Physica A, 175, 1991, pp. 87-113.

[38] Chelton, D. B., P. A. Bernal, and J. A. McGowan, "Large-scale interannual physical and biological interaction in the California Current", J. Mar. Res., 40, 1982, pp.1095-1125.

[39] Chen, H., and Sh. Chen, "Probability distribution of a stochastically advected scalar field", Phys. Rev. Letters, 63, N24, 1989, pp. 2657-2660.

[40] Cheney, R. E., J. G. Marsh, and B. D. Beckley, "Global mesoscale variability from collinear tracks of Seasat altimeter data", J. Geophys. Res., 88, 1983, pp. 4343-4354.

[41] Chereskin, T. K., "Direct evidence for an Ekman balance in the California Current", J. Geophys. Res., 100, 1995, pp. 18261-18269.

[42] CLIVAR, "A Study of Climate Variability and Predictability", Scientific Plan, World Climate Research Programme, CLIVAR Scientific Steering Group, WCRP-89, WMO/TD No. 690, ICSU, WMO, UNESCO, August 1995.

[43] Csanady, G.T., "Turbulent Diffusion in the Environment", Geophysics and Astrophysics Monographs, Reidel, Dordrecht, 1973.

[44] Davis, R., "Predictability of sea surface temperature and sea level pressure anomalies over the North Pacific Ocean", J. Phys. Oceanogr., 6, 1976, pp. 249-266.

[45] Davis, R., "Predictability of sea level pressure anomalies over the North Pacific Ocean", J. Phys. Oceanogr., 8, 1978, pp. 233-240.

[46] Davis, R. E., "On relating Eulerian and Lagrangian velocity statistics: single particles in homogeneous flows", J. Fluid Mech., 114, 1982, pp. 1-26.

[47] Davis, R. E., "Drifter observations of coastal surface currents during CODE: The statistical and dynamical views", J. Geophys. Res., 90, 1985, pp. 4756-4772.

[48] Davis, R. E., "Modeling eddy transport of passive tracers", J. Marin. Res., 45, 1987, pp. 635-666.

[49] Davis, R. E. "Observing the general circulation with floats", Deep-Sea Res., 38, 1991, pp. 5531-5571.

[50] Davis, R. E., R. De Szoeke, and P. Niiler, "Variability in the upper ocean during MILE, Part II, Modeling the mixed layer response", Deep-Sea Res., 28A, 1981, pp. 1453-1475.

[51] Dennis, J. E. and R. B. Schnabel, "Numerical Methods for Unconstrained Optimization and Nonlinear Equations", Printice Hall, Englewood Cliffs, NJ, 1983.

[52] Dever, E. P. and S. J. Lentz, "Heat and salt balances over the northern California shelf in winter and spring", J. Geophys. Res., 99, 1994, pp. 16001-16017.

[53] Dobrovolski, S. G., "Global Climatic Changes in Water and Heat Transfer - Accumulation Processes", Developments in Atmospheric Science, 21, Elsevier, Amsterdam, 1992.

[54] Eckart, C. "An analysis of the stirring and mixing process in incompressible fluids", J. Mar. Res., 7, 1948, pp. 265-275.

[55] Elsberry, R. L. and N. T. Camp, "Oceanic thermal response to strong atmospheric forcing, I Characteristics of forcing events", J. Phys. Oceanogr., 8, 1978a, pp. 206-214.

[56] Elsberry, R. L. and N. T. Camp, "Oceanic thermal response to strong atmospheric forcing, II The role of one-dimensional processes", J. Phys. Oceanogr., 8, 1978b, pp. 215-224.

[57] Elsberry, R. L. and R. W. Garwood, "Sea surface temperature anomaly generation in relation to atmospheric storms", Bull. Amer. Meteorol. Soc., 59, 1978, pp. 786-789.

[58] Elsberry, R. L. and S. D. Raney, "Sea surface temperature response to variations in atmospheric wind forcing", J. Phys. Oceanogr., 6, 1976, pp. 881-887.

[59] Emery, W. J., "The role of vertical motion in the heat budget of the upper northeastern Pacific Ocean", J. Phys. Oceanogr., 6, 1976, pp. 299-305.

[60] Emery, W. J., "Global summary: Review of eddy phenomena as expressed in temperature measurements", In: Eddies in Marine Science, Ed. A. R. Robinson, Springer-Verlag, NY, 1983, pp. 354-375.

[61] Emery, W. J. and K. Hamilton, "Atmospheric forcing of interannual variability in the northeast Pacific Ocean", J. Geophys. Res., 90, 1985, pp. 857-868.

[62] Emery, W. J., A. C. Thomas, M. J. Collins, W.R. Crawford, and D. L. Mackas, "An objective method for computing advective surface velocities from sequential infrared satellite images", J. Geophys. Res., 91, 1986, pp. 12865-12878.

[63] Enfield, D. B. and J. S. Allen, "On structure and dynamics of monthly mean sea level anomalies along the Pacific coast of North and South America", J. Phys. Oceanogr., 10, 1980, pp. 557-578.

[64] Favorite, F. and D. R. McLain, "Coherence in transpacific movements of positive and negative anomalies of sea surface temperature, 1953-60", Nature, 244, 1973, pp. 139-143.

[65] Fedorov, K., "The Physical Nature and Structure of Oceanic Fronts", Lecture Notes on Coastal and Estuarine Studies, 19, Springer-Verlag, Berlin, 1986.

[66] Fedorov, K. N. and A. I. Ginzburg, "The Subsurface Layer of the Ocean", Hydrometeoizdat, Leningrad, 1988, p.304 (in Russian).

[67] Fedorov, K. N. and A. G. Ostrovskii, "Climatically Significant Physical Parameters of the Ocean", In: Time Series of Ocean Measurements, 3, Intergovernmental Oceanographic Commission, Tech. Ser. 31, UNESCO, Paris, 1986, pp.9-31.

[68] Fedoruk, M. V., "Asymptotics. Integrals and Series", Nauka, Moscow, 1987 (in Russian).

[69] Fiaderio, M. E. and G. Veronis, "Obtaining velocities from tracer distributions", J. Phys. Oceanogr., 14, 1984, pp. 1734-1746.

[70] Frankignoul, C., "Sea surface temperature anomalies, planetary waves, and air-sea feedbacks in the middle latitudes", Rev. Geophys., 23, 1985, pp. 357-390.

[71] Frankignoul, C., C. Duchine, and M. Cane, "A statistical approach to testing equatorial ocean models with observed data", J. Phys. Oceanogr., 19, 1989, pp. 1191-1208.

[72] Frankignoul, C. and K. Hasselmann, "Stochastic climate models, Part II, Application to sea-surface temperature anomalies and thermocline variability", Tellus, 29, 1977, pp. 289-305.

[73] Frankignoul, C. and P. Müller, "Quasi-geostrophic response of an infinite -plane ocean to stochastic forcing by the atmosphere", J. Phys. Oceanogr., 9, 1979, pp. 104-127.

[74] Frankignoul, C. and R. W. Reynolds, "Testing a dynamical model for midlatitude sea surface temperature anomalies", J. Phys. Oceanogr., 13, 1983, pp. 1131-1145.

[75] Friedman, A., "Partial Differential Equations of Parabolic Type", Prentice Hall, Englewood Cliffs, NJ, 1964.

[76] Friedman, A., "Stochastic Differential Equations and Applications", v.1, Academic Press, NY, 1975.

[77] Fu, L-L., T. Keffer, P. P. Niiler, and C. Wunsch, "Observations of mesoscale variability in the western North Atlantic: A comparative study", J. Mar. Res., 1982, pp. 809-848.

[78] Furutsu, K., "On the statistical theory of electromagnetic waves in a fluctuating media", J. Res. NBS, D-67, 1963, p.303-324.

[79] Garcia-Ortiz, J. M. and A. Ruizde-Elvira, "An analysis of North Pacific SST anomalies by means of a linear thermodynamic stochastic two-dimensional model", Tellus, 47, 1995, pp. 118-131.

[80] Gardiner, C. W., "Handbook of Stochastic Methods for Physics, Chemistry and the Natural Sciences", Springer-Verlag, NY, 1985.

[81] Ghil, M. and P. Malanotte-Rizzoli, "Data assimilation in meteorology and oceanography", Advances in Geophysics, Academic Press, 33, 1991, pp. 141-266.

[82] Gikhman, I. and V. Skorokhod, "Theory of Random Processes", Springer-Verlag, Berlin, 1984.

[83] Gill, A. E., "Eddies in relation to climate", In: Eddies in Marine Science, Ed. A. R. Robinson, Springer-Verlag, Berlin, 1983, pp. 441-445.

[84] Girdyuk, G. V. and S. P. Malevskii-Malevich, "Method for computation of the ocean outgoing irradiance", Main Geoph. Observatory Trudy, 297, 1973, pp. 10-20 (in Russian).

[85] Gulev, S. K., "Influence of space-time averaging on the ocean-atmosphere exchange estimates in the North Atlantic midlatitudes", J. Phys. Oceanogr., 24, 1994, pp. 1236-1255.

[86] Gurbatov, S., Malakhov, A. and Saichev, A. "Nonlinear Random Waves and Turbulence in Nondispersive Media: Waves, Rays and Particles", Manchester U. Press, Cambridge, 1991.

[87] Haidvogel, D. B., A. B. Robinson, and C. G. H. Rooth, "Eddy-induced dispersion and mixing", In: Eddies in Marine Science, Ed. A. R. Robinson, Springer, Berlin, 1983, pp. 481-491.

[88] Hajek, J., "Asymptotically most powerful rank order test", Ann. Math. Stat., 33, 1962, pp. 1124-1147.

[89] Halliwell, G. R., P. Cornillon, and D. A. Byrne, "Westward-propagating SST anomaly features in the Sargasso Sea, 1982-88", J. Phys. Oceanogr., 21, 1991a, pp. 635-649.

[90] Halliwell, G. R., Y. J. Ro, and P. Cornillon, "Westward-propagating SST anomalies and baroclinic eddies in the Sargasso Sea", J. Phys. Oceanogr., 21, 1991b, pp. 1664-1680.

[91] Hanawa, K., and I. Hoshino, "Temperature and mixed layer in the Kuroshio Region over the Izu Ridge". J. Mar. Res., 46, 1988, pp. 683-700.

[92] Hanawa, K., Y. Yoshikawa, and T. Watanabe, "Composite analyses of wintertime wind stress vector fields with respect to SST anomalies in the western North Pacific and the ENSO events, Part I, SST composite", J. Meteor. Soc. Japan, 67, 1989a, pp. 385-400.

[93] Hanawa, K., Y. Yoshikawa, and T. Watanabe, "Composite analyses of wintertime wind stress vector fields with respect to SST anomalies in the western North Pacific and the ENSO events, Part II, ENSO composite", J. Meteor. Soc. Japan, 67, 1989b, pp. 833-844.

[94] Hasselmann, K., "Feynman diagrams and interaction rules of wave-wave scattering processes", Rev. Geophys. Space Phys, 4, 1966, pp. 1-32.

[95] Hasselmann, K., "Stochastic climate models, Part I, Theory", Tellus, 28, 1976, pp. 473-485.

[96] Hasselmann, K., "An ocean model for climate variability studies", Prog. Oceanog., 11, 1982, pp. 69-92.

[97] Hellerman, S. and M. Rosenstein, "Normal monthly wind stress over the world ocean with error estimates", J. Phys. Oceanogr., 13, 1983, pp. 1093-1104.

[98] Herterich, K., and K. Hasselmann, "Extraction of mixed layer advection velocities, diffusion coefficients, feedback factors and atmospheric forcing parameters from the statistical analysis of North Pacific SST anomaly fields", J. Phys. Oceanogr., 17, 1987, pp. 2145-2156.

[99] Hickey, B. M., "The California Current System - Hypotheses and Facts", Prog. Oceanogr., 8, 1979, pp. 191-279.

[100] Holland, W. R., "Eddy-resolving numerical models of large-scale ocean circulation", In: Eddies in Marine Science, Ed. A. R. Robinson, Springer, Berlin, 1983, pp. 379-403.

[101] Horel, J. D., "On the annual cycle of the tropical Pacific atmosphere and ocean", Mon. Weather Rev., 110, 1982, pp. 1863-1878.

[102] Hsiung, J., and R. E. Newell, "The principal nonseasonal modes of variation of global sea surface temperature", J. Phys. Oceanogr., 13, 1983, pp. 1957-1967.

[103] Huebner, M., R. Z. Khasminskii and B. L. Rozovskii, "Two examples of parameter estimation for SPDE's", In: A Festschrift in Honour of Copanath Kallianpur, Springer-Verlag, Berlin, 1993, pp. 34-49.

[104] Huebner, M. and B. L. Rozovskii, "On asymptotic properties of ML estimator for pararabolic stochastic PDE's", Probab. Theory and Related Topics, 103, 1995, pp.143-164.

[105] Huebner, M."Parameter estimation for stochastic differential equations", Thesis, Univ. of South. Cal., 1993.

[106] Huyer, A. and P. M. Kosro, "Mesoscale surveys over the shelf and slope in the upwelling region near Point Arena, California", J. Geophys. Res., 92, 1987, pp. 1655-1681.

[107] Huyer, A. and R. L. Smith, "The signature of El Niño off Oregon, 1982-1983", J. Geophys. Res., 90, 1985, pp. 7133-7142.

[108] Ibragimov, I. A. and R. Z. Khasminskii, "Statistical Estimation (Asymptotic Theory)", Springer-Verlag, NY, 1981.

[109] Isemer, H.-J. and L. Hasse, "The Bunker Atlas of the North Atlantic Ocean, Vol. 2: Air-Sea Interactions", Springer-Verlag, Berlin, 1987, p. 252.

[110] Isichenko, M.B. "Percolation, statistical topography, and transport in random media", Reviews of Modern Physics, 64, 1992, N4, pp. 961-1043.

[111] Iwasaka, N., and K. Hanawa, "Climatologies of marine meteorological variables and surface fluxes in the North Pacific computed from COADS", Tohoku Geoph. J. (Sci. Rep. Tohoku U., Ser. 5), 33, 1990, pp. 185-239.

[112] Iwasaka, N., K. Hanawa, and Y. Toba, "Analysis of SST anomalies in the North Pacific and their relation to 500-mb height anomalies over the Northern Hemisphere during 1969-1979", J. Meteor. Soc. Japan, 65, 1987, pp. 103-113.

[113] Jacobs, G. A., W. J. Emery and G. H. Born, "Rossby waves in the Pacific ocean extracted from Geosat altimeter data", J. Phys. Oceanogr., 1993, 23, pp. 1155-1175.

[114] Kantha, L. H. and C. A. Clayson, "An improved mixed layer model for geophysical applications", J. Geophys. Res., 99, 1994, pp. 25235-25266.

[115] Kawai, H., "Hydrography of the Kuroshio extension", In: Kuroshio, Its Physical Aspects, Eds. H. Stommel and K. Yoshida, U. Tokyo Press, 1972, pp. 235-352.

[116] Kawamura, H., K. Mizuno and Y. Toba, "Formation of a warm-core ring in the Kuroshio-Oyashio frontal zone - December 1981 - October 1982", Deep Sea Res, 33, 1986, pp. 1617-1640.

[117] Kawamura, R., "Relation between atmospheric circulation and dominant sea surface temperature anomaly patterns in the North Pacific during the Northern Winter", J. Meteorol. Soc. Japan, 62, 1984, pp. 910-916.

[118] Kawamura, R., "A rotated EOF analysis of global sea surface temperature variability with interannual and interdecadal scales", J. Phys. Oceanogr., 24, 1994, pp. 707-715.

[119] Keldysh, M. V., "On the eigenvalues and eigenfunctions of some classes of the non-self-adjoint equations", Sov. Math. Dokl., 127, 1951, pp.11-14.

[120] Kelly, K. A., "An inverse model for near-surface velocity from infrared images", J. Phys. Oceanogr., 19, 1989, pp. 1845-1864.

[121] Kelly, K. A. and P. T. Strub, "Comparison of velocity estimates from advanced very high resolution radiometer in the coastal transition zone", J. Geophys. Res., 97, 1992, pp. 9653-9668.

[122] Kessler, W. S., "Observations of long Rossby waves in the northern tropical Pacific", J. Geophys. Res., 95, 1990, pp. 5183-5217.

[123] Kitaigorodskii, S. A., "On the computation of the thickness of the wind-mixing layer in the ocean", Izv. Akad. Nauk SSSR, Geophys. Ser., 3, 1960, pp. 425-431. (English edition pp. 284-287).

[124] Kitaigorodskii, S. A., "Physics of Air-Sea Interactions, Hydrometeoizdat, Leningrad", 1970, p.304. (in Russian)

[125] Klyatskin, V. I., "Statistical description of diffusing tracers in random velocity fields", Uspekhi Fizicheskikh Nauk, Russ. Acad. Sci., 37, 1994, pp. 5-22.

[126] Klyatskin, V. I., W. A. Woyczynski, and D. Gurarie, "Short-time correlation approximations for diffusing tracers in random velocity fields: a functional approach", In: Stochastic modeling in physical oceanography, Eds. R. Adler, P. Muller, B. L. Rozovskii, Birkhauser, Boston, 1996, pp. 221-270. .

[127] Komorowski, T., "Application of the parametrix method to diffusions in a turbulent Gaussian environment", Stochastic processes and their applications, 1996 (to appear)

[128] Kondo, J., "Air-sea bulk transfer coefficients in diabatic conditions", Bound.-Layer Meteor., 9, 1975, pp. 91-112.

[129] Kondo, J., and A. Miura, "Surface heat budget of the western Pacific for May 1979", J. Meteor. Soc. Japan, 63, 1985, pp. 633-646.

[130] Koshlyakov, M. N., "Eddies in the open ocean", In: Synoptic Eddies in the Ocean, Eds. V. M. Kamenkovich, M. N. Koshlyakov, and A. S. Monin, D. Reidel Publishing Co., Dordrecht, 1986, pp. 265-376.

[131] Kozlov, S. M., "Reductibility of quasiperiodic operators and homogenization", Trans. Moscow Math. Soc., 46, 1983, pp. 99-123.

[132] Kraichnan, R. M., "The structure of isotropic turbulence at very high Reynolds numbers", J. Fluid Mech., 5, 1959, pp. 497-543.

[133] Kraichnan, R. M., "Convection of a passive scalar by a quasi-uniform random straining field", J. Fluid Mech., 64, 1974, pp. 737-762 .

[134] Kraus, E. B., "Atmosphere-Ocean Interaction", Claredon Press, Oxford, 1972.

[135] Kraus, E. B. and C. Rooth, "Temperature and steady state vertical heat flux in the ocean surface layers", Tellus, 13, 1961, pp. 231-238.

[136] Kraus, E. B. and S. Levitus, "Annual heat flux variations across the tropic circles", J. Phys. Oceanogr., 16, 1986, pp. 1479-1486.

[137] Kraus, E. B. and J. S. Turner, "A one-dimensional model of the seasonal thermocline, II", Tellus, 19, 1967, pp. 98-106.

[138] Kubo, R., "Stochastic Liouville equation", J. Math. Phys., 4, 1963, pp. 174-183.

[139] Kudryavtsev, V. N. and A. V. Soloviev, "Slippery near-surface layer of the ocean arising due to a daytime solar heating", J. Phys. Oceanogr., 20, 1990, pp. 617-628.

[140] Kutoyants, Yu. A., "Identification of Dynamical Systems with Small Noise", Kluwer, Dordrecht, 1994.

[141] Kutsuwada, K. and T. Teramoto, "Monthly maps of surface wind stress fields over the North Pacific during 1961-1984", Bull. Ocean Res. Inst., U. Tokyo, 24, 1987.

[142] Lanzante, J. R., "A rotated eigenanalysis of the correlation between 700-mb heights and the sea surface temperatures in the North Pacific and Atlantic", Mon. Weather Rev., 112, 1984, pp. 2270-2280.

[143] Lappo, S. S. and S. K. Gulev, "Introduction to analysis of the energetically active regions of the northern part of the Atlantic Ocean", In: Study of the Ocean-atmosphere Interaction Processes, Hydrometizdat, Moscow, 1984, pp. 5-30 (in Russian).

[144] Large, W. G., J. C. McWilliams, and S. Doney, "Oceanic vertical mixing: A review and a model with a nonlocal boundary layer parameterization", Rev. Geophys., in press.

[145] Large, W. G., J. C. McWilliams, and P. P. Niiler, "Upper ocean thermal response to strong autumnal forcing of the Northeast Pacific", J. Phys. Oceanogr., 16, 1986, pp. 1524-1550.

[146] Larin, D. A.,"Concerning the estimation acuracy problem for monthly mean values of meteorological variables based on measurements of the opportunity ships", Meteorology and Hydrology, N8, 1982, pp. 109-114 (in Russian).

[147] Latif, M., and N. E. Graham, "How much predictive skill is contained in the thermal structure of an oceanic GCM?", J. Phys. Oceanogr., 22, 1992, pp. 951-962.

[148] LeCam, L., "Asymptotic Methods in Statistical Decision Theory", Springer-Verlag, NY, 1986.

[149] Ledvina, D. V., G. S. Young, R. A. Miller, and C. W. Fairall, "The effect of averaging on bulk estimates of heat and momentum fluxes for the tropical western Pacific Ocean", J. Geophys. Res., 98, 1993, pp. 20211-20217.

[150] Lee, J. H. and G. Veronis, "Determining velocities and mixing coefficients from tracers", J. Phys. Oceanogr., 19, 1989, pp. 487-500.

[151] Lipcombe, T. C., A. L. Frenkel, and D. ter Haar, "On the convection of a passive scalar by a turbulent Gaussian velocity field", J. Statist. Physics, 63, 1991, pp. 305-313.

[152] Lipert, A. and P. Muller, "Direct atmosphering forcing of geostrophic eddies, Part II, Coherence maps", J. Phys. Oceanogr., 25, 1995, pp. 106-121.

[153] Lomakin, A. F. and K. A. Rogachev, "Connection of sea surface temperature anomalies of the North Pacific with atmospheric processes in the transient seasons", Meteorology and Hydrology, N11, 1983, pp. 51-58 (in Russian).

[154] Lundgren, T. S., "Trubulent pair dispersion and scalar diffusion", J. Fluid Mech., 111, 1981, pp. 27-57.

[155] Magaard, L., "On the potential energy of baroclinic Rossby waves in the North Pacific", J. Phys. Oceanogr., 13, 1983, pp. 38-42.

[156] Majda, A. J., "The random uniform shear layer: an explicit example of turbulent diffusion with broad tail probability distributions", Phys. Fluids A5 , 1993, pp. 963-1969.

[157] Markus, A. S. and V. I. Matsaev," On the asymptotics of the spectrum of operators, close to normal", Funct. Anal. Appl., 13, 1979, pp. 93-94.

[158] Marquardt, D. W., "An algorithm for least-square estimation of nonlinear parameters", SIAM J. Appl. Math., 11, 1963, pp. 431-441.

[159] Martin, P. J., "Simulation of the mixed layer at OWS November and Papa with several models", J. Geophys. Res., 90, 1985, pp. 903-916.

[160] McCreary, J. P., "Eastern tropical response to chnaging wind systems: With application to El Nino", J. Phys. Oceanogr., 6, 1976, pp. 632-645.

[161] McLeish, W., "Spatial spectra of ocean surface temperature", J. Geophys. Res., 75, 1970, pp. 6872-6877.

[162] McNally, G. J. and W. B. White, "Wind driven flow in the mixed layer observed by drifting buoys during autumn-winter in the midlatitude North Pacific", J. Phys. Oceanogr., 15, 1985, pp. 684-694.

[163] Michaelsen, J., "A statistical study of large-scale, long-period variability in North Pacific sea surface temperature anomalies", J. Phys. Oceanogr., 12, 1982, pp. 694-703.

[164] Miller, A. J., "Large-scale ocean-atmosphere interactions in a simplified coupled model of the midlatitude wintertime circulation", J. Atmos. Sci., 49, 1992, pp. 273-286.

[165] Miller, A. J., D. R. Cayan, T. B. Barnett, N. E. Graham, and J. M. Oberhuber, "Interdecadal variability of the Pacific Ocean: Model response to observed heat flux and wind stress anomalies", Climate Dyn., 9, 1994, pp. 287-302.

[166] Minster, J. F. and C. Brossier, "Seasonal variations in sea level variability", CNES French Space Agency, AVISO Altimetry Newsletter, N4, 1996, pp. 12-13.

[167] Mizuno, K. and W. B. White, "Annual and interannual variability in the Kuroshio Current system", J. Phys. Oceanogr., 13, 1983, pp. 1847-1867.

[168] Molchanov, S. A. and L. I. Piterbarg, "Heat propagation in random flows", Russ. J. Math. Phys., 1, 1994, pp.353-376.

[169] Monin, A. S. and R. V. Ozmidov, "Turbulence in the Ocean", Reidel, Hingham, MA, 1985.

[170] Monin, A. S. and A. M. Yaglom, "Statistical Fluid Mechanics; Mechanics of Turbulence", MIT Press, Cambridge, MA, 1975.

[171] Muller, P. and C. Frankignoul, "Direct atmospheric forcing of geostrophic eddies", J. Phys. Oceanogr., 11, 1981, 287-308.

[172] Namias, J., "Recent seasonal interactions between North Pacific waters and the overlying atmosphere circulation", J. Geophys. Res., 64, 1959, pp. 631-646.

[173] Namias, J., "Large-scale air-sea interactions over the North Pacific from summer 1962 through the subsequent winter", J. Geophys. Res., 68, 1963, pp. 6171-6186.

[174] Namias, J., "Microscopic association between mean monthly sea surface temperature and overlying winds", J. Geophys. Res., 70, 1965, pp. 2307-2318.

[175] Namias, J., "Seasonal interactions between the North pacific ocean and the atmosphere during thle 1960's", Mon. Weather Rev., 97, 1969, pp. 173-192.

[176] Namias, J., "Experiments in objectively predicting some atmospheric and oceanic variables for the winter of 1971-72", J. Appl. Meteorol., 11, 1972, pp. 1164-1174.

[177] Namias, J., "Thermal communication between the sea surface and the lower atmosphere", J. Phys. Oceanogr., 3, 1973, pp. 373-378.

[178] Namias, J., "Negative ocean-air feedback systems over the North Pacific in the transition from warm to cold seasons", Mon. Weather Rev., 104, 1976, pp. 1107-1121.

[179] Namias, J. and R. M. Born, "Temporal coherence in North Pacific sea surface temperature patterns", J. Geophys. Res., 75, 1970, pp. 5952-5955.

[180] Namias, J. and D. R. Cayan, "Large-scale air-sea interactions and short period climatic fluctuations", Science, 214, 1981, pp. 869-876.

[181] Nesterov, E. S., "On effect of storms on formation of the temperature anomalies in the ocean in the fall season", Meteorology and Hydrology, N5, 1984, pp. 84-87 (in Russian).

[182] Nesterov, E.S., "On the influence of the dynamical atmospheric forcing on the SST anomaly formation", Meteorology and Hydrology, N11, 1991, pp. 63-75 (in Russian).

[183] Neumann, G., and W. J. Pierson, "Principles of Physical Oceanography", Prentice-Hall, Englewood Cliffs, NJ, 1966, p.545.

[184] Niiler, P. P., "Deepening of the wind-mixed layer", J. Mar. Res., 33, 1975, pp. 405-422.

[185] Niiler, P., "Global Drifter Program: Measurements of velocity, SST and atmospheric pressure", Intl. WOCE Newslett., Number 20, September 1995, WOCE Office at NERC IOS Deacon Lab., Wormley, UK, 1995, pp. 3-6.

[186] Niiler, P. P., and E. B. Kraus, "One-dimensional models of the upper-ocean". In Modeling and Prediction of the Upper Layer of the Ocean, Ed. E. B. Kraus, Pergamon, NY, 1977, pp. 143-172.

[187] Nitani, H., "Beginning of the Kuroshio", In: Kuroshio, Its Physical Aspects, Eds. H. Stommel and K. Yoshida, U. Tokyo Press, 1972, pp. 129-163.

[188] Norton, J. G. and D. R. McLain, "Diagnostic patterns of seasonal and interannual temperature variation off the west coast of the United States: Local and remote large-scale atmospheric forcing", J. Geophys. Res., 99, 1994, pp. 16019-16030.

[189] Nosko, V. P., "Local structure of the random Gaussian fields in vicinity of high specks", Doklady Acad. Sci. USSR, 189, 1969, pp. 714-717 (in Russian).

[190] Nosko, V. P., "Limit distributions of the excursion characteristics above high level in the homogeneous random Gaussian field", In: 4th International Vilnus Conference of the Probability theory and Mathematical Statistics, Abstracts Volume 2, Vilnus, 1985, pp. 269-271.

[191] Novikov, E.A., "Functionals and the random-force method in turbulence theory", Sov.Phys. JEPT, 20, N5, 1964, pp. 1290-1311.

[192] Okubo, A., "Oceanic diffusion diagrams", Deep-Sea Res., 18, 1971, pp. 789-802.

[193] Okubo, A. "Diffusion and Ecological Problems: Mathematical Models", Springer-Verlag, Berlin, 1979.

[194] Ollitrault, M., "The general circulation at the subtropical North Atlantic, Near 700 m depth, revealed by the TOPOGULF SOFAR floats", Intl. WOCE Newslett., Number 20, September 1995, WOCE Office at NERC IOS Deacon Lab., Wormley, UK, 1995, pp. 15-18.

[195] Ortiz-Bevia, M. J. and A. Ruizde-Elvira, "A cyclostationary model of sea surface temperatures in the Pacific Ocean", Tellus, 37A, 1985, pp. 14-23.

[196] Osborne, M. R., "Nonlinear least squares - The Levenberg algorithm revisited", J. Austral. Math. Soc., 19, 1976, pp. 343-357.

[197] Ostrovskii, A. G., "Essentials of the sea surface temperature variability in the North Atlantic", Abstract of Ph. D. thesis, Moscow University, Moscow, 1983, p. 23.

[198] Ostrovskii, A. G., "Signatures of stirring and mixing in the Japan Sea surface temperature patterns in autumn 1993 and spring 1994", Geophys. Res. Lett., 22, 1995, pp. 2357-2360.

[199] Ostrovskii, A. G., "Advection, diffusion, and feedback as estimated with maximum likelihood method from time dependent distributions of passive tracer", 1995, unpublished manuscript.

[200] Ostrovskii, A. G. and L. I. Piterbarg, "Autoregression model of sea surface temperature anomaly field in the North Atlantic", Oceanology, 25, 1985, pp. 333-334 (English transl.).

[201] Ostrovskii, A. G. and L. I. Piterbarg, "Diagnosis of the seasonal variability of water surface temperature anomalies in the North Pacific", Meteorology and Hydrology, N12, 1985, pp. 51-57 (in Russian).

[202] Ostrovskii, A. G. and L. I. Piterbarg, "On adaptation of a numerical model of SST anomalies to real data", In: Numerical Modeling of Hydrophysical Fields and Processes in the Ocean, Eds. A. S. Monin and D. G. Seidov, Nauka, Moscow, 1986, pp. 133-140 (in Russian).

[203] Ostrovskii, A. and L. Piterbarg, "Inversion for heat anomaly transport from sea surface temperature time series in the northwest Pacific", J. Geophys. Res., 100, 1995, pp. 4845-4865.

[204] Papanicolaou, G. and S. R. S. Varadhan, "Boundary value problems with rapidly oscillating coefficients". In: Random Fields, colloq. Math. Soc. Janos Bolyai, 27. Eds. J. Fritz, J.L. Kebowitz, D. Szasz, D., Amsterdam, North Holland, 1981, pp.835-875.

[205] Philander, S. G. H., "El Niño, La Niqa and the Southern Oscillation", Academic, San Diego, 1990, p. 287.

[206] Phillips, O. M., "The Dynamics of the Upper Ocean", The University Press, Cambridge, 1969, p. 287.

[207] Phythian, R. "Dispersion by random velocity fields", J. Fluid Mech., 67, 1975, pp. 145-153.

[208] Piterbarg, L. I., "On the stochastic nature of generation of the large scale sea surface temperature anomalies", Doklady Acad. Sci. USSR, 282, 1985, pp. 1473-1477 (in Russian).

[209] Piterbarg, L. I., "Formation of features of the sea surface temperature field by the action of synoptic atmospheric processes", Izv. Acad. Sci. USSR., Atmosph. Ocean. Phys., 23, 1987, pp. 48-53 (English transl.).

[210] Piterbarg, L. I., "The Dynamics and Prediction of the Large-Scale SST anomalies (Statistical Approach)", Hydrometeoizdat, Leningrad, 1989, (in Russian).

[211] Piterbarg, L. I. and A. G. Ostrovskii, "Dynamical stochastic model of long-term sea surface temperature variability", Doklady Acad. Sci. USSR, 276, 1984, pp. 1467-1470 (in Russian).

[212] Piterbarg, L. I. and B. L. Rozovskii, "Maximum likelihood estimators in the equations of physical oceanography", In: Stochastic modeling in physical oceanography, Eds. R.Adler, P.Muller, B.L.Rozovskii, Birkhauser, Boston, 1996, pp. 398-421.

[213] Price, J. F., R. A. Weller, and R. R. Schudlich, "Wind-driven ocean currents and Ekman transport", Science, 238, 1987, pp. 1534-1538.

[214] Qiu, B., and T. M. Joyce, "Interannual variability in the mid- and low-latitude western North Pacific", J. Phys. Oceanogr., 22, 1992, pp. 1063-1079.

[215] Qiu, B., and K. A. Kelly, "Upper ocean heat balance in the Kuroshio extension region", J. Phys. Oceanogr., 23, 1993, pp. 2027-2041.

[216] Qiu, B., and K. A. Kelly, and T. M. Joyce, "Mean flow and variability in the Kuroshio extension from Geosat altimetry data", J. Geophys. Res., 96, 1991, pp. 18490-18507.

[217] Reed, R., "On estimating insolation over the ocean", J. Phys. Oceanogr., 7, 1977, pp. 482-485.

[218] Reynolds, R. W., "Sea surface temperature anomalies in the North Pacific Ocean", Tellus, 30, 1978, pp. 97-103.

[219] Reynolds, R. W., "A real-time global sea surface temperature analysis", J. Climate, 1, 1988, pp. 75-86.

[220] Reynolds, R. W. and T. M. Smith, "Improved global sea surface temperature analyses", J. Climate, 7, 1994, pp. 929-948.

[221] Reznik, G. M., "Generation of eddies by direct forcing by the atmosphere", In: Synoptic Eddies in the Ocean, Eds. V. M. Kamenkovich, M. N. Koshlyakov, and A. S. Monin, D. Reidel Publ., Dordrecht, 1986, pp. 153-171.

[222] Rhines, P. B., "ATLAST. A World-Ocean Atlas of Hydrography, Nutrients and Chemical Tracers", U. Washington, Tech. Rep. 91-1, 1992.

[223] Richards, K. J., Y. Jia, and C. F. Rogers, "Dispersion of tracers by ocean gyres", J. Phys. Oceanogr., 25, 1995, pp. 873-887.

[224] Richardson, L.F., "Atmospheric diffusion shown on a distance-neighbor graph", Proc. Roy. Soc., 110, 1926, pp. 709-727.

[225] Rienecker, M. M., C. N. K. Mooers, D. E. Hagan, and A. R. Robinson, "A cool anomaly off northern California: An investigation using IR imagery and in situ data", J. Geophys. Res., 90, 1985, pp. 4807-4818.

[226] Roberts, P. H., "Analytical theory of turbulent diffusion", J. Fluid Mech., 11, 1961, pp. 257-283.

[227] Robinson, I. S., "Satellite Oceanography. An Introduction for Oceanographers and Remote-sensing Scientists", Ellis Horwood Series in Marine Science, Ellis Horwood Ltd., Halsted Press of John Wiley and Sons, Chichester, 1985, p. 455.

[228] Roden, G. I., "On North Pacific temperature, salinity, sound velocity and density fronts and their relation to the wind and energy flux fields", J. Phys. Oceanogr., 5, 1975, pp. 557-571.

[229] Rogachev, K. A., "Prognostic connections for temperature anomalies of the upper layer of the North Pacific", Doklady Acad. Sci. USSR, 274, 1984, pp. 1197-1200 (in Russian).

[230] Saffman, P. G., "An application of the Weiner-Hermite expansion to diffusion of a passive scalar in a homogeneous turbulent flow", Phys. Fluids, v.12, 1969, pp. 1786-1798.

[231] Saichev, A. I. and W. A. Woyczynski, "Probability distributions of passive tracers in randomly moving media", In: Stochastic models in geosystems, Eds. Molchanov, S. A. and W. Woyczynski, IMA Volumes, Springer-Verlag, Berlin, 1996.

[232] Saino, T., "Particle dynamics in the western subarctic Pacific - Implications from CZCS imagery", In: Proceedings of Pacific Ocean Remote Sensing Conference PORSEC'92, Vol. 1, Okinawa, pp. 330-335.

[233] Sarkisian, A. S. (1969): "Theory and Computation of Ocean Currents", Springfield, Jerusalem, 1969.

[234] Seager, R., S. E. Zebiak, and M. A. Cane, "A model of the tropical Pacific sea surface temperature climatology", J. Geophys. Res., 93, 1988, pp. 1265-1280.

[235] Semenov, D. V., "Averaging of differential equations of parabolic type with random coefficients", Math. Notes, 45, N5, 1989, pp. 123-126.

[236] Semtner, A. J., and R. M. Chervin, "Ocean general circulation from a global eddy-resolving model", J. Geophys. Res., 97, pp. 5493-5550.

[237] Sennichael, N., C. Frankignoul, and M. A. Cane, "An adaptive procedure for tuning a sea surface temperature model", J. Phys. Oceanogr., 24, 1994, pp. 2288-2305.

[238] Shum, C. K., R. A. Werner, D. T. Sandwell, B. H. Zhang, R. S. Nerem, and B. D. Tapley, "Variations of global mesoscale eddy energy observed from Geosat", J. Geophys. Res., 95, 1990, pp. 17865-17876.

[239] Simpson, J. J., "Large-scale thermal anomalies in the California Current during the 1982-1983 El Niño", Geophys. Res. Lett., 10, 1983, pp. 937-940.

[240] Simpson, J. J., "Response of the southern California Current system to the mid-latitude North Pacific coastal warming events of 1982-1983 and 1940-1941", Fish. Oceanogr., 1, 1992, pp. 57-79.

[241] Sinai, Ya. G. and Yachot V., "Limiting probability distributions of a passive scalar in a random velocity field", Phys. Rev. Lett., 63, N18, 1992, pp. 1962-1964.

[242] SSL II, "Scientific Subroutine Library", Fujitsu Ltd., Communication and Electronics, Tokyo, 1987.

[243] Sverdrup, H. U., M. W. Johnson, and R. H. Fleming, "The 0ceans, their Physics Chemistry and General Biology", Printice-Hall, NY, 1946.

[244] Tabata, S., "Oceanic time-series measurements from Station P and along Line P in the northeast Pacific Ocean", In: Papers Presented at the Meeting on Time Series of Ocean Measurements , Tokyo, 11-15 may 1981, WCP-21, 1981, pp. 171-192.

[245] Taft, B. A., "Characteristics of the flow of the Kuroshio south of Japan", In: Kuroshio, Its Physical Aspects, Eds. H. Stommel and K. Yoshida, U. Tokyo Press, 1972, pp. 165-216.

[246] Tai, C. T., and W. B. White, "Eddy variability in the Kuroshio Extension as revealed by Geosat altimetry: Energy propagation away from the jet, Reynolds stress, and seasonal cycle", J. Phys. Oceanogr., 20, 1990, pp. 1761-1777.

[247] Tapley, D., "Variations of global mesoscale eddy energy observed from Geosat", J. Geophys. Res., 90, 1990, pp. 17865-17876.

[248] Taylor, G. I., "Diffusion by continuous movements", Proc. London Math. Soc., 20, 1921, pp. 196-211.

[249] Taylor, G. I., "Dispersion of soluable matter in solvent flowing slowly through a tube", Proc. Roy. Soc. A, 219, 1953, pp. 186-203.

[250] "The NOAA/NASA Pathfinder Program", A Publication of the U. Corporation for Atmospheric Research pursuant to NOAA Award Number NA27GP0232-01, NOAA, NASA, 1994.

[251] Tourre, Y. M. and W. B. White, "ENSO signals in global upper-ocean temperature", J. Phys. Oceanogr., 25, 1995, pp. 1317-1332.

[252] Trenberth, K., "Mean annual poleward energy transports by the oceans in the Southern Hemisphere", Dyn. Atmosph. Oceans, 4, 1979, pp. 57-64.

[253] Tsuchiya, M., "On the Pacific upper-water circulation", J. Mar. Res., 40, Suppl., 1982, pp. 777-799.

[254] Turner, J. S. and E. B. Kraus, "A one-dimensional model of the seasonal thermocline, I, A laboratory experiment and its interpretation", Tellus, 19, 1967, pp. 88-97.

[255] Tutubalin, V. N., "A variant of the local limit theorem for products of random matrices", Theor. Probab. Applic., 22, N2, 1977, pp. 203-214.

[256] Van den Bos, A., "Alternative interpretation of maximum entropy spectral analysis", IEEE Trans. Inf. Theory, 17, 1971, pp. 493-494.

[257] Van der Hoven, I., "Power spectrum of horizontal wind speed in the frequency range from 0.0007 to 900 cycles per hour", J. Meteor., 14, 1957, pp. 160-164.

[258] Vonder Haar, T. H. and A. H. Oort, "New estimate of annual poleward energy transport by northern hemisphere oceans", J. Phys. Oceanogr., 3, 1973, pp. 169-172.

[259] Wallace, J. M., C. Smith, and Q. Jiang, "Spatial patterns of atmosphere-ocean interaction in the northern winter", J. Climate, 3, 1990, pp. 990-998.

[260] Weare, B. C., A. Navato, and R. E. Newell, "Empirical orthogonal analysis of Pacific sea surface temperature", J. Phys. Oceanogr., 6, 1976, pp. 671-678.

[261] Weare, B. C., "Uncertainties in estimates of surface heat fluxes derived from marine reports over the tropical and sutropical oceans", Tellus, 41A, 1989, pp. 357-370.

[262] White, W. B. and A. E. Walker, "Time and depth scales of anomalous subsurface temperature at ocean weather stations P, N, and V in the North Pacific", J. Geophys. Res., 79, 1974, pp. 4517-4522.

[263] White, W. B., G. A. Meyers, J. R. Donguy, and S. E. Pazan, "Short-term climatic variability in the thermal structure of the Pacific Ocean during 1979-1982", J. Phys. Oceanogr., 15, 1985, pp. 1080-1094.

[264] Willebrand, J., "Temporal and spatial scales of the wind field over the North Pacific and North Atlantic", J. Phys. Oceanogr., 8, 1978, pp. 1080-1094.

[265] Willebrand, J., S. G. H. Philander, and R. C. Pacanowski, "The oceanic response to large-scale atmospheric disturbances", J. Phys. Oceanogr., 10, 1980, pp. 411-429.

[266] Woodruff, S. D., R. J. Slutz, R. L. Jenne, and P. M. Steurer, "A comprehensive ocean-atmosphere data set", Bull. Amer. Meteor. Soc., 68, 1987, pp. 1239-1250.

[267] Woodruff, S. D., S. J. Lubker, K. Wolter, S. J. Worley, and J. D. Elms, "Comprehensive Ocean-Atmosphere Data Set (COADS) Relese 1a: 1980-92," Earth Syst. Monitor, 4, 1993, pp. 1-8.

[268] Wunsch, C., "The North Atlantic general circulation west of 50 W determined by inverse methods", Rev. Geophys. Space Phys., 16, 1978, pp. 583-620.

[269] Wunsch, C., "Can a tracer field be inverted for velocity?", J. Phys. Oceanogr., 15, 1985, pp. 1521-1531.

[270] Wunsch, C. and J.-F. Minster, "Methods for box models and ocean circulation tracers: Mathematical programming and nonlinear inverse theory", J. Geophys. Res., 87, 1982, pp. 5647-5662.

[271] Wyrtki, K., L. Magaard, and J. Hager, "Eddy energy in the oceans", J. Geophys. Res., 81, 1976, pp. 2641-2646.

[272] Yasuda, I., K. Okuda, and M. Hirai, "Evolution of a Kuroshio warm-core ring - variability of the hydrographic structure", Deep-Sea Res., 39, Suppl. 1, 1992, pp. S131-S161.

[273] Yoshida, K. and T. Kidokoro, "A subtropical counter-current in the North Pacific - An eastward flow near the subtropical convergence", J. Oceanogr. Soc. Japan, 23, 1967, pp. 88-91.

[274] Zambianchi, E. and A. Griffa, "Effects of finite scales of turbulence on dispersion estimates", J. Mar. Res., 52, 1994, pp. 129-148.

[275] Zwiers F., and H. von Storch, "Regime dependent auto-regressive time series modelling of the southern oscillation", Max-Plank-Institute fur Meteorologie, Rep. N44, Hamburg, 1989.

INDEX